MPS

Using Statistics in Industry

CW01403633

IAN GARRY

The Manufacturing Practitioner Series

Using Statistics in Industry

Quality Improvement Through Total Process Control

Gordon Betteley, Neville Mettrick
Edward Sweeney, David Wilson

Prentice Hall

New York   London   Toronto   Sydney   Tokyo   Singapore

First published 1994 by
Prentice Hall International (UK) Ltd
Campus 400, Maylands Avenue
Hemel Hempstead
Hertfordshire, HP2 7EZ
A division of
Simon & Schuster International Group

Typeset in 10/12 pt Times
by Mathematical Composition Setters Ltd, Salisbury

Printed and bound in Great Britain by
T.J. Press (Padstow) Ltd

Library of Congress Cataloging-in-Publication Data

Using statistics in industry : quality improvement through total
   process control / Gordon Betteley ... [et al.].
      p.   cm. − (Manufacturing practitioner series
   Includes bibliographical references and index.
   ISBN 0-13-457862-7
      1. Process control−Statistical methods.   2. Quality control−
Statistical methods.   I. Betteley, Gordon.   II. Series.
TS156.8.U83   1994
658.5′015195−dc20   93-35786
                                                              CIP

British Library Cataloguing in Publication Data

A catalogue record for this book is available from
the British Library

ISBN 0−13−457862−7

---

1 2 3 4 5 98 97 96 95 94

Contents

Preface

For many years there has been a gulf between theoretical and practical technology. Nowhere is this more so than in the mathematical sciences, where traditional teaching based on academic texts seems to be poles apart from popular understanding.

During the last few years there has been an increasing recognition of the need to bridge this gap, so that all available knowledge can be brought to bear on everyday problems and the pursuit of continuous improvement. To this end, universities and industrial companies have taken joint initiatives which enable learning to relate to common experience.

Particular initiatives include those taken by the Open University and the University of Warwick. In the early 1980s, these institutions started work with the Rover Group and others on schemes that now extend to many companies which are household names in the United Kingdom: including, for example, Rolls Royce, British Airways, Lucas, British Aerospace, Short Brothers and Westland Helicopters.

The schemes are aimed at all levels of employee in commercial and manufacturing industry from executives to shopfloor team leaders in marketing, design, purchasing, production, sales and service. Some schemes lead to a master's degree (for example, the OU IMTM course and Warwick's IGDS course), while others qualify successful students for the award of a diploma (for example, Warwick's IMDS course). In addition, material developed for these schemes provides a basis for participating companies' in-house courses within the framework of vocational training.

The schemes vary in content and depth, but they all have the one subject of statistics in common and, through teaching based on real occurrences in industry, all develop the subject from practical applications rather than classical mathematics.

This book has been written to accompany the courses and to be used as a reference text at the workplace. It covers basic methods, but offers more advanced techniques to those who have an interest or a need. Its approach is based on the premise that statistical techniques do not exist as an end in themselves, but rather as a means to an end. This approach does not deny the fact that an understanding of theory is essential if practical application is to be carried out in a meaningful way.

The book sets out the subject of statistics for people who have no aspirations to becoming mathematicians. It might be read as a whole from beginning to end (certainly by engineering students taking a degree course) or only in those parts relevant to a particular task or problem. To guide the casual reader, references indicate related text in other parts of the book. However it is read, previous knowledge should not be necessary – except, of course, for a grasp of the three Rs!

In more detail, it is expected that useful reading for an executive is merely the overview of the subject in Chapter 1 and the treatment of process control in Chapter 9. However, line managers should have sufficient appreciation of the whole subject to understand its results and to be able to coach their staff in its use.

At the team leader level, there should be the ability to use the ideas and methods described in Chapters 2, 3, 9, 10 and 13. These chapters discuss the basic principles of data collection and presentation for process monitoring, control and improvement.

The remaining chapters explore the subject in greater depth and perhaps are for production and design engineers or similar specialists in other areas. Whoever it is, any working area should have at least one person with the ability to use the data investigation techniques that are covered in Chapters 5, 8, 11 and 12. Also among these people, there should be an understanding of some less common aspects of probability presented in Chapters 4, 6 and 7.

Generally, the book focuses on techniques that are appropriate to process control, which is the wide-ranging activity described in Chapter 9.

In particular, it offers discussion and treatment of process control itself and topics such as probability plotting (Chapter 8) and capability assessment (Chapter 9) that are not readily available in other texts.

Readers should note that the terms and symbols used are widely accepted. However, they occasionally differ from the conventions of some publications which cover other techniques using similar mathematical principles.

1 Introduction

■ 1.1 The context

Since the late 1970s, the industrialized nations of the Western world have recognized a need to change the way in which their manufacturing and commercial industries are managed. This recognition has been triggered by a continuing and declining share of world markets.

The principle reason for the decline is increased competition in most market sectors from sources around the world, in particular from the Far East. Developing nations are becoming serious contenders in markets, including home markets, that were previously captive to the West. The overthrow of communism in Eastern Europe presented opportunities, but the other side of the coin is a further increase in competition.

Competition has been stimulated by the removal of trade barriers. The creation of a single market in the European Economic Community during 1992 is just one example.

The challenge, of the need to change, has forced Western industries to put themselves under scrutiny in a manner never witnessed before. Many companies have identified *quality* as the major issue in the quest for immediate survival and for longer-term profitability and growth.

Quality

Quality is a concept and industry has not been well served by those who have attempted some narrow definition: for example, 'fitness for use' or 'conformance to requirements'. These 'buzz-word' definitions have caused some to think that they have achieved quality because they have performed as specified − no thought that the specification might be flawed.

Quality is the same in any language. In the United Kingdom, English words (like the constitution) are accepted through legal precedent rather than through 'expert' definition. The *Shorter Oxford English Dictionary* describes quality through 23 different accepted understandings which, taken as a whole, represent common perception.

Companies that have recognized this perception no longer attempt to define quality. Rather, they define what must be done in order to achieve quality. For example, the Rover Group uses the following definition:

> Quality achievement means satisfying customer requirements, continually improving products and services, and making a profit. This demands team-working relationships founded on trust, respect, fairness and the highest standards of honesty and integrity.

This approach summarizes *total quality management* or TQM for short. It is codified in the International Standard ISO 9004, which is reproduced in British Standard BS 5750 Part 0 Section 2.

The 'gurus'

The development of TQM is associated with a number of 'gurus' who have made contributions over the years. Among them are Juran, Taguchi and Ishikawa, who are mentioned later in the book.

Perhaps the most significant of the gurus, in the context of TQM, is the American statistician W. Edwards Deming. Shortly after the Second World War he visited Japan, where his views about quality were listened to and put into practice. His philosophy is expressed in his *fourteen points for management*, which are summarized in Figure 1.1. A recurrent theme in his philosophy is an insistence that the use of statistics is essential to all involved in the achievement of quality.

So far as this book is concerned, the important guru is Walter Shewhart, who pioneered the use of statistics for process control in the 1920s.

Control philosophies

The United States of America is the source of two distinct control philosophies. Both use all the statistical theory that is described in other chapters of this book, and both are aimed at the achievement of quality. However, at a working level there are important practical differences.

The *quality control* philosophy has been used by Western-based industries since the 1920s. It is linked to a style of management called *scientific*, where the results of tasks are contractually specified and are the basis of payment for work done. Scientific management developed from the work of F.W. Taylor and gave rise to time and motion studies, the use of tolerances and the science of industrial engineering.

1	**Constancy of purpose**	*Provide for long-term needs*
2	**New philosophy**	*Meet customer expectations*
3	**Statistical evidence**	*Eliminate go/no-go inspection*
4	**Minimal total cost**	*End lowest tender contracts*
5	**Continuous improvement**	*Seek out and solve problems*
6	**Task training**	*All employees must know their job*
7	**Leadership**	*The role of management*
8	**Drive out fear**	*Uninhibited two-way communication*
9	**Break down barriers**	*Eliminate departmental self-interest*
10	**Eliminate exhortations**	*Make reasonable requests*
11	**Eliminate targets**	*The resort of helpless management*
12	**Encourage pride in work**	*Accountability for quality*
13	**Education**	*Ability to adapt to change*
14	**Executive obligation**	*Harness everybody, including themselves*

Figure 1.1   Fourteen points for management

The quality control philosophy implies that there is an acceptable range for material, dimensional and performance features of products or services. It infers that achievement within the range is the same as achievement of quality, and controls are designed to detect and correct any nonconformance with specification.

The *process control* philosophy developed in Japan after Deming's visit in 1946 and since the 1970s has been adopted by most of the world's major companies. The style of management associated with this philosophy is described as *human*. The distinctive features of scientific and human management are summarized in Figure 1.2.

SCIENTIFIC MANAGEMENT	HUMAN MANAGEMENT
Fair day's work for fair day's pay	**Limits to money as a motivator**
Standards set by specialists	**People need some independence**
Assumption of the perfect inspector	**Self-checking**
Autocratic organization	*Democratic organization*
Work compartmentalized by skills	*Multi-skill teamworking*
Work objectives limited to tasks	*Customer satisfaction objective*

Figure 1.2   Management styles

3

Process control requires the customers' economic optimum to be the suppliers' target, but recognizes that there is variability in all products and services. This requirement leads to a view of product specification which is different to that in quality control. The principal controls are designed towards maintaining processes on target, rather than within tolerance. Other controls are used to prevent customer dissatisfaction and to improve continuously both the process and the product, through a reduction of the variability which represents waste. In process control, *customers* are the people and machines at the next processing stage, the ultimate product or service users and all in-between.

■ 1.2 Statistics and numbers

Whatever the control philosophy of an organization, *statistics* have a crucial role. Statistical methods are used consciously or unconsciously in many situations, especially in presentations of information from numerical data.

Actual presentations often owe as much to art as they do to mathematical theory, but in their preparation, statistical methods will have been used to design data collection techniques and the presentation itself. Statistical methods have applications in process control, product control, problem solving and last, but not least, in improvement of manufacturing economics and customer satisfaction.

Anyone who derives information from data is acting as a statistician. However, individuals are sometimes not as aware as they might be about the tools of the statistician's trade. The job of getting information from data is tackled less well than it might be. This book aims to help the individual through practical methods, ideas and supporting theory.

Numbers are all around us. In almost everything that we do, whether personally or in our occupations, numbers and an understanding of numerical data are of fundamental importance.

'What time is it?', 'How much is that?', 'How far is it to there?' are examples of simple questions which are asked every day. In each, the demand for a numerical response is implicit. During our daily work or career, the effective use of numbers is often crucial to achievement or success.

Without considering those professions that depend upon numbers for their very existence, from the artist to the zoologist, it is hardly possible to conceive a pursuit which is not heavily reliant upon numerical information in one way or another. Indeed, it is impossible to imagine a world without the concept of numbers and the associated analysis of numerical information.

The undoubted involvement we all have with numbers is reflected in the fact that mathematics as a discipline, with its many and varied branches, has existed from as far back as historians have been able to document. In today's society, children are introduced to ideas about and analysis of numbers from a very early age. Despite this, concern is frequently aired about the lack of grasp of the basic concepts of mathematics and statistics by people of all ages and in all walks of life.

Industry has long realized that the use of statistical methods can bring benefits to many aspects of commercial life. Nowhere is this more true than in the continuous quest for extraordinary customer satisfaction through quality management. First, there was a recognition that the traditional approach to the achievement of quality in products and services through inspection had failed. Then came a further recognition, that variation in manufacturing and commercial processes exists and contributes to the inability of companies to satisfy their customers' expectations. These recognitions suggested the way forward.

Statistical method is the natural tool to use to tackle variation because statistical analysis is the only basis for attempting to understand variability. Western industry has awakened to the use of statistical methods and allied techniques in the face of ever-increasing competition from all parts of the world. However, if this awakening is to become something which genuinely changes the way in which business is conducted, then people must become more conversant with the language of statistics. Also, there must be a greater understanding of available statistical techniques and more competence in effectively applying the results of statistical analyses.

■ References to further reading on the topics of Chapter 1

Deming (1986) discusses the fourteen points and many other aspects of quality and management. Walton (1989) provides a useful introduction to Deming's approach. Mitra (1993) discusses some philosophies and their impact on quality. Shewhart (1931) is of historical interest.

See Bibliography for details of titles and publishers.

2 | Basics

■ 2.1 Steps in statistical analysis

A statistical analysis is like any other analysis, it depends upon a logical development of thought from raw data. The principal steps are outlined below. A particular piece of analysis might consist of all or only a few of the steps; experience or well-developed theories, based on previously available data, often enable some steps to be bypassed.

Step 1: Determine objective

Too many people in too many organizations spend too much time collecting data for no reason that is apparent to them. Records are created virtually for their own sake and little or no use is made of them.

> During the late 1970s and early 1980s, there were some manufacturing companies which used a greater weight of paper than metal or other product materials, in a misguided attempt to mitigate the effects of consumer legislation. The recorded data was intended to prove that regulations were being observed; it was not used in the development of improved products, which was the purpose of the legislation.

- What needs to be controlled?
- What problem needs to be solved?
- What opportunities need to be sought?

Step 2: Determine data source

Data that yields information relevant to the matter at hand must be identified. Often, qualitative techniques (Chapter 13) will help to pinpoint appropriate data. Then, the data can be collected by the statistician or by somebody else;

it might be the output of an existing activity or it might have to be specially obtained.

Step 3: Explore data and derive theory

The data must be examined to check that it is valid and to develop an explanation, theory or hypothesis of what it portrays. All sorts of methods are available for the development of hypotheses; they tend to be of two general types: sketched pictures and calculations using simple formulae (Chapter 3).

Step 4: Test the theory

When a theory has been proposed, it must be tested. In some situations there will be independent corroborating information. In other situations, the theory is a 'stake in the ground' for comparison with later information. In both circumstances, more data is needed to test the theory.

A theory cannot be tested on the same data set that was used to derive the theory, although the original data set could be split to create a test sample not used in the derivation.

Step 5: Use the theory

A statistical theory is a basis for prediction of what new data should look like if nothing changes. Any later discrepancy is a signal for investigation.

The fit between predictions and new data can be inexact in some respects without invalidating the theory. However, any discrepancy might mean that the theory needs rethinking, or that there has been a process change whose cause needs to be identified.

The purpose of the theory is to help decision making about the course of future action.

Step 6: Recheck

Remember that the world can change. Theories might hold only at certain times and in certain places. It is worth checking periodically that the theory still works.

■ 2.2 Eyeballing

When a statistician has collected, produced or been presented with a set of data, the first thing to be done is to sit back and examine the data visually.

Any conclusions drawn from a statistical analysis can be only as reliable as the data upon which they are based. So, before contemplating lengthy number crunching, it is well worth spending a fair amount of time studying, in order to be satisfied with the quality of the data.

A visual inspection, accompanied by a rough histogram, plot or scatter diagram, often reveals peculiarities, patterns or trends which throw considerable light on the subject, and, in many cases, no further analysis is necessary. If these things are not spotted, they might lead to violations of the assumptions behind subsequent analysis, overstatement of accuracy and false conclusions.

Being able to pick out relevant patterns or peculiarities is, to some extent, a matter of experience – getting a feel for that sort of thing – but there are certain things that should be looked for as a matter of course, before embarking on more sophisticated analysis. In particular, attention should be given to the following:

- Unlikely numbers.
- Data precision.
- Patterns and trends.
- Data type and homogeneity.

These items are considered in the remaining sections of this chapter. The details are by no means exhaustive, but should help to save time and effort spent in unnecessary and inappropriate analysis.

■ 2.3 Unlikely numbers

Figure 2.1 contains a data set that provides some examples of unlikely numbers. The data has been abstracted from computer print-outs of motor vehicle emissions test results. The vehicle identification number (VIN) was manually entered, while the revolutions per minute (RPM) and air fuel ratio (AFR) were automatically entered.

Typing or recording errors

There are several things to look for here:

- Successive numbers being identical.
  Reference *A* in Figure 2.1 – it is unlikely that the same vehicle has been tested twice.
- Numbers differing by a factor of 10.
  Reference *B* in Figure 2.1 – identification numbers should have six digits.
- Transpositions of digits.
  Reference *C* in Figure 2.1 – 702 should be 207.

VIN	RPM	AFR
204998	681	14.60
205004 ⌐	752 ⌐	13.64
205004 ⌐ A	678	13.64
205060	752 ⌐	16.36
20507 ⌐	684	14.63
20538 ⌐	688	14.03
205471	692	14.21
205475	752 ⌐	606.80 ——— F
20550 ⌐	719	15.77
205502	634	15.33
20554 ⌐ B	759	14.65
205541	713	16.18
207204	676	15.57
207221	752 ⌐	14.07
702253 ——— C	758	14.79
207325	752 ⌐	13.95
207476	715	14.29
207813	752 ⌐ D	11.58
207907	713	14.58
207915	724	13.77
207919	9 ⌐	13.52
208172	12 ⌐	15.24
208226	9 ⌐	21.43 ——— G
208268	4 ⌐	15.88
208269	12 ⌐	13.83
208286	2 ⌐ E	16.90
208327	695	15.28
208402	703	14.03
208991	717	13.76
209049	757	14.80

Figure 2.1 A data set

Conversions

Multiple occurrences of unusual combinations of digits, particularly if they differ from others by a more or less constant factor, often indicate that some sort of conversion has taken place.

The most common conversion is, of course, from metric to imperial units. Measurements that are quoted to an unusually large number of significant figures also suggest a conversion. If possible, any analysis should use the unconverted figures.

Repeated numbers

In a large range of values there may be certain numbers that appear an unexpected number of times, even though a conversion has not been applied.

Reference *D* in Figure 2.1 – the sixfold appearance of 752 is unexpected.

It is particularly important to look for and note this sort of peculiarity in data from equipment with a digital readout and where data is automatically fed into a computer. This situation is difficult to tackle and often requires close investigation of the measuring and recording equipment.

Outliers

Possibly the easiest type of peculiarity to spot is an outlier. This is a value which does not appear to belong with the remainder of the data. Sometimes and wrongly, outliers are called 'fliers' and are swept under a convenient carpet! Statistical tests for outliers (Section 11.12) can be used but, before they are, questions should be asked and answered when a particular value looks different to the rest:

- Why is it different?
- What is meant by 'different'?
- Is it providing information about the matter in hand?
- Is it indicative of some previously unconsidered matter?

The data at Reference E in Figure 2.1 arose because the revolution counter was not connected and at Reference F in Figure 2.1 because the engine being tested was incomplete.

An outlier can be dealt with in one of three ways:

- **Accommodation**: Analyse with and without the presence of an outlier. If the conclusions are the same, the theory is acceptable.
- **Incorporation**: Develop a new theory when the data including the outlier does not fit the assumed theory.
- **Rejection**: In some cases there is no alternative but to discard the outlier and perform the appropriate analysis on the remainder of the data. This is the usual course of action when the matter in hand is control of a manufacturing process. Note that discarding the outlier for purposes of the analysis does not mean ignoring it: its cause should be determined and appropriate action taken.

If an outlier goes undetected, it can have drastic effects on the subsequent analysis. Statistics which are particularly susceptible to the influence of outliers are measures of process variability and the correlation coefficient. Rough plots of the data are helpful: they can indicate distortions to statistics that are likely to be caused by the presence of odd points.

Reference G in Figure 2.1 is an outlier that was confirmed by a plot.

■ 2.4 Data precision

Data from subjective or objective measurement is the basis of statistics. It follows that data must be precise: in other words, it must be accurate within known limits.

The process of measurement is influenced by the same factors that influence every other process: facilities, methods, people and so on. The statistician must be satisfied that these factors are under control before making potentially expensive statistical pronouncements.

Facilities

It is important that a measured value has the same meaning at different times and in different places: for example, a measurement of one millimetre made today in Longbridge, Birmingham (England), should be the same length as one millimetre measured 12 months hence in Austin, Texas (United States of America).

The key to precision of objective measurement facilities lies in international standards for calibration such as set out in British Standard 5781 or the NATO standard AQAP6.

Precision in subjective measurement requires well-defined and maintained comparators that sufficiently guide individuals' interpretations.

Methods

In addition to being calibrated, equipment must be appropriate, suitably positioned and properly used. Some implications for precision of improper positioning or use are illustrated in Figure 2.2.

Figure 2.2(a) shows the bending of light as it passes through thick glass. When viewed from some positions, the readings will be false if the scale is on the outside and the level is on the inside.

Figure 2.2(b) might be a motor vehicle speedometer well below eye level – it could cause unintended attention from traffic police!

People

Individuals' perceptions influence their ability to discriminate between scale markings such as those illustrated in Figure 2.3. When using the scale Figure 2.3(a), there is a preference for reporting in whole-millimetre rather than half-millimetre values; also, there is a preference for the bolder markings of scale Figure 2.3(b).

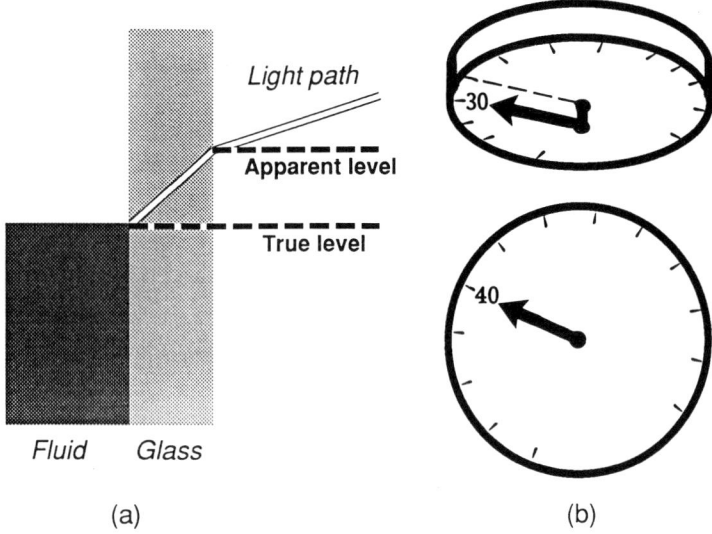

Figure 2.2 Equipment position

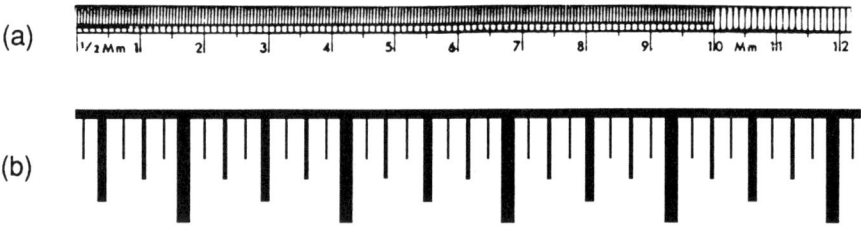

Figure 2.3 Scales

These preferences are not confined to purely visual measurements. In general terms: any set of numbers which appear to be accurate to (say) the nearest 0.1 mm might on closer inspection be seen to contain a high proportion of values ending in .5 and .0, which suggests a lower degree of precision than 0.1.

Where the range of values is large compared with the measuring interval, the distribution of final-digit values should be expected to be uniform. For example, if the measuring interval is 1 in a fairly large set of three-digit whole numbers, there should be approximately equal numbers of 0s, 1s, 2s … 9s among the last digits. A histogram (see Figure 2.4) of the frequencies of recorded values or final digits can be used to help decide whether a distribution is different from that which is expected. If there is still an element of doubt, exact probabilities could be calculated or a goodness-of-fit test such as the chi-squared test (Section 11.3) could be performed.

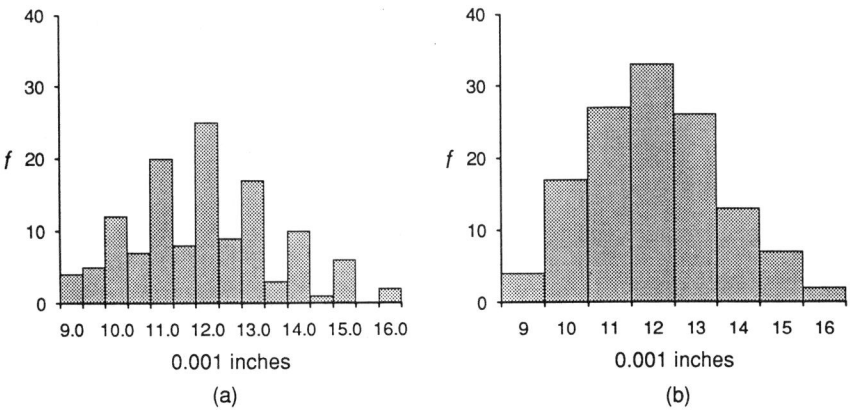

Figure 2.4 Measurement precision

Figure 2.4 illustrates motor vehicle engine tappet clearance data from measurements that were made using a feeler gauge set with 0.0005″ increments. Figure 2.4(a) is of the recorded data and shows measurement preference for multiples of 0.0010″. Figure 2.4(b), of the same data but rounded to multiples of 0.0010″, is a better reflection of the precision of this measurement process.

More generally, actual or full data should be used in any calculations, but the reported results should be rounded to reflect the true precision or accuracy of measurement.

Figure 2.5 illustrates an example where results reflecting true precision were important.

The data used in Figure 2.5 was obtained during an investigation of gas leakage from motor vehicle engine combustion chambers. Other work had highlighted gas leakage as a cause of reduced engine performance.

The investigator pressurized combustion chambers and recorded pressure loss in a given time from readings of a dial gauge with widely spaced scale markings. Values between the markings were estimated.

The recorded data is presented in Figure 2.5(a); it tends to confirm the preconceived hypothesis 1: that there was a single leak source.

The histogram in Figure 2.5(a) led to the measurement precision being questioned; it was realized that value estimations were not reliable; values were rounded and the histogram in Figure 2.5(b) was constructed.

Figure 2.5(b) led to hypothesis 2: that there were at least two leak sources. Further investigation eventually proved the second hypothesis.

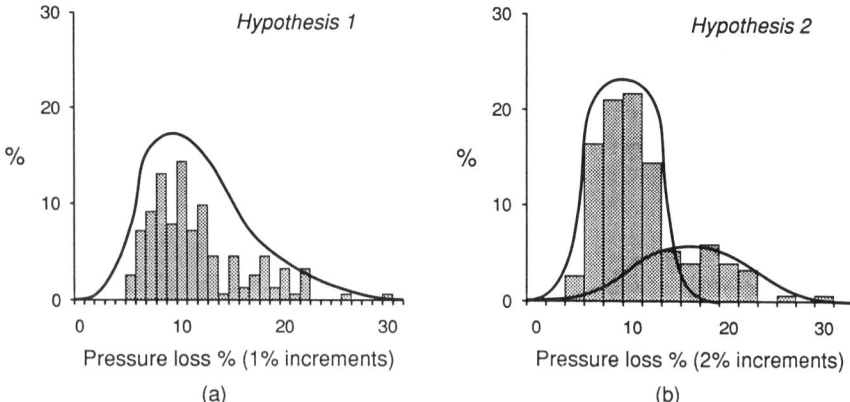

Figure 2.5 Effect of measurement precision

■ 2.5 Patterns and trends

It is sometimes possible to spot patterns and trends in written data by eye, but a picture or plot is usually more informative and more reliable. Plots often highlight previously unseen movement in data and give more clear expression of relationships between variables. Figures 2.6 to 2.10 provide some examples.

Occasionally, patterns and trends in the plotted data might not be obvious to the inexperienced eye, but they are worth a try before indulging in long-winded calculations.

Even to the experienced eye, interpretation demands knowledge of the process to which the data refers and an understanding of its implications in practical and statistical terms.

Interpretation of the statistics is the matter that is aimed at in the remainder of this book. So far as this section is concerned, altering the way in which data is tabulated or presented can be quite an eye-opener. A record of product defects ordered by time of product manufacture can provide a different picture from that given by the same data ordered by time of defect detection. The example in Figure 2.6 is another illustration of the effect of altering presentation.

Figure 2.6 is constructed from audit data of motor vehicle quality. The purpose of the data presentation is to identify trends in a production factory's quality performance.

The number of vehicles audited each week is small and is therefore unlikely to be representative of total production. To overcome this difficulty, weekly reports combine results from several weeks. The example illustrates how different treatments of the data can lead to different conclusions.

(a)

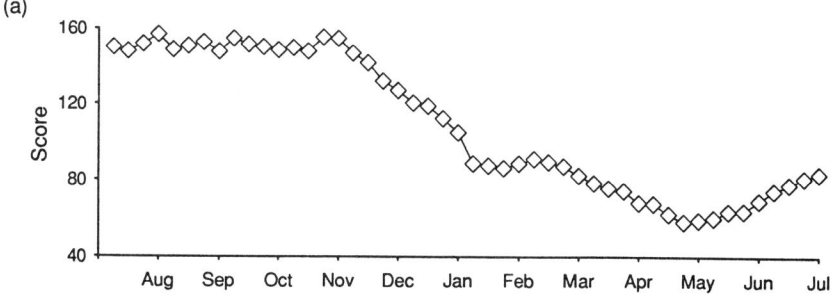

(b)

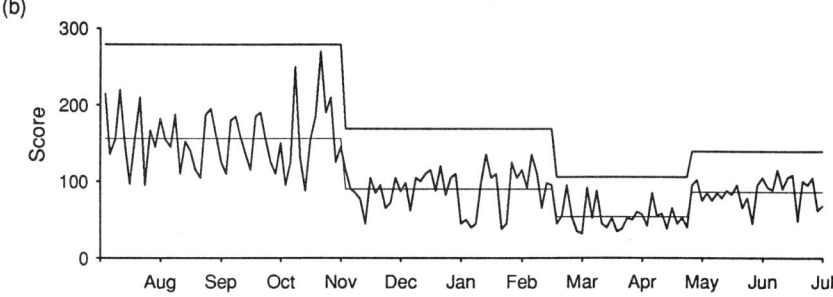

Figure 2.6 A trend and a pattern for the same data

Figure 2.6(a) shows a 'rolling' or 'moving' average. A plot is made each week of the last 30 vehicles audited; it shows a steady decrease in values from November to May with a discontinuity during January and a steady increase from May. These changes might be associated with management intervention by way of new model introduction or holidays.

The plot in Figure 2.6(b) is of the same data but for individual vehicles. Different to the smoothed trend of the rolling average, this pattern shows step changes in values during November, February and April. It suggests that the rolling average trend might be misleading if there is need to determine causes of improvement or deterioration – perhaps changes are due to other management intervention, such as variation in standards or personnel.

Figure 2.7 is a picture, over a period of time, of the variability of shaft alignment in a gearbox.

In Phase 1, control of the alignment was in the hands of the on-line machine operator, who periodically checked his own work with go/no-go gauges based upon specification. He called in the machine setter when out-of-specification items were produced. The plotted data is from off-line audit measurements.

In Phase 2, similar control was exercised. The different pattern was because a 'new-broom' supervisor insisted upon 'good housekeeping'

15

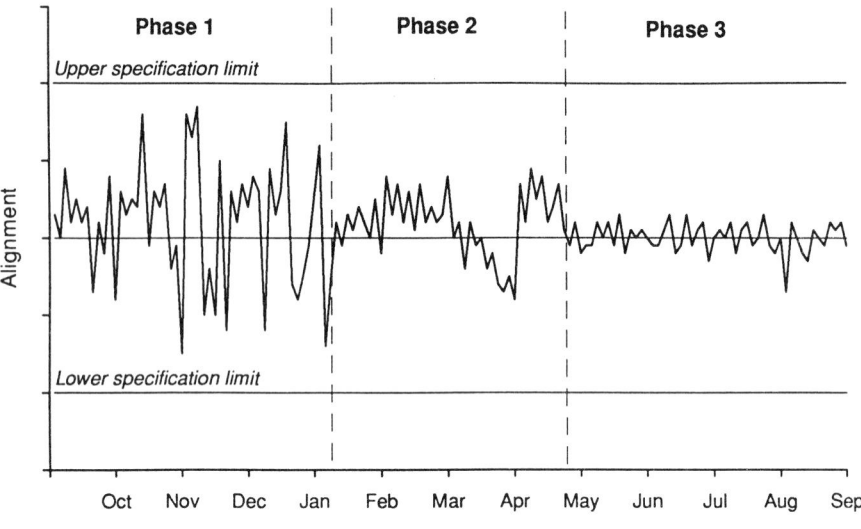

Figure 2.7 Interpretation of patterns

(keep machine clean, cover machines when they are not being used, etc.) and upon reacting to statistical interpretations of the off-line measurements.

In Phase 3, the go/no-go gauges and off-line measurement had been scrapped. The operator periodically measures his own work, plots the data on a control chart and calls in the setter when his interpretation of the plot indicates a process change. This phase brought consistent shaft alignment, a quieter gearbox and greater customer satisfaction – better quality, less cost and no further need for tolerances.

The data used in Figure 2.8 is of blemishes in motor vehicle paint finish. The sketch summarizes measurements from samples of 100 vehicles per week. The information suggests improvements in average levels of blemishes and in process controls that give reduced process variability.

In an attempt to increase the production rate of a machined component, cycle times were reduced where it was easy to do so – on seven out of the eight operations that made up the machining process. The attempt failed. The reason for failure is discernible from a study of the available data, but is shown much more dramatically on the sketch of the data in Figure 2.9. The sketch emphasizes the dangers of piecemeal efforts to improve processes; it also focuses attention on the area to be tackled if there is to be improvement.

The histograms in Figure 2.10(a) illustrate the changes that occur in the

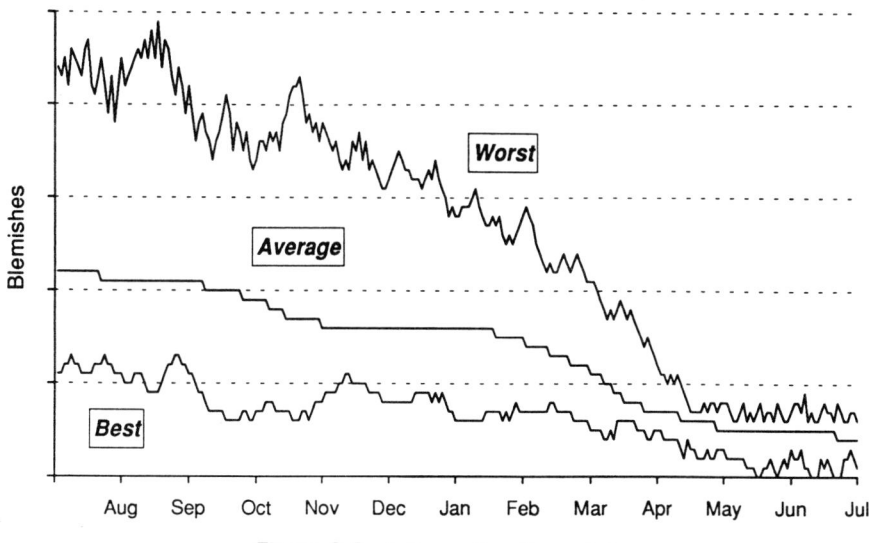

Figure 2.8 Information from data

Operation	1	*2*	3	4	5	6	7	8	Mean
Time before	6.2	*7.3*	8.2	8.5	9.4	9.8	10.2	9.6	8.65
Time after	6.1	*11.4*	7.9	7.4	9.1	9.7	10.1	8.9	8.83

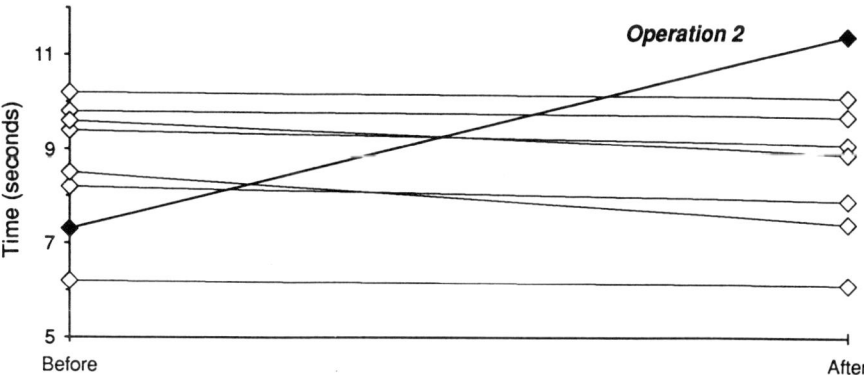

Figure 2.9 Highlighting significant information

bearing diameter of a gear during carbo-nitriding. Allowance can be made at 'soft' machining stages for typical or average changes. The question is: do these changes occur predictably in individual components? In other words, will small end up as small, large as large, etc?

An answer is in Figure 2.10(b), where the scatter diagram shows that change is unpredictable (all the points would be close to a line if change were predictable). The interpretation, for the studied component and

17

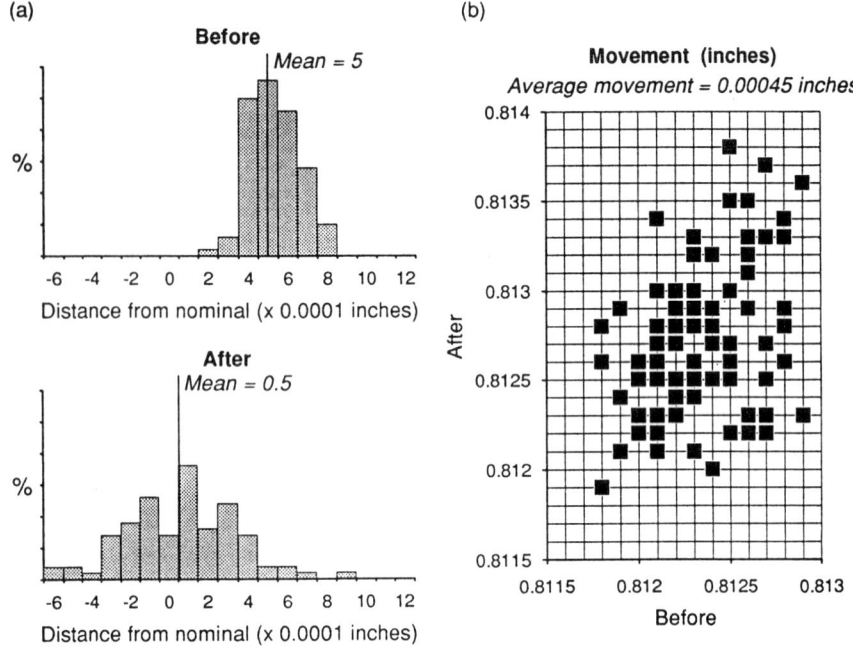

Figure 2.10   Investigating data relationships

process, is that size when soft does not determine size when hard, hence the need for critical dimensions to be machined after heat treatment.

■ 2.6 Data types

While the basic steps of assessing accuracy and simply analysing data are being taken, the type of data should be recognized or determined, in preparation for application of models and methods that are described in later chapters.

The distinction between continuous and discrete data types is important because their statistical descriptions are based upon slightly different mathematical models.

Continuous data

Data of measured quantities (length, mass and time, or compounds of these such as volume, energy, acceleration) is called continuous or *continuously variable data*. Measurements of these variables can be recorded only to the accuracy of the measurement process and therefore they will be rounded.

Temperatures are often reported as the nearest whole degree, but the precise value could be any value within a range and it could have any number of digits after the decimal point.

Discrete data

Other data (called *discrete data*) involves counts which might be of *attributes* (such as good, bad, accept, reject, high or low) or of *occurrences* (such as accidents, live births or aircraft movements). The count might be expressed by number or by proportion (per cent, per unit, per area and so on). Counts are restricted to certain values and are usually whole numbers.

For example, the raw data of blemishes used to construct Figure 2.8 can be only whole numbers. However, the individual plots might not be: they will depend on the scale, which in this case can be units, units per vehicle or units per square metre.

■ 2.7 Data homogeneity

The use and interpretation of many statistics start from the assumption that data is homogeneous: in other words, that successive samples are taken from the same mix of data. Also when successive samples are compared, their data must have common units. If sample sizes are the same, data having the same units can be compared; otherwise the data must be converted.

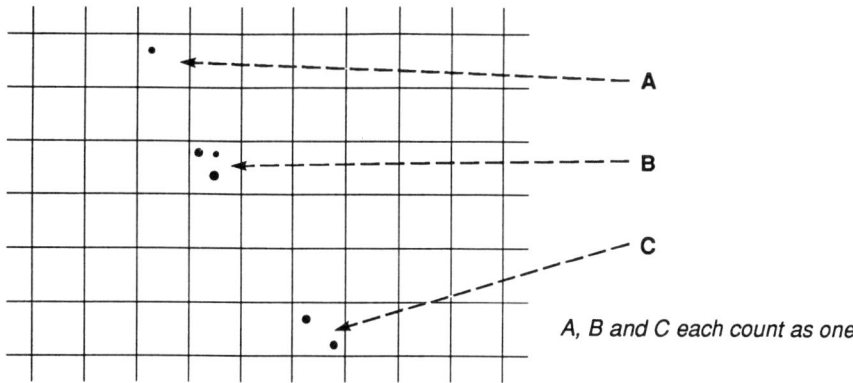

Figure 2.11 Definition of a blemish

For example, the raw data of Figure 2.6(a) is not directly comparable from month to month because some months include a holiday shut-down and therefore sample sizes vary. The data has been converted to the average score per vehicle audited during a month.

Homogeneity can be affected by the measurement process. It is important that there are clear conventions, especially for attribute and subjective data collection.

For example, when is an unwanted mark a blemish? Should two almost inseparable marks be counted as one blemish? Figure 2.11 illustrates a grid that defines a blemish as the presence of one or more unwanted marks in a grid square.

■ References to further reading on the topics of Chapter 2

Chatfield (1983) starts with a clear non-technical discussion of statistics for readers with an engineering background.
See Bibliography for details of titles and publishers.

<table>
<tr><td></td><td>

3

</td><td>

Data Presentation

</td></tr>
</table>

■ 3.1 Introduction

The main purpose of statistical analysis is to extract and concisely present all useful information from a set of data. Invariably this data exists initially as a list of numbers which often is best used when it has been arranged in a table. The process of arranging data is described in Section 3.8.

After arrangement, it is usual to draw a picture to gain an impression of the way that the data is distributed or to pick out the relevant points shown by the data. In other words, a suitable diagram is drawn for the values in the table. Many different types of diagram can be used to illustrate data. The more important ones are explained in this chapter.

Later, it is often necessary to carry out some calculations, so that the main characteristics of the data are explained by two or three numbers. This will be particularly important if comparisons are to be made with other sets of data.

■ 3.2 A set of data

A set of data relating to car sales for one company over one year is shown in Table 3.1.

Table 3.1 Numbers of cars sold

	Small	Medium	Large	Total
UK	92 645	83 012	108 613	284 270
EEC except UK	42 290	30 022	20 858	93 170
USA	0	0	3 317	3 317
Japan	12 087	0	538	12 625
Rest of world	2 113	2 219	3 107	7 439
Total	149 135	115 253	136 433	400 821

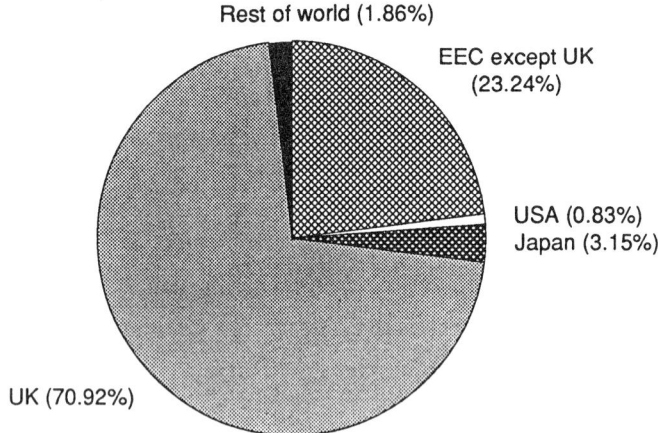

Figure 3.1 Total car sales

■ 3.3 Pie chart

Figure 3.1 is a pie chart. It illustrates some information from the data of Table 3.1 in a pictorial form. The area of a segment of the circle represents the proportion of cars sold in a particular region. The segments are drawn by dividing the angle of $360°$ at the centre of the circle into the correct proportions.

■ 3.4 Pictogram

A diagram that is more simple than the pie chart is shown in Figure 3.2. It can be used if only approximate values are required. In this picture it is difficult

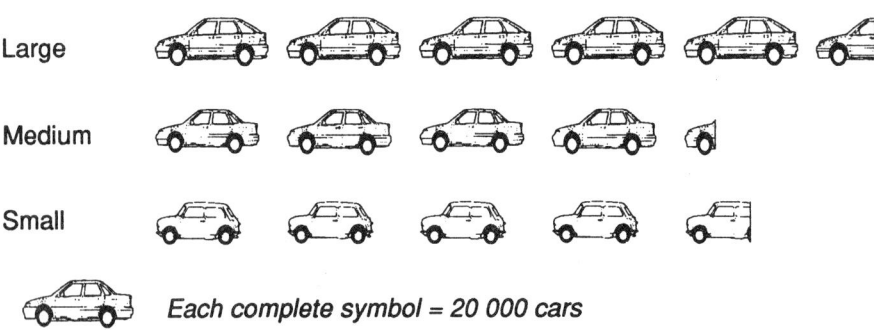

Figure 3.2 Cars sold in the UK

to know how many actual cars are represented by a fraction of a car in the diagram.

■ 3.5 Bar chart

The information illustrated in the pictogram is presented in another way by the vertical bar chart in Figure 3.3. A bar chart is often used when the illustrated categories are not numerical, in this case 'small', 'medium' and 'large' cars.

A bar chart with the rectangles drawn horizontally makes it easier to label the categories, as shown in Figure 3.4.

■ 3.6 Pareto analysis

Pareto analysis makes use of an empirical relationship between value and quantity. The relationship is empirical because it has not been proved mathematically; however, it fits the facts of many everyday observations. For example, a comparatively small percentage of fault types account for a comparatively large percentage of total fault incidence; a comparatively small percentage of stocked items account for a comparatively large percentage of total stock.

The technique can be used to suggest priorities for action. It is named after an Italian economist called Vilfredo Pareto, who first expressed the relationship, commonly called the 80/20 rule, which often appears as the statement

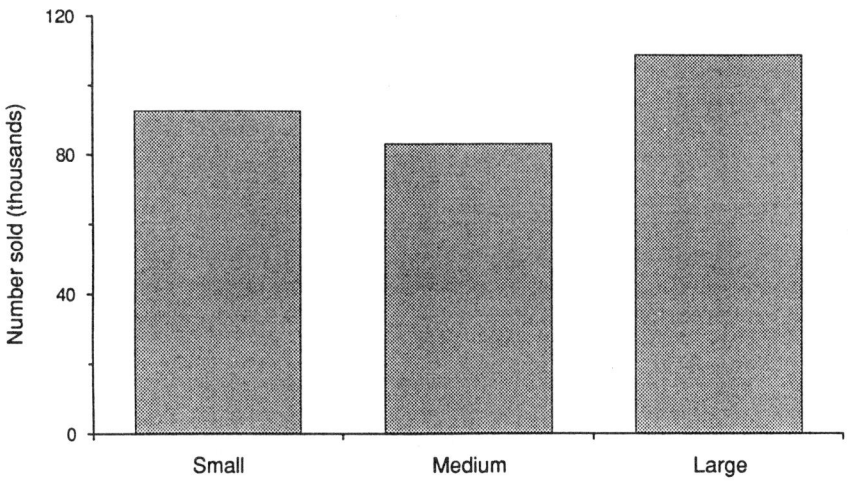

Figure 3.3 Cars sold in the UK

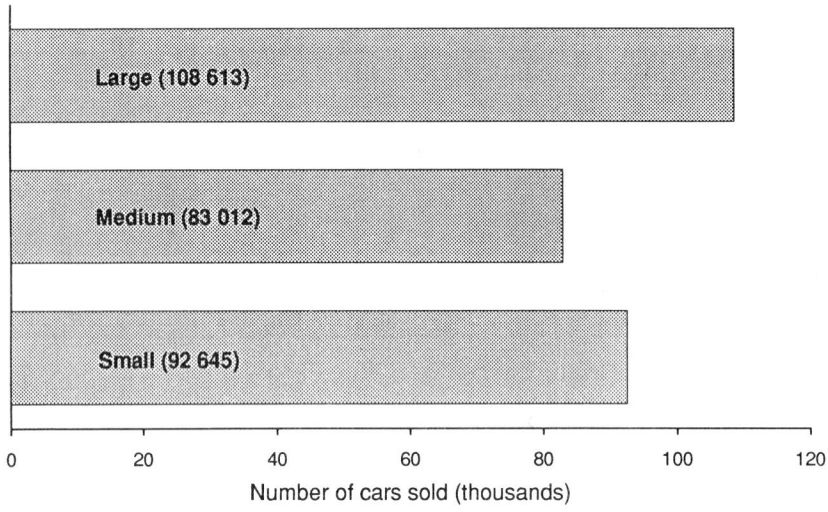

Figure 3.4 Cars sold in the UK

that approximately 80% of total fault costs or fault incidence are accounted for by approximately 20% of fault types.

Figure 3.5 is an analysis of complaints dealt with by a motor vehicle dealer's workshop. The complaints have been categorized and presented in order of magnitude on a bar chart. The ordering of the data is determined by the purpose of the analysis: for example, Figure 3.5(a) orders matters that affect immediate customer satisfaction and Figure 3.5(b) orders matters relevant to business management (time and hence cost). Also, starting with the largest category, a curve has been drawn to represent the progressive cumulative sum of complaints.

The analysis highlights items where improvement action should give substantial pay-back: for example, in Figure 3.5(a), on alignments and bulbs. When action to reduce these categories is successful, more priority items are highlighted.

> It might happen, in some other analysis, that the 80/20 rule is not followed (say 20% of fault types account for only 50% of the total fault incidence). In this situation it is sometimes helpful to review the categorization of faults for reasons similar to those described below.

Continued use of the technique often results in a long list of 'one-off' faults (ultimately, 20% of fault types might account for only 20% of the total fault incidence). This is the time when faults categorized by description or effect provide little help to prioritizing problem-solving activities. At this stage, it is the process which produces faults that needs to be addressed. The Pareto

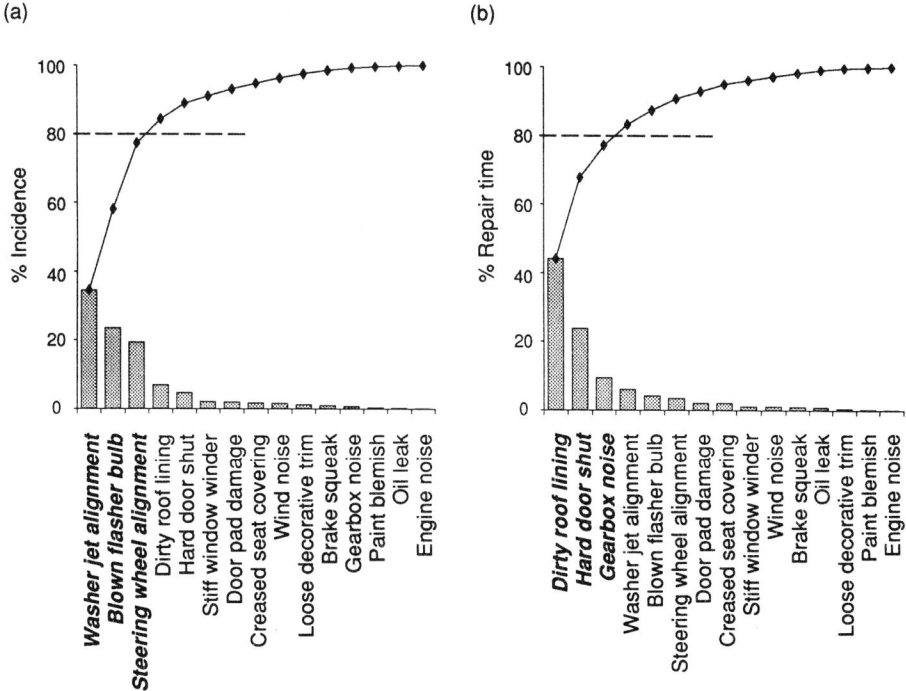

Figure 3.5 Pareto analysis of workshop data

analysis might be usefully applied again when the data has been recategorized by fault cause (in terms of people, material, method, environment and facilities).

■ 3.7 Histogram of discrete data

The histogram is one of the most common diagrams that is used when recordings are numerical values. Table 3.2 lists the number of faults per car in 120 vehicles audited and Figure 3.6 is a histogram for this data.

The value plotted on the horizontal axis is called the *observation* (x) and is the reading or measurement that was recorded. The *frequency* (f) of that observation is plotted on the vertical axis. Although this is the more usual case, the axes are sometimes interchanged, particularly when presented on a

Table 3.2   Faults in 120 vehicles

Number of faults	0	1	2	3	4	5	6	7	8	9
Number of cars	2	4	9	19	23	27	17	11	5	3

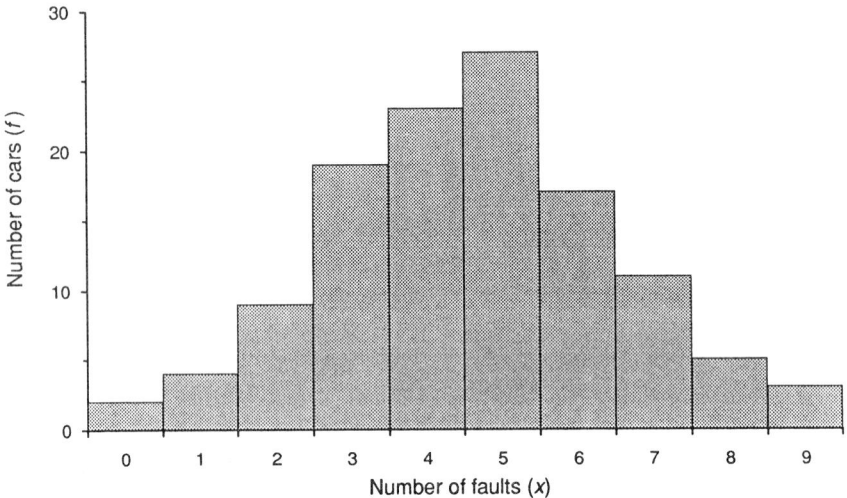

Figure 3.6 Number of faults in cars

computer. Note that the zero does not have to lie at the intersection of the axes; and also that the observations are the values at the centre of each rectangle.

In addition to the representation in Figure 3.6 discrete data is presented sometimes on a bar chart as in Figure 3.7(a) and sometimes on a line chart as in Figure 3.7(b). In fact, the line chart is what should be mentally recorded when looking at the other two diagrams.

The reason that the histogram may be preferred in many cases to the line chart is that, for certain purposes explained in Section 3.9, frequencies should be represented by areas. Also with the line chart, it is too easy to inadvertently 'lose' the top of a line.

Unless each class corresponds to just one x value (as in Table 3.2), grouping values will always result in a slight loss of information. This loss is offset by a better understanding of the shape of the overall distribution.

■ 3.8 Tally chart and frequency table

In all the diagrams so far in this chapter, the data has first been arranged in tables so that diagrams can be drawn easily. The process of arranging data in tables is described as *grouping the data in classes*. The process is illustrated using the data shown in Figure 3.8. These are measurements of dimensions of 50 gear wheel coupling teeth (coupling teeth are sometimes called dog-teeth; they provide means of gear selection). The figures as listed are actually the

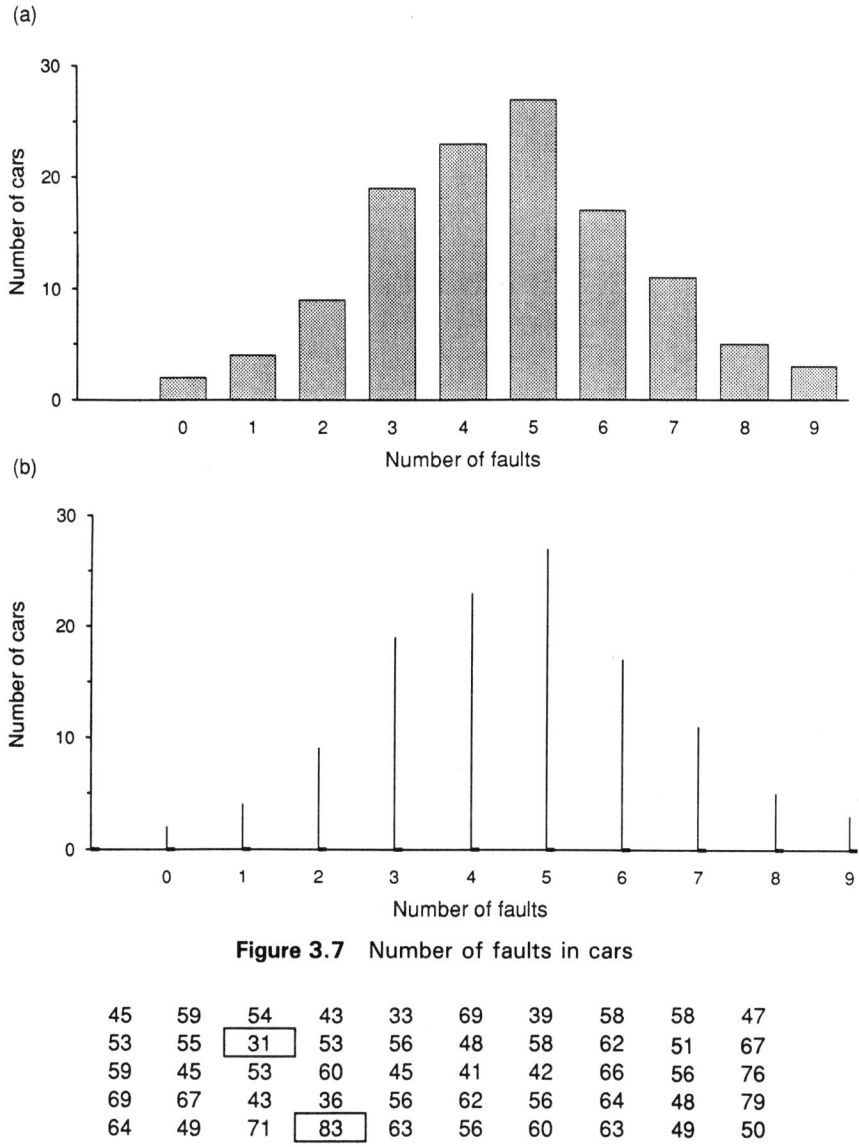

Figure 3.7 Number of faults in cars

45	59	54	43	33	69	39	58	58	47
53	55	31	53	56	48	58	62	51	67
59	45	53	60	45	41	42	66	56	76
69	67	43	36	56	62	56	64	48	79
64	49	71	83	63	56	60	63	49	50

Figure 3.8 Fifty measurements of coupling teeth

number of hundredths of a millimetre greater than 79 mm, so that the first measurement is 79.45 mm, the second is 79.53 mm and so on.

To obtain most information from a picture of data and especially if further calculations are to be carried out on the data, it is best to group the bulk of the data into between 8 and 15 classes. There is no harm in other groupings – the actual number is a matter of convenience.

27

Important information might be lost if there are fewer than 8 classes. If there is a suitable spread and quantity of data, the target should be about 12 classes; this target is linked to the characteristics of the normal distribution (Section 5.4).

The table might become unwieldy with more than 15 classes, although larger numbers often help when there is unusual data such as outliers (Section 2.3). Calculations of class width should be made from the bulk of the data: in other words, any unusual numbers are discarded.

- First, find the smallest and largest values. In Figure 3.8, where there are no unusual numbers, these values are 31 and 83, which have been boxed. The difference between the two values is $83 - 31 = 52$.
- Then calculate the class width for 10 classes: in other words, divide the difference by 10 (52/10). This gives 5.2, which is an inconvenient number. However, 5.0 is a convenient number to work with, is very close to 5.2 and so is chosen as the class width. The number of classes need not be exactly 10 (in fact in this case there are 11), but usually that is not important.

The next step is to construct a tally chart for the data. Classes are written down so that they cover all the data. Whatever the number of classes, they should all be of equal width even if there is to be no data in some classes.

A tally chart is illustrated in Table 3.3, where the class width is 5.0.

- It is necessary to leave an apparent gap between the *upper limit* of one class and the *lower limit* of the next class. This is necessary in order that values go into only one class. If both values were the same then it would not be clear into which class a reading should go.
- The real limits, called *class boundaries*, have the upper boundary of one class equal to the lower boundary of the next and in fact could be used in place of the class limits in the table.

Table 3.3 Tally chart for data in Figure 3.8

Classes	Tally	Frequency (f)
30–34	//	2
35–39	//	2
40–44	////	4
45–49	₩ ///	8
50–54	₩ /	6
55–59	₩ ₩ /	11
60–64	₩ ///	8
65–69	₩	5
70–74	/	1
75–79	//	2
80–84	/	1
	Total	50

- For example, the first class with limits of 30–34 could have been written as boundaries of 29.5–34.5, the second as 34.5–39.5 and so on. Note that it is necessary to introduce an extra decimal place to do this but the boundaries are used in any case in the histogram that follows (Figure 3.9). Also, note that the class width of 5.0 is the difference between upper and lower boundaries rather than limits.

Having fixed the class limits, go through the data one number at a time and place a *tally mark* against the class in which a reading falls. It is usual to use each fifth tally in a class to cross out the previous four, and hence the chart is sometimes called a five-bar gate diagram.

The final column in the tally chart is a count of the tally marks in each class: in other words, the class frequency (f). The frequency column is a table of arranged data that can be used in a diagram.

■ 3.9 Histogram of continuous data

A histogram for the arranged continuous data in Table 3.3 is shown in Figure 3.9. Notice that the rectangles are all of the same width. This means that, because it is areas that represent frequencies and the heights of the rectangles are proportional to the areas, the heights also represent frequencies.

Occasionally, histograms are drawn with rectangles of varying widths. The method is illustrated below. It is somewhat tedious and often does not give

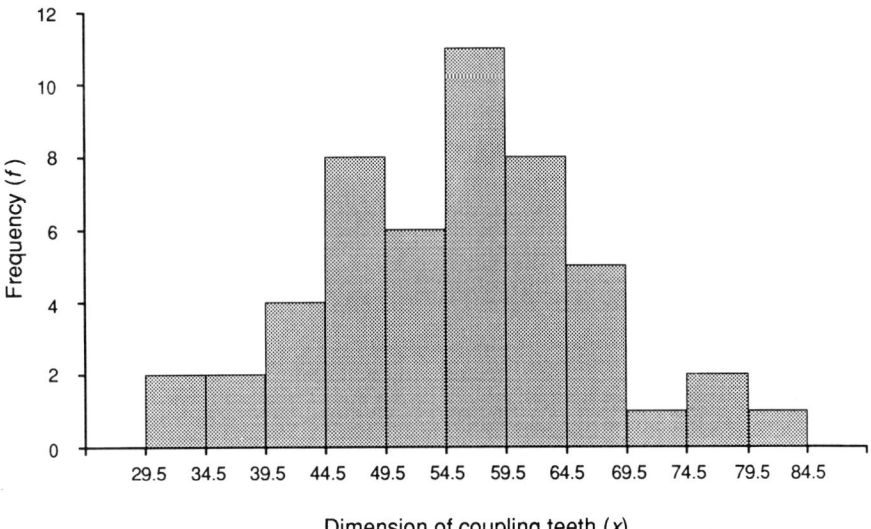

Figure 3.9 Histogram for continuous data

a clear picture; a more satisfactory way of representing such data is explained in Section 6.7. However, when a histogram is used, it should be remembered that *areas must be used to represent frequencies*. The height of each rectangle in the histogram is calculated as:

$$\frac{\text{frequency}}{\text{class width}} \times \text{standard width}$$

where standard width can be the width of any class but is usually taken as the smallest class width. For example, the data in Table 3.4 is the only readily available information on deaths during one year in a particular community. Standard width has been taken as 1 and the resulting histogram is drawn in Figure 3.10.

■ 3.10 Ogive

An *ogive* or cumulative frequency graph is a plot of values (x) against their *cumulative frequency* (F). It is the basis of several statistical techniques: in particular, in determinations of the median (Section 3.15) and in checking whether data arises from a particular distribution (Section 8.2).

Before an ogive can be drawn, a table of cumulative frequency must be constructed. Cumulative frequency in the table is the frequency less than a class boundary. It is sometimes helpful to number the classes in a table from 1 at the class with lowest values: in other words, the class below 1 contains no

Table 3.4 Data for histogram in Figure 3.10

Age group (class)	Deaths (frequency)	Class width	Rectangle height
<1	5	1	5.0
≥1 to <2	4	1	4.0
≥2 to <3	3	1	3.0
≥3 to <4	4	1	4.0
≥4 to <5	2	1	3.0
≥5 to <10	12	5	2.4
≥10 to <20	30	10	3.0
≥20 to <35	25	15	1.7
≥35 to <50	10	15	0.7
≥50 to <65	30	15	2.0
≥65 to <70	12	5	2.4
≥70 to <75	8	5	1.6
≥75 to <80	5	5	1.0
≥80 to <81	1	1	1.0
≥81 to <82	4	1	4.0
≥82 to <83	3	1	3.0
≥83 to <84	1	1	1.0
≥84 to <85	1	1	1.0

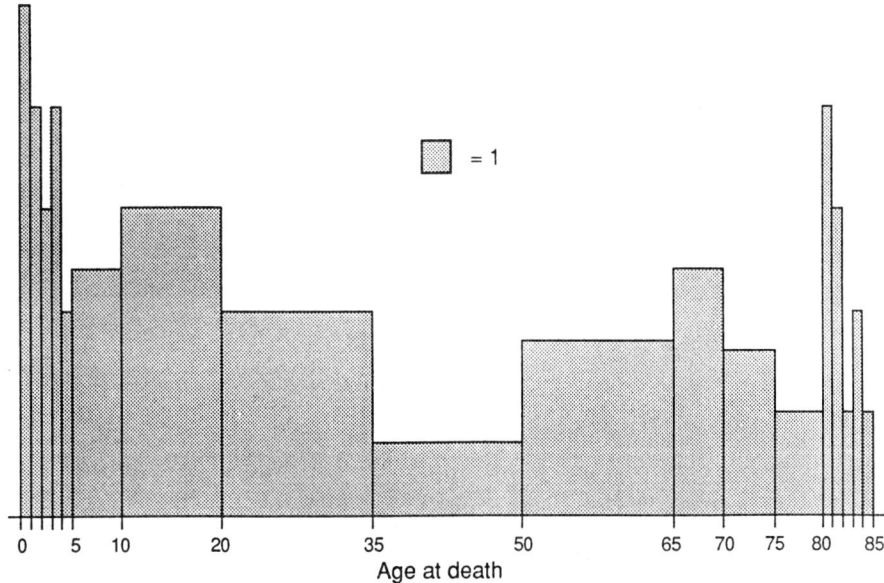

Figure 3.10 Histogram with classes of different widths

values and is numbered 0. The cumulative frequency of Class 0 is designated F_0. For example and using the data in Table 3.3:

the frequency less than 29.5 is $F_0 = 0$, by definition,
the frequency less than 34.5 is $F_1 = F_0 + f_1 = 0 + 2 = 2$,
the frequency less than 39.5 is $F_2 = F_1 + f_2 = 2 + 2 = 4$.

Continued summing of fs gives Fs for the cumulative frequency shown in Table 3.5. From this table, an ogive is plotted in Figure 3.11. It has been superimposed on the histogram in Figure 3.9 so that the relationship can be seen. The height of the ogive at any value of x represents the area of the histogram to the left of that value.

■ 3.11 Theoretical distributions

So far, this chapter has presented several ways of describing the distribution of actual data in pictorial form, the most useful of these being the histogram. All the statistical methods that are considered later in the book are based upon theoretical distributions. This section emphasizes some points already made and also illustrates the link between actual and theoretical distributions. The starting point is actual data whose distribution can be described by a histogram.

Table 3.5 Data for ogive in Figure 3.11

Class	0	1	2	3	4	5	6	7	8	9	10	11
Class boundary (x)	29.5	34.5	39.5	44.5	49.5	54.5	59.5	64.5	69.5	74.5	79.5	84.5
Cum frequency (F)	0	2	4	8	16	22	33	41	46	47	49	50

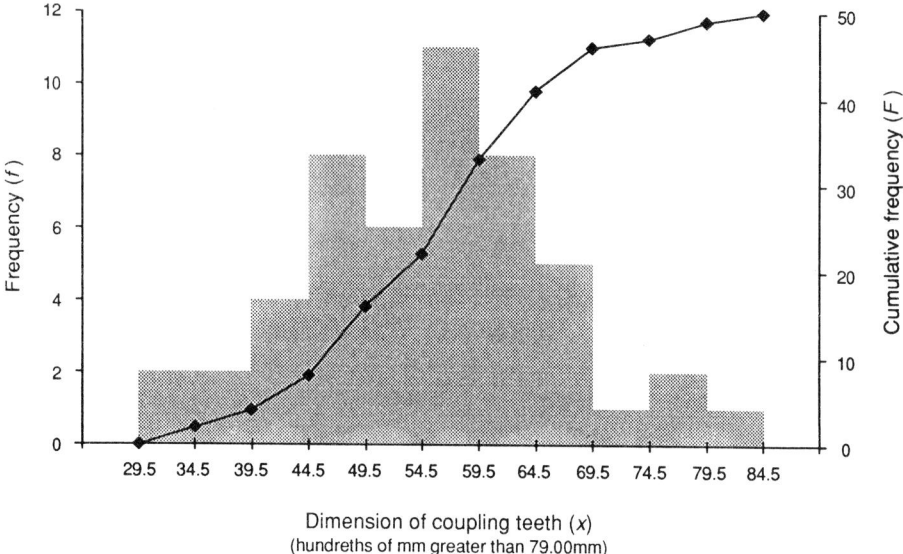

Dimension of coupling teeth (x)
(hundreths of mm greater than 79.00mm)

Figure 3.11 Ogive for continuous data

The data in Figure 3.12 is a record of the last two digits in measurements of flange separation on 125 crankshafts. The separation is nominally 15.0050 inches and was measured on each crankshaft to the nearest ten-thousandth of an inch. Recording the complete measurement seemed unnecessarily tedious! So 73 (the first value in Figure 3.12) actually represents 15.0073 inches, 63 represents 15.0063 inches and so on.

Even when the digits in common (15.00) have been taken off the beginning of each measurement, the mass of detail in Figure 3.12 still masks the shape of the distribution.

Figure 3.13 is a histogram of the individual (that is, not grouped) values in Figure 3.12. The histogram does not reveal any coherent pattern because

```
73 63 58 30 54 72 69 79 53 44 68 59 50 56 60 50 81 62 52 52 33 51 71 18 57
53 54 52 87 68 31 29 44 54 43 61 42 61 37 59 40 57 46 35 45 65 58 51 48 50
50 66 44 15 58 82 51 61 54 29 38 59 72 80 40 43 26 56 33 68 64 55 57 48 78
42 82 65 82 56 72 58 63 49 39 28 58 47 26 60 66 89 55 34 63 38 78 37 64 44
55 48 27 99 70 12 44 72 46 61 57 53 54 74 66 28 60 83 41 32 51 59 43 68 52
```

Figure 3.12 Flange separations for 125 crankshafts

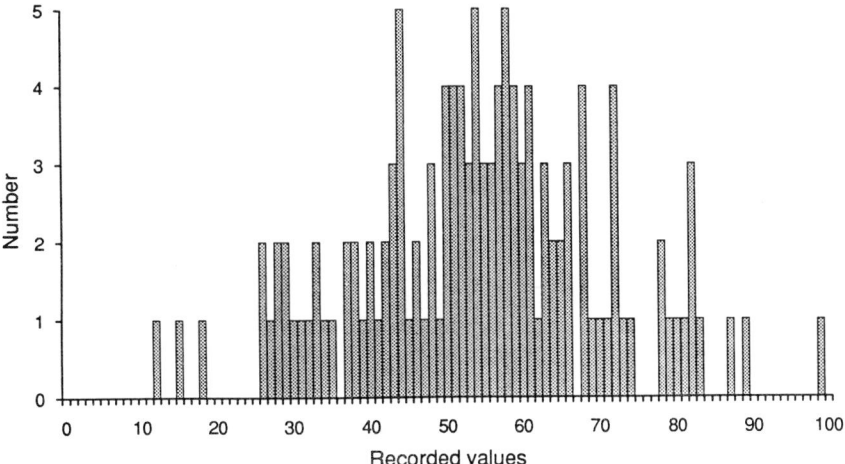

Figure 3.13 Histogram of the flange separation data in Figure 3.12

there are too many increments: in other words, it is spread over too many classes. With this amount of data (only 125 measurements), the histogram should have been spread over only 8 to 12 classes, in order to get a reasonable idea of the measurement distribution.

Using the frequency table method, the data has been grouped in Table 3.6 and presented again in Figure 3.14 as a histogram of grouped values that gives a better idea of the shape of the measurement distribution.

The histogram is drawn with rectangles of equal width, so that the heights of the rectangles represent the relative percentages of values falling in the chosen classes and, with suitable scaling, the frequency of occurrence within the classes.

After the initial sample of 125 crankshafts had been taken, more data was collected. Figure 3.15 is a histogram that shows the distribution of separations

Table 3.6 Frequency table of the data in Figure 3.12

Class	Limits	Tally chart	f	%
0	below 10		0	0
1	10 to 19	///	3	2.4
2	20 to 29	₦₦ //	7	5.6
3	30 to 39	₦₦ ₦₦ //	12	9.6
4	40 to 49	₦₦ ₦₦ ₦₦ ₦₦ /	21	16.8
5	50 to 59	₦₦ ₦₦ ₦₦ ₦₦ ₦₦ ₦₦ ₦₦ ////	39	31.2
6	60 to 69	₦₦ ₦₦ ₦₦ ₦₦ ///	23	18.4
7	70 to 79	₦₦ ₦₦ /	11	8.8
8	80 to 89	₦₦ ///	8	6.4
9	90 to 99	/	1	0.8
10	above 99		0	0

33

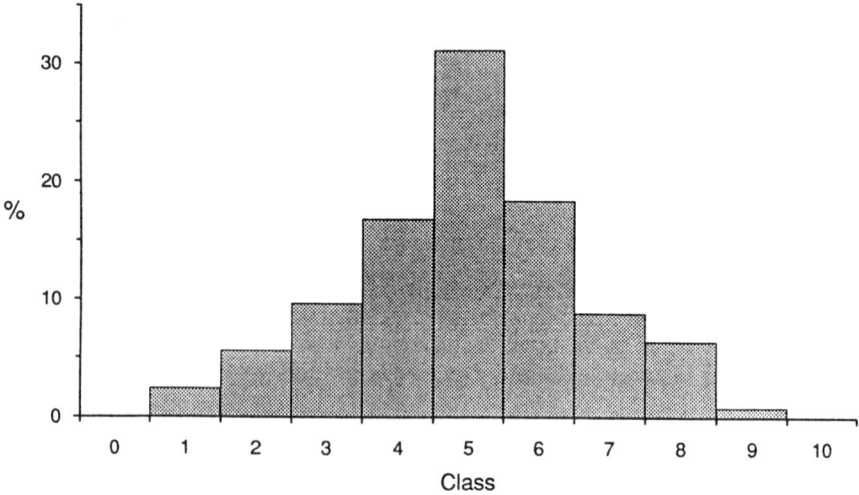

Figure 3.14 Histogram of the grouped data in Table 3.6

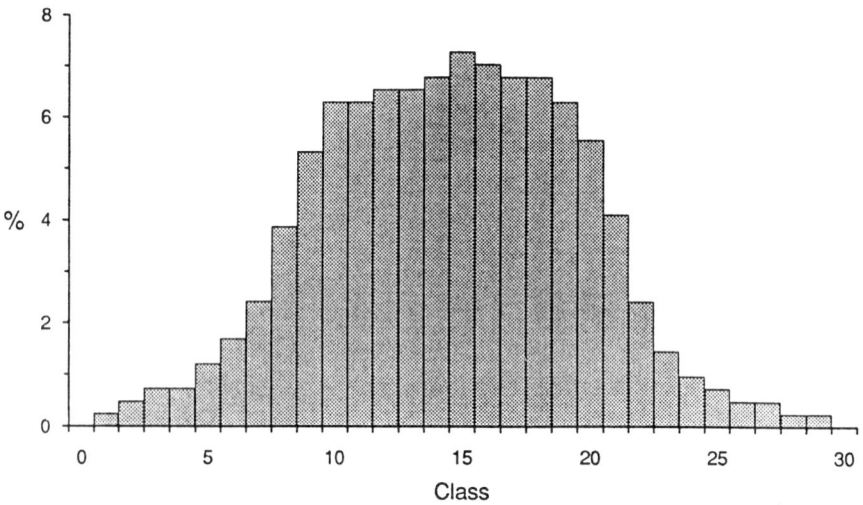

Figure 3.15 Histogram of flange separations for 4350 crankshafts

on 4350 crankshafts, grouped into 31 classes. Although the number of classes has been increased, there is still a sufficient number of values in each class to give the histogram a fairly regular outline.

As more data is collected, the number of classes can increase further and the width of their rectangles in the histogram can be reduced. The result is a clearer and more accurate picture of the shape of the data distribution.

As the process of data collection continues and if the data itself is continuous (as in this example), the number of classes can approach the theoretical extreme where there is an infinite amount of data, the width of the rectangles in a histogram is infinitely small and the outline of the histogram has become smooth. The theoretical extreme is illustrated by the frequency density curve

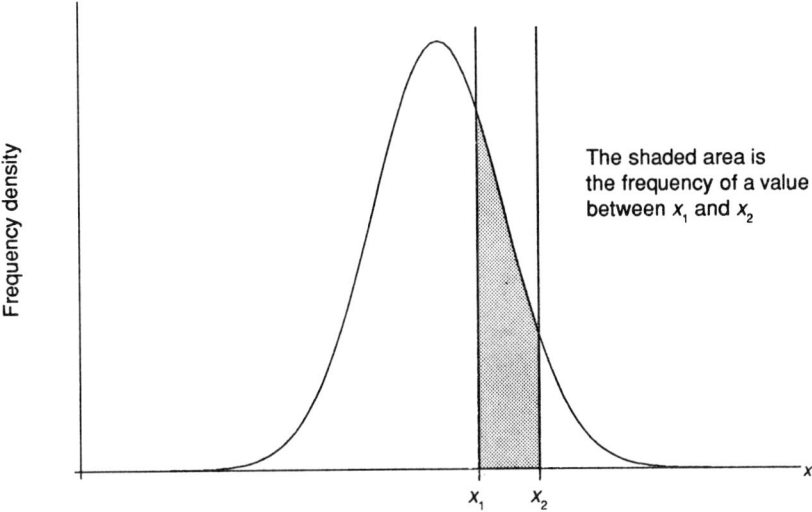

Figure 3.16 Frequency density curve

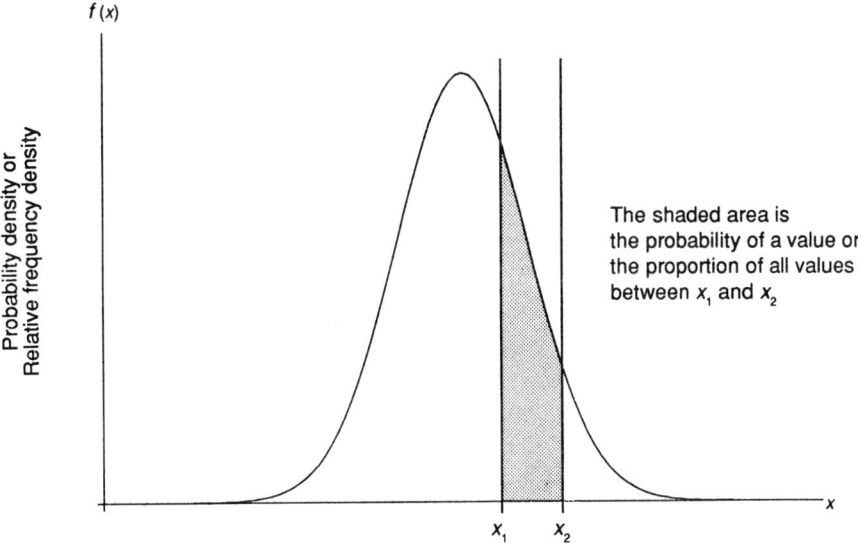

Figure 3.17 Probability density or relative frequency curve

in Figure 3.16 and by the probability density (or relative frequency) curve in Figure 3.17. The probability density curve differs from the frequency density curve in that the total area under the curve is defined as one unit.

The frequency density and the probability density curves are theoretical distributions. They can have many shapes in manufacturing and other engineering situations, but the most important is the bell-shaped type of distribution that is illustrated in Figures 3.16 and 3.17. Chapter 5 is devoted to considering this type of distribution in detail.

■ 3.12 Mathematical distributions

All lines can be described by a mathematical formula as well as pictorially. The straight line in Figure 3.18 is described by the formula

$$f(x) = cx$$

where c is a constant that determines the slope of the line.
The exponential curve in Figure 3.18 is described by the formula

$$f(x) = \lambda \exp[-\lambda x]$$

where λ is a constant and $\exp[\ldots]$ is a way of writing $e(\ldots)$ when e is a constant that is called the base of natural logarithms and is approximately equal to 2.7183. The symbol λ is the Greek lower case letter 'lambda'. The complete Greek alphabet is given in Appendix A.

Theoretical distributions have lines or curves that can be described in a similar way (Section 4.14). Chapters 5 and 6 consider the principal distributions and their formulae which are the basis of industrial statistical methods.

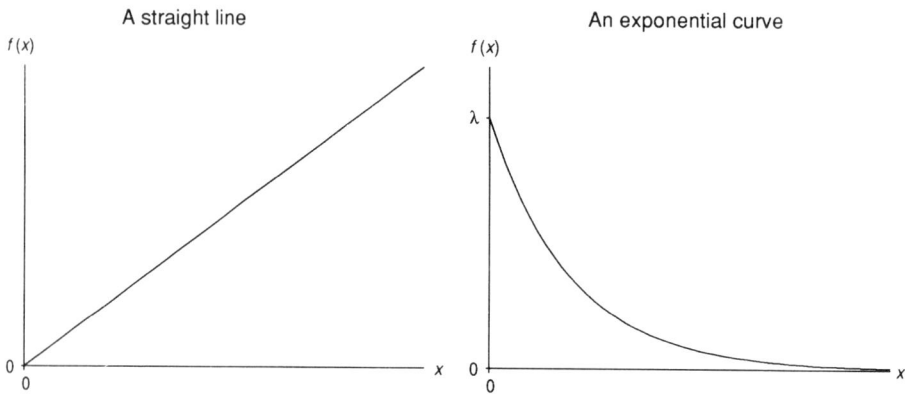

Figure 3.18 Lines and curves

■ 3.13 Summary measures of the 'middle' of a distribution

Although diagrams indicate the overall shape of a distribution, it is often necessary to describe the distribution in mathematical terms, especially if comparisons are to be made with other distributions. Even if the distribution consists of only a few values, it may be necessary still to summarize them by one or two special numbers.

One summary that is almost always required is an estimate of the 'middle' of the distribution. Sometimes, this summary is called a measure of the *location* or *central tendency* of a distribution. For example, it would help to answer questions such as these:

- For about how many miles will a particular car tyre stay within the law?
- What is the average annual cost of servicing a television set?
- What salary can be expected by a 28-year-old engineer in manufacturing industry?

These questions lead to other questions. For example:

- Does one make of tyre last longer than another?
- Is it worth paying for the retailer's insurance package?
- Do engineers earn more or less than people in other professions?

Significance tests (Sections 11.7 to 11.10) are used to answer such questions and they require estimates of the middle of sets of data.

Three commonly used ways of estimating the middle of a distribution are the *mode*, the *median* and the *mean*; they are described below through the following example.

> Nine similar cars have been checked for faults and the number of faults found per car was 2, 5, 3, 6, 4, 3, 8, 5 and 3. How many faults might be expected in other similar cars?

Mode

The mode is defined as *the value with the maximum frequency*: in other words, the value which occurs most often. It does not have a standard designation but $\hat{x}$ is commonly used.

> In the example, there are three 3s, two 5s and one of each of the other four numbers (4, 5, 6, 8); therefore the mode is $\hat{x} = 3$.

Median

The median is defined as *the middle value when the data is arranged in order of magnitude*. It is denoted by $\tilde{x}$.

> Rearranging the numbers, in order of magnitude, gives 2, 3, 3, 3, 4, 5, 5, 6 and 8. The middle number is 4; thus the median is $\tilde{x} = 4$.

Mean

The most common measure is the mean, which is defined as the *arithmetic average* and is denoted by $\bar{x}$. It is found by summing all the values and dividing by their number. For example:

$$\bar{x} = \frac{2 + 5 + 3 + 6 + 4 + 3 + 8 + 5 + 3}{9} = \frac{39}{9}$$

$$= 4.33 \text{ (to two decimal places).}$$

The mean is the value that most people would quote when asked for such things as the average number of miles they drive each week or the average amount they spend on food.

■ 3.14. Calculations of mean, mode and median

Ungrouped data

The mathematical general description of data similar to the example in the previous section, which consists of just a few values, is *ungrouped data* and is written as $x_1, x_2, x_3, \ldots x_n$. The *mean* of these values is expressed as

$$\bar{x} = \frac{x_1 + x_2 + x_3 + \cdots x_n}{n} = \frac{\Sigma x}{n}$$

where the symbol Σ is the Greek capital letter 'sigma' and denotes 'sum of'. Other special uses of Greek and English letters are given in Appendix A.

The *median* is the middle value when all values are arranged in order of magnitude. If there is an even number of values, there will not be a middle value. In this case the median is the mean of the middle two values. For example, given 2, 4, 5, 6, 9, 11, 35 and 99: $\tilde{x} = (6 + 9)/2 = 7.5$.

In this example, the mode cannot be defined because none of the values is repeated.

Multiple frequencies

When there are a large number of observations, it is very time consuming to treat them as ungrouped data. For ease in calculations, they can be dealt with as *grouped data* as described below; or, if there are just a few different observations each of which occurs a number of times, they can be treated as *multiple frequencies*. In other words, if x_1 occurs f_1 times, x_2 occurs f_2 times and so on, then the mean is expressed as

$$\bar{x} = \frac{f_1 x_1 + f_2 x_2 + f_3 x_3 + \cdots}{f_1 + f_2 + f_3 + \cdots} = \frac{\Sigma\, fx}{\Sigma\, f}$$

Notice that this expression uses the same principle as for ungrouped data: values are summed and then divided by their number.

For example, the number of cars checked for faults has increased to 25 from the 9 in the previous section. The results are now presented in a table of multiple frequencies where the number of faults per vehicle is denoted by x (the observation) and the number of cars having x by f (the frequency).

x	f	fx	F	
0	1	0	1	$F_0 = \qquad f_0 = 1$
1	0	0	1	$F_1 = F_0 + f_1 = 1 + 0 = 1$
2	3	6	4	$F_2 = F_1 + f_2 = 1 + 3 = 4$
3	4	12	8	$F_3 = F_2 + f_3 = 4 + 4 = 8$
4	6	24	14	$F_4 = F_3 + f_4 = 8 + 6 = 14$
5	5	25		
6	3	18		
7	3	21		
Total	25	106		$\bar{x} = \dfrac{\Sigma\, fx}{\Sigma\, f} = \dfrac{106}{25} = 4.24$

The *mean*, as calculated above, is 4.24 but, being sensible about the number of decimal places, it is probably better quoted as 4.2. Note that this mean value cannot actually occur since the data is discrete. However, that does not matter; it is a theoretical value which is used only to indicate the middle of the distribution.

Because 4 occurs most often (6 times), the *mode* is 4.

So far as the *median* is concerned, the data could be arranged in order of magnitude as below. The median would then be the middle one, in this case the 13th value which is 4.

$$\Downarrow$$
$$0,2,2,2,3,3,3,3,4,4,4,4,4,4,5,5,5,5,5,6,6,6,7,7,7$$

A quicker method of finding the median is to use the cumulative frequency (F) discussed in Section 3.10. In this case, F is the frequency less than or equal to x. F is calculated by summing f until the midpoint is passed.

Strictly, the midpoint is passed at $(n + 1)/2$ where n = total frequency $= \Sigma f$ but $n/2$ is often used. For the data in the table above, $n/2 = 25/2 = 12.5$ and $F = 14$ is the first value greater than this. Reading across to the corresponding observation gives $x = 4$ as the median.

Grouped data

Calculation of the mean for grouped data is illustrated below, using Table 3.3. The x values are taken as the midpoints of each class. The mean is then found in the same way as for multiple frequencies, using $\bar{x} = \Sigma fx / \Sigma f$.

Classes	x	f	fx
30–34	32	2	64
35–39	37	2	74
40–44	42	4	168
45–49	47	8	376
50–54	52	6	312
55–59	57	11	627
60–64	62	8	496
65–69	67	5	335
70–74	72	1	72
75–79	77	2	154
80–84	82	1	82
Total		50	2760

$$\bar{x} = \frac{\Sigma fx}{\Sigma f} = \frac{2760}{50} = 55.20$$

Rounded, the mean is 55.2

The graphical methods in the next section can be used to find the median and the mode. These graphical methods do not give very accurate values, but they are good enough for most practical purposes. The methods already described in this section give more precise values. The empirical relationship between mode, median and mean which is described in Section 3.16 also can be used to determine values in some situations.

■ 3.15 Determination of median and mode from diagrams

Median from ogive

In order to find the median it is necessary to determine where the middle value lies. For grouped values, a graphical method using the ogive usually gives sufficient accuracy.

The ogive in Figure 3.11 is reproduced in Figure 3.19. Since there are 50 observations in this distribution, a line drawn at the 25th as in Figure 3.19 gives the median as 55.9.

Mode from histogram

The histogram can be used to find a value for the mode of grouped data. The method is illustrated using the histogram in Figure 3.9, which is reproduced in Figure 3.20.

The rectangle with the largest frequency is selected and lines are drawn between *AB* and *CD*, as shown in Figure 3.20. These lines intersect at the mode, which is 57.6 in this case.

■ 3.16 An empirical relationship

It has been found, for most practical purposes in situations where a distribution is skew and has one mode, that the formula

mode = (3 × median) − (2 × mean)

gives a sufficiently accurate value for the mode when mean and median have been determined from a large number of measurements. It must be

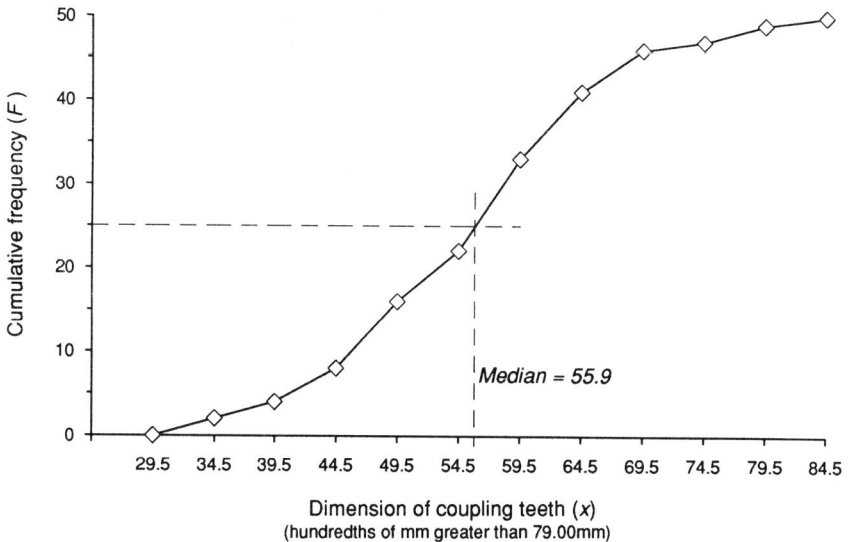

Figure 3.19 Median found from an ogive

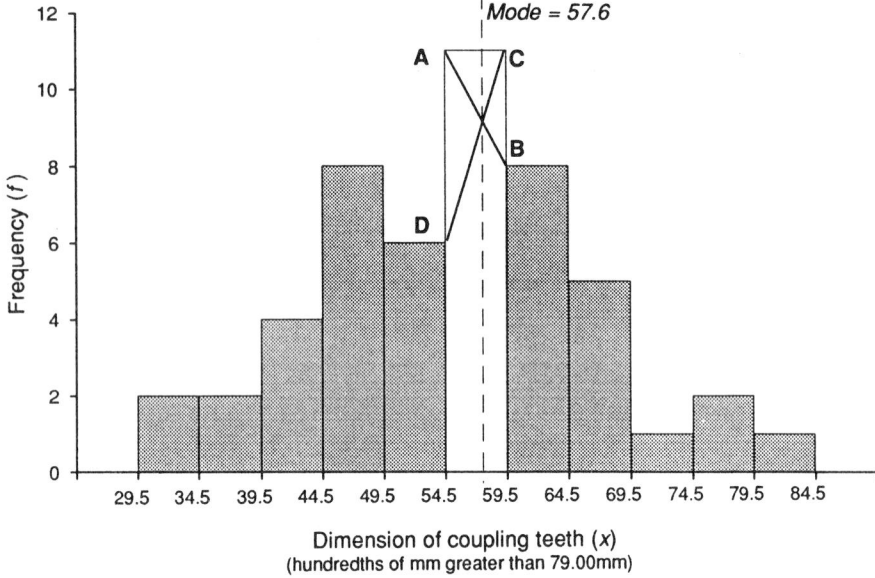

Figure 3.20 Mode found from a histogram

emphasized that this relationship is empirical: in other words, there is no mathematical reason why it should be true.

■ 3.17 Comparison of mean, mode and median

The mean, mode and median are illustrated for a symmetrical and a non-symmetrical distribution in Figure 3.21.

When the distribution is symmetrical, all three occur at what is obviously the middle of the distribution. The effect of a 'tail' on a distribution is to pull the median away from the mode and the mean even further. Note that, in both situations, the median has 50% of the distribution (as indicated by 50% of the area) on each side of its value. Also, the mean is always the value of x at the centre of gravity of a uniform solid with the shape of the distribution.

The advantages and disadvantages of each summary value are listed below.

Mode

Advantages:

■ Easy to calculate.
■ Not affected by extreme values.

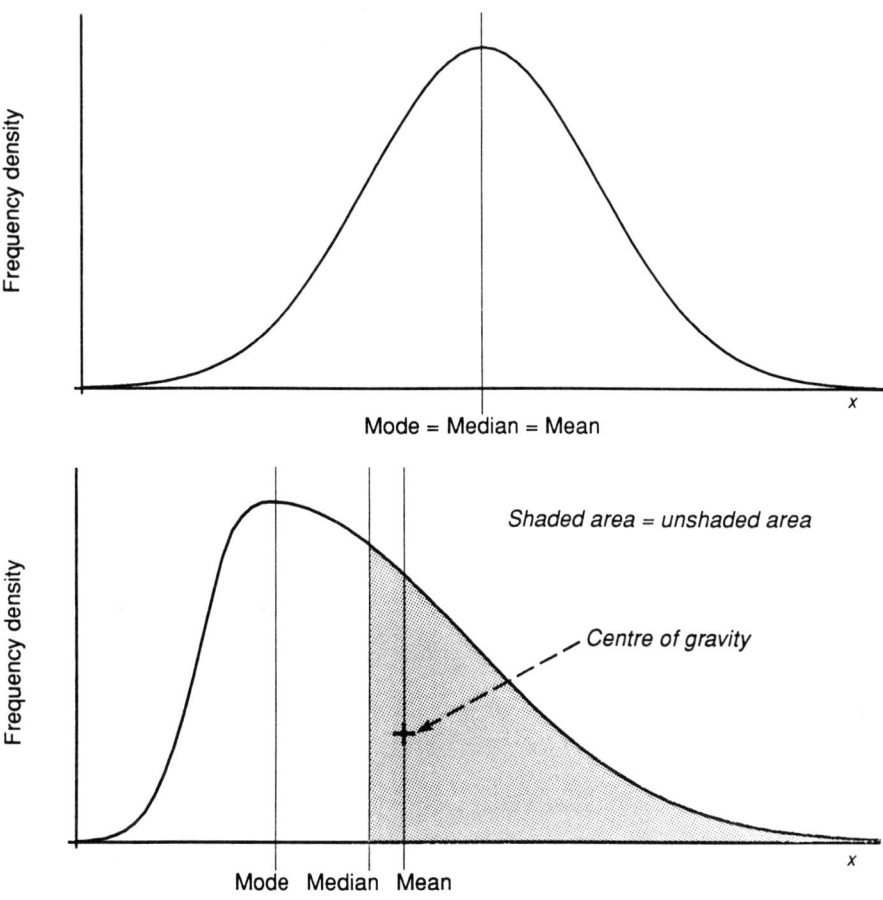

Figure 3.21 Illustration of mean, median and mode

Disadvantages:

- May be well away from the 'middle'.
- Difficult to handle in mathematical equations.
- Distribution could have more than one mode.
- Does not use all the available data.

Median

Advantages:

- Easy to determine.
- Not affected by extreme values.

43

- Can seem logically correct since 50% above and 50% below.
- Only requires 50% of the distribution in order to calculate it.

Disadvantages:

- Can be misleading in a distribution with a long tail.
- Difficult to handle in mathematical equations.
- Does not use all the available data.

Mean

Advantages:

- Easily understandable and in common use.
- Uses all values in the distribution.
- Easy to handle in mathematical equations.

Disadvantages:

- May be affected by extreme values.
- Need to know all values of the distribution.
- May not be representative of the data in that all values may be well away from the mean.

Although the mean is used extensively in statistical work, there are times when the mode or median is preferred. The mode is best thought of as the value occurring most often, rather than as an estimate of the middle of the distribution. For example:

- The record of accidents per day occurring at a road junction is summarized as

Number of accidents per day	0	1	2	3	4
on number of days	80	53	32	4	1

  It might might be of comfort to some that the most likely case is that of no accidents.
- People who design improved products usually follow the indications of potential customer surveys. In practice, this amounts to following the mode.

The median tends to be used in salary negotiations, rather than the mean, probably because it seems logically correct to quote a value which has 50% of salaries above and below it. Also, the median is used in manual charting for process control (Section 10.6), mainly because it is easily calculated and understood by anyone given the job of plotting results on a control chart.

■ 3.18 Samples and populations

One expression used widely in statistics is *a sample of size n is taken from a population*. What does this mean?

The expression *sample of size n is taken* means that *n* values have been obtained. The *population* is the basic distribution from which they are taken. In Figure 3.22, five values x_1, x_2, x_3, x_4 and x_5 are shown. These constitute a sample of size 5.

If the sample is to be truly representative of the population, then the values need to be chosen at *random* and hence the expression *random sample* is often used. Most people assume that samples are representative and therefore chosen at random, unless it is stated to the contrary. Because of this, much thought needs to be given to the sampling procedure, to make sure that the values are as random as possible before using statistics derived from them.

> A worker taking components from a tray will invariably end up with a higher proportion of defectives than exist in the whole tray. He or she will have chosen them unconsciously.

However, in taking, say, five consecutive components from a machine, the assumption of randomness is usually correct for the sample as a whole and is therefore representative of production at that moment in time.

Unless otherwise stated in this book, sample values are assumed to be *independent*. That is to say, the population is considered to be very large so that the distribution does not change as readings are taken. This is different from taking cards from a pack of cards and not replacing the chosen card; in this case, the distribution changes as each card is taken and therefore the value of the second card depends to some extent on the value of the first one, the third depends on the first two, and so on.

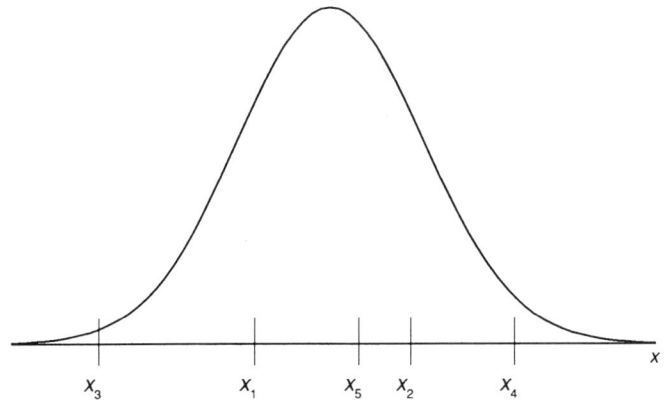

Figure 3.22 A sample and a population

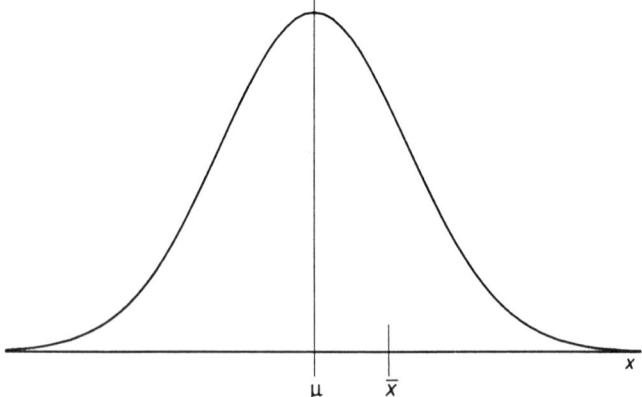

Figure 3.23 Sample and population means

The assumption of a large population is reasonable in most manufacturing situations; 5 cars can be assumed to constitute a sample from the production during a shift and so on.

The *population mean* is denoted by μ and the *sample mean* is denoted by $\bar{x}$. Usually these will differ, as shown for example in Figure 3.23.

In most cases the population mean (μ) is the value that is required but frequently it is unknown and the sample mean ($\bar{x}$) is used as a best estimate. In general, a numerical characteristic of a population is called a *parameter* and is denoted by a Greek letter. A numerical value which describes a sample is called a *statistic* and is denoted by an English letter.

> For example, if a random sample of size 5 is taken from the flange separations set out in Figure 3.12, and suppose that the sample mean, $\bar{x}$ turns out to be 15.0053 inches, then that is the best estimate of the population mean (which is the true mean) until other readings are taken. It might be that the population mean, μ, is 15.0058, but there is no way of knowing that. Of course, the larger the sample, the more accurate should be the estimate.

It is important that $\bar{x}$ is recognized as an *estimate* of μ. For most of this chapter, it is assumed that calculations are made on samples, and so $\bar{x}$ is used to denote the mean.

■ 3.19 Measures of the spread of a distribution

The mode, median or mean is used to give a measure of the middle of a distribution. Just as important and, for process and quality control often more important, is a measure of the spread of a distribution.

Quoting the price range of a weekend bargain break for two as £45 to £55 gives a very different picture from quoting £30 to £70, but the average or middle price is the same.

Usually, if the setting of a machine is wrong, it can be adjusted. However, the spread which indicates the variation of values is inherent in the machine and cannot be changed merely by turning a knob. Most of the effort required for continuous improvement is in identifying the causes of and reducing this variation.

This variation needs to be assessed and the more important measures are described below.

Range

Range is illustrated in Figure 3.24 and is defined as

Maximum value – Minimum value.

For the values 3, 2, 7, 6 and 11, the range is $11 - 2 = 9$.

The range is easily calculated and is widely used as a measure of variation. However, as a single estimate of the spread of a large distribution, it is unsatisfactory. For example:

Imagine the distribution of the market value of cars owned by people living in a single street. Perhaps there are 150 cars, varying in value from £300 to £25 000, giving a range of £25 000 – £300 = £24 700. But suppose that one of the residents is left a large sum of money and buys a luxury car worth £72 000. Look at the effect on the range. It is now £72 000 – £300 = £71 700, nearly three times the previous figure.

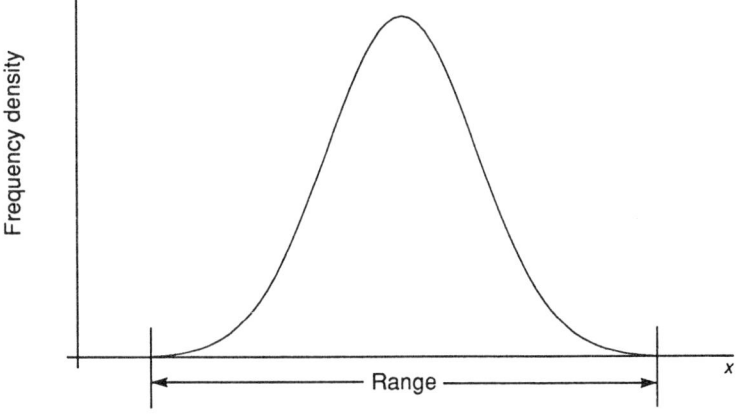

Figure 3.24 The range of a distribution

47

A measure which is so influenced by one value is unreliable for many applications. It must be stressed that range can be acceptable under the conditions imposed for process or quality control charting (Chapter 10). In general, however, something is required which takes all values in the distribution into account.

There are a number of such measures; those that are useful to industry are described in the following paragraphs.

Population variance (σ^2)

If the individual values in a whole population are known – in other words, the complete distribution is known – the *population variance* is designated σ^2 and is defined as *the mean square difference of the values from the population mean*.

This is illustrated in Figure 3.25 for the four values x_1, x_2, x_3 and x_4.

A value y in Figure 3.25 is the difference between x and the population mean, μ. From the definition, the variance is the mean of the squared y values, which is expressed mathematically as

$$\frac{y_1^2 + y_2^2 + y_3^2 + y_4^2}{4}$$

or, in terms of the original four xs, since $y = x - \mu$,

$$\sigma^2 = \frac{\Sigma(x - \mu)^2}{4}$$

and more generally, for N values

$$\sigma^2 = \frac{\Sigma(x - \mu)^2}{N}.$$

Clearly, the wider the spread of the data, the larger the values of y and hence σ^2. Therefore, it is a measure of the spread of the data.

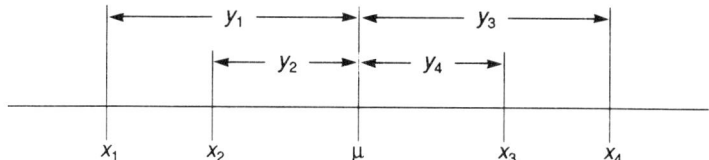

Figure 3.25 Illustration of population variance

Population standard deviation (σ)

This is simply *the square root of the variance.*

Sample variance (s^2)

The formulae used for calculation of population parameters are based on theoretical distributions, but most calculations in industry are based on samples. The formulae used to obtain statistics from samples are slightly different from those used for populations.

The *sample variance* is

$$s^2 = \frac{\Sigma(x - \bar{x})^2}{n - 1}$$

where $\bar{x}$ is the sample mean and there are n values in the sample. Note that sample variance is used as an estimator of σ^2. If sample variance is defined with a divisor n instead of $n - 1$, the estimate of σ^2 will, on average, tend to be too small by an amount called the *bias*. Dividing by $n - 1$ instead of n makes s^2 an *unbiased* estimator of σ^2: that is to say, although s^2 will sometimes be greater than the true σ^2 and sometimes less, it will be correct on the average. For practical purposes, the difference between $n - 1$ and n as divisor is negligible for large sample sizes (above about 30).

In practice, calculations are made easier by using a formula obtained by expanding $(x - \bar{x})^2$, which gives

$$s^2 = \frac{\Sigma x^2 - n\bar{x}^2}{n - 1}.$$

Sample standard deviation (s)

Just as population standard deviation is the square root of population variance, so sample standard deviation is the square root of sample variance.

The advantage of using standard deviation rather than variance is that the measuring units will be the same as the original data and the mean (the variance has units squared, since it is based on y^2).

Although it is of little help to try to picture exactly what a numerical value of standard deviation represents, it is useful to note that, if the standard deviation is doubled, then the spread of the data is doubled; and if it is halved, the spread is halved. Also, in the normal distribution (Chapter 5) which often occurs in practice, the range of observed data is usually about six standard deviations.

Just as the sample mean, $\bar{x}$, is an estimate of the population mean μ, so the sample variance s^2 is an estimate of the population variance σ^2.

In the remainder of this chapter $\bar{x}$ and s^2 are used. In other words, the assumption is made that all data arises as a sample, which is the usual case in manufacturing.

■ 3.20 Calculations of variance and standard deviation

Ungrouped data

When the standard deviation is required for a small sample, the calculation is that for ungrouped data.

> The calculation is illustrated by the following example, where individual cars in a sample of four had 3, 7, 4 and 5 faults respectively.
>
> | x | x^2 | | |
> |-----|-------|--|--|
> | 3 | 9 | $\bar{x} = \Sigma x/n$ | $= 19/4 = 4.75$ |
> | 7 | 49 | variance | |
> | 4 | 16 | $s^2 = (\Sigma x^2 - n\bar{x}^2)/(n-1)$ | $= (99 - 4[4.75]^2)/3$ |
> | 5 | 25 | | $= 2.917$ |
> | 19 | 99 | standard deviation | |
> | | | $s = \sqrt{2.917}$ | $= 1.71$ |

Most scientific calculators have keys which give the mean and standard deviation at the press of a key once the original data has been fed into the calculator. The relevant keys are often marked $\bar{x}$ for the mean and σ_{n-1} for the sample standard deviation ($\sigma_{n-1} = s$).

Multiple frequencies

When there are a number of values for each x, the formulae need modifying so that calculations are simplified. Again suppose that x occurs f times. The formula that defines sample variance in the previous section is replaced by

$$s^2 = \frac{\Sigma f(x - \bar{x})^2}{n - 1}$$

and the more useful formula for calculations, also in the previous section, is replaced by

$$s^2 = \frac{\Sigma fx^2 - n\bar{x}^2}{n - 1}.$$

In both the above two formulae, $n = \Sigma f$.

The division by $(n - 1)$ rather than by n is not so important for larger samples because it makes very little difference to the answer. Sometimes the divisor n is used, but in the following examples the more correct divisor $(n - 1)$ is retained.

In Section 3.14, the mean was calculated for data of the number of faults in 25 cars. The same data is used below to illustrate calculation of standard deviation for the same data.

Compared with the calculation of the mean, this calculation requires an extra column, fx^2 which is simply calculated by multiplying fx by x.

x	f	fx	fx^2
0	1	0	0
1	0	0	0
2	3	6	12
3	4	12	36
4	6	24	96
5	5	25	125
6	3	18	108
7	3	21	147
	25	106	524

$\bar{x} = \Sigma\, fx / \Sigma\, f$ $= 106/25 = 4.24$

variance
$s^2 = (\Sigma fx^2 - n\bar{x}^2)/(n - 1)$ $= (524 - 25\,[4.24]^2)/24$
$= 3.107$

standard deviation
$s = \sqrt{3.107}$ $= 1.76$

Grouped frequencies

The formula and method used for grouped frequencies is the same as that for multiple frequencies, but, as in the calculation of the grouped data mean (Section 3.14), x is taken as the middle of the class.

The illustration below follows from the example in Section 3.14.

Classes	x	f	fx	fx^2
30–34	32	2	64	2048
35–39	37	2	74	2738
40–44	42	4	168	7056
45–49	47	8	376	17672
50–54	52	6	312	16224
55–59	57	11	627	35739
60–64	62	8	496	30752
65–69	67	5	335	22445
70–74	72	1	72	5184
75–79	77	2	154	11858
80–84	82	1	82	6724
		50	2760	158440

$\bar{x} = \Sigma\, fx / \Sigma\, f$ $= 2760/50$
$= 55.20$

variance
$s^2 = (\Sigma fx^2 - n\bar{x}^2)/$ $= (158\,440 - 50$
$(n - 1)$ $[55.2]^2)/49$
$= 124.2449$

standard deviation
$s = \sqrt{146.735}$ $= 11.15$

Many scientific calculators can be used to obtain the mean and standard deviation of this type of data, provided x and f values are entered correctly.

■ 3.21 Other measures of the shape of a distribution

Measures of middle and spread together provide a summary of a distribution which will be adequate for most purposes. However, there are situations which require other measures to be considered, in particular when testing for outliers (Section 11.12). The features which need to be considered are as follows:

■ Departure from symmetry, which is called *skewness*.
■ Whether the distribution is peaked or flat-topped, which is called *kurtosis* (a Greek word meaning bulging or convexity).

Many different names and symbols have been and are used for measures of skewness and kurtosis. The descriptions below are sufficient, but they refer only to the coefficients C_s and C_k. Their importance lies in the fact that, if two distributions have almost equal values of each of the four measures $\bar{x}$, s, C_s and C_k, then they will be very similar in overall dimensions and shape.

Skewness

The most simple measures of skewness are based on measures of middle and spread. That is

$$\text{skewness} = \frac{\text{mean} - \text{mode}}{\text{standard deviation}} \text{ or } \frac{3(\text{mean} - \text{median})}{\text{standard deviation}}.$$

The value of both these measures will have the following characteristics:

■ It will be zero for a symmetric distribution.
■ It will be positive for a distribution with a longer tail on the right.
■ It will be negative for a distribution with a longer tail on the left.

These three cases are illustrated in Figure 3.26. Larger numerical values indicate greater skewness and hence less symmetry in the distribution.

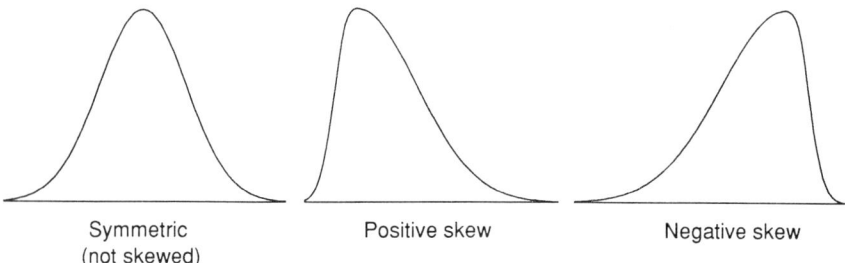

Symmetric (not skewed)   Positive skew   Negative skew

Figure 3.26 Skewness

When a detailed analysis of skewness is required (Section 11.12) a more complicated measure is used. Several possibilities have been proposed, but the most widely used is the *coefficient of skewness*, C_s. For ungrouped data, where n is the sample size and $\bar{x}$ is the sample mean, this is defined as

$$C_s = \frac{\sqrt{n}\, \Sigma(x - \bar{x})^3}{[\Sigma(x - \bar{x})^2]^{3/2}}$$

The term in square brackets in the denominator is related to the sample variance s^2. Hence if s^2 is already known, this term can be evaluated from the expression

$$[\Sigma(x - x)^2] = (n - 1)s^2.$$

Raising this to the power 3/2 is equivalent to finding its square root and then cubing the result. The reason for this slightly complicated calculation is that it makes C_s a 'dimensionless number', which is a number that is independent of the units in which x is measured.

Kurtosis

Kurtosis is measured by a similar dimensionless number which is called the *coefficient of kurtosis*, C_k. This is always positive and is defined as

$$C_k = \frac{n\, \Sigma(x - \bar{x})^4}{[\Sigma(x - \bar{x})^2]^2}$$

■ A flat-topped distribution will tend to have a low value of C_k and is often said to be *platykurtic* (literally 'flat bulging', as in Figure 3.27).
■ A sharp-peaked distribution will tend to have a high value of C_k and is often said to be *leptokurtic* (literally 'thin bulging', as in Figure 3.27).

A high C_k does not necessarily imply a sharp peak, nor does a low C_k imply a flat top, because curves with different outlines may have the same C_k. In fact, C_k is influenced more by the shape of the tails than values near the middle of

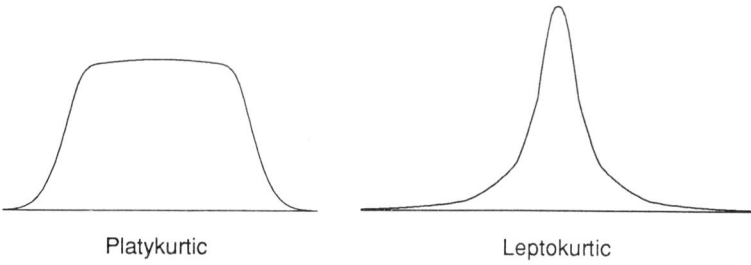

Platykurtic   Leptokurtic

Figure 3.27   Kurtosis

the distribution. When the tails end abruptly as if the distribution has 'shoulders' then C_k is low, whereas long-tailed distributions result in a large C_k.

A practical engineering use of the idea of kurtosis arises in monitoring the condition of bearings. Vibration in bearings is measured by a kurtosis meter which displays more peaked results (in other words, kurtosis increases) as wear increases.

■ References to further reading on the topics of Chapter 3

Chatfield (1983) and Oakland (1986) give simple introductions to the material covered in this chapter. Walpole and Myers (1993) give a slightly more mathematical treatment.

See Bibliography for details of titles and publishers.

■ Self-assessment questions on Chapter 3

1. The number of breakages per day in a laboratory were recorded for 500 days as follows.

Number	0	1	2	3	4	5	6	7	8
Frequency	98	122	108	82	57	26	5	0	2

   Represent the distribution by a histogram and calculate the sample mean and sample variance.
2. Find the mode, median and standard deviation for the data in question 1.
3. (a) The number of faulty components in 5 consecutive batches received from a supplier were 4, 3, 7, 3 and 6. Find the mean, median and standard deviation of these values.
   (b) Another 3 batches contained 3, 8 and 2 faulty components. Find the mean, median, mode and standard deviation for all 8 values.
4. A chemical test was carried out on 12 pieces of steel to find traces of cobalt. The percentage of cobalt found in each was 0.072, 0.069, 0.070, 0.072, 0.076, 0.075, 0.065, 0.068, 0.071, 0.069, 0.073 and 0.072. Find the mean, range and standard deviation for this sample.

5. The following data results from an analysis of access times (in micro-seconds) to a computer disc system during the runs of a particular computer programme.

Access time	20–24	25–29	30–34	35–39	40–44	45–49	50–54	55–59	60–64	65–69
Frequency	11	17	19	18	23	28	16	10	6	2

Find the mean and the standard deviation for this sample.
6. Draw a histogram for the data in question 5 and find the mode.
7. Draw an ogive for the distribution in question 5 and find the median.
8. Fifty employees in a factory were timed for how long it took each one to assemble a new component. Times taken, rounded to the nearest minute were as follows.

25	27	35	24	41	26	31	20	28	26
35	45	38	29	42	30	29	36	22	32
34	46	27	49	21	43	51	27	39	28
30	35	40	30	37	29	40	29	36	31
27	31	23	33	29	28	28	26	29	34

(a) Construct a frequency distribution table from this data using 8 classes of equal width.
(b) Estimate the mean and standard deviation of this sample using the grouped frequency formulae on the frequency table.
(c) Find the exact sample mean and standard deviation using the appropriate keys on a calculator.

4 | Probability and Probability Distributions

■ 4.1 Introduction

The words *probability* and *chance* are used often by the media and in everyday conversations. Before the FA Cup Final, reporters write about the probability of a particular team winning rather than a statement that they believe the team will win. Weather reports give a 'chance of rain', whereas not so long ago it was either 'going to rain' or it 'would remain dry'. But what exactly is the meaning of these references to probability or chance?

In a manufacturing process, if 2% of components are defective in a long run, it can be said that the proportion of defectives being produced is 0.02. If one component is selected at random from the run, it can be said that the probability of it being defective is 0.02. Also, it can be said that there is a 1 in 50 chance of selecting a defective component. In other words, *the proportion in a long run is equivalent to the probability or chance in a single selection.*

This will be true only if the process conditions are the same when each selection is made. If a bent coin is tossed many times and overall gives 70% 'heads', the probability of a head in a single toss of that coin is 0.7. In manufacturing, it is assumed that process conditions remain approximately the same, at least over short periods. It is to check this assumption and to detect change that so much effort is put into statistical monitoring of process parameters, product conformance and quality. The information obtained offers objective probabilities, which are based on an association of data from successive events with data from previous events.

There are occasions when events cannot be associated with previous events because the conditions are not sufficiently similar. For examples: the probability that a new model of car will be an outstanding success; the probability that red will be the preferred fashion colour next year; the probability that a new invention will revolutionize air transport sometime in the next three years; the probability that the reader is involved in an accident during the next 12 months. The examples are all of events which are one-off situations in this ever-changing world, and there may be some argument about the validity of any numerical values attached to the term 'probability'. In these cases, probabilities have to be thought of as the *degree of belief* in the event or the *confidence* that the event will occur. They will differ from person to person and therefore are called *subjective probabilities*.

Throughout this chapter, the examples that explain and illustrate probability are based on 'selections' which are made at random.

■ 4.2 The ratio rule

The ratio rule is explained and its use is illustrated through the examples below. It applies to events when *all possible outcomes are equally likely* and is expressed mathematically by

$$P(E) = \frac{m}{n}$$

where $P(E)$ is the probability the event occurs,

> m is the number of different outcomes which result in the event occurring,
> n is the total number of different outcomes.

A large batch of semi-conductor devices in several boxes contains 10% faulty units. Using the description introduced in the previous section, if one device is chosen from the whole batch, the probability that it is defective is 0.1. If the device is chosen from only one box, the probability that it is defective is calculated using the ratio rule.

Figure 4.1 represents a box containing 8 devices where G identifies a good unit and D a defective unit. Each of the devices is equally likely to be the first taken out of the box.

There are 8 possible choices including 2 defectives, i.e., $n = 8$ and $m = 2$ when defective probability is being considered. The ratio rule is used to calculate the probability that the first taken is defective:

$$P(\text{defective unit}) = \frac{2}{8} = 0.25.$$

Also, because there are 6 good units out of 8, $m = 6$, $n = 8$ and

$$P(\text{good unit}) = \frac{6}{8} = 0.75.$$

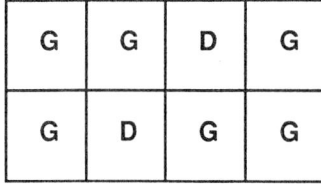

Figure 4.1   Box of semi-conductor devices

Similarly:

- The probability of picking an ace when a card is chosen from a pack is 4/52.
- The probability of obtaining a six when a 6-sided cubical die is thrown is 1/6.

4.3 Range of probabilities

If a device is chosen from a batch in which all are defective, $m = 0$ and

$$P(\text{good unit}) = \frac{0}{n} = 0.$$

On the other hand, if there are no defectives, $m = n$ and

$$P(\text{good unit}) = \frac{n}{n} = 1$$

The examples above illustrate that it is impossible to have probabilities outside the values 0 to 1. These extremes are expressed mathematically by

$$0 \leqslant P(E) \leqslant 1.$$

At the lower end, $P(E) = 0$ is the situation where the event is impossible. $P(E) = 1$ is the situation where the event is certain.

At the upper end, sometimes probabilities are close to 0 or 1. The probability that the reader is in a car accident next week is (hopefully) nearly 0; the probability that his or her car starts tomorrow morning is (again hopefully) nearly 1.

4.4. Special addition rule

Addition rules are used to calculate probabilities when outcomes are associated with more than one event.

Table 4.1 Sales of cars

Price (£000s)	Price group			Total
	1	2	3	
	5–10	10–20	>20	
Male buyers	16	23	12	51
Female buyers	22	8	2	32
Total buyers	38	31	14	83

The *special addition rule* applies when the events cannot occur together. In this situation the events are said to be *mutually exclusive*. The special addition rule is illustrated for two mutually exclusive events (*a* and *b*) in the example below and is expressed mathematically as

$P(a \text{ or } b) = P(a) + P(b)$.

Table 4.1 summarizes car sales from a showroom by vehicle price and sex of customer. Suppose that a decision is made to contact a single customer to find his or her reaction to the sales procedure and the performance of the car purchased. The customer's response might be influenced by the type of vehicle that was bought. If that customer is chosen at random from the 83, it could be useful to know the combined probability of the customer buying from price group 1 or from price group 2.

One way of calculating this is to use the special addition rule because, in this case, a car cannot be in both group 1 and group 2. The probability that the car was in group 1 is 38/83 (38 of the 83 cars sold were in group 1) and the probability that the car was in group 2 is 31/83. Therefore

$P(\text{group 1 or group 2}) = P(\text{group 1}) + P(\text{group 2})$
$= 38/83 + 31/83 = 69/83$.

An alternative way of calculating, in this case, is to treat groups 1 and 2 as a single group and use the more simple ratio rule for which, $m = 69$, $n = 83$ and hence

$P(E) = m/n = 69/83 = 0.831$ (to 3 decimal places).

■ 4.5 General addition rule

However, a different approach is needed in a slightly different situation where it would be useful to find the combined probability that the customer is male or the car is from group 1. The requirement looks similar to that in the previous section, but the special addition rule is not appropriate, as illustrated by the example below.

P(owner is male or car was from group 1)
= *P*(owner is male) + *P*(car was from group 1)
= 51/83   +   38/83
= 89/83 = 1.07.

Something is wrong! From the range of probabilities, probability cannot be greater than 1.

If the ratio rule is used, m is the number of sales in the category of men owners or group 1 cars = 12 + 23 + 16 + 22 = 73 and n = 83.
Hence $P(E) = m/n = 73/83 = 0.880$, which is the correct answer.

Table 4.2 illustrates why the first answer was wrong. The 16 men who bought cars in group 1 have been included twice, once in group 1 and again as male buyers. The special addition rule can be used only as stated: that is, if the two events cannot both occur together. In this example they do occur together.

The *general addition rule* is not so limited; it covers all situations and should be used in this case as shown below. The mathematical expression of the general addition rule for two events (a and b) is

$$P(a \text{ or } b) = P(a) + P(b) - P(ab)$$

where $P(ab)$ is the probability that a and b both occur.

■   In the case when it is impossible for a and b to occur together, $P(ab) = 0$ and the general addition rule becomes the special addition rule.

Applying the general addition rule to the example where the answer was wrong:

P(owner is male or the car is from group 1)
= *P*(owner is male) + *P*(car is from group 1)
   − *P*(owner is male and car is from group 1)
= 51/83 + 38/83 − 16/83 = 73/83

which is the same answer as that found using the ratio rule.

The two addition rules are pictured in Figure 4.2, where the circles represent the probabilities $P(a)$ and $P(b)$ and their intersection represents $P(ab)$.

Table 4.2   Male buyers and price group 1 cars

	Price group			Total
	1	2	3	
Male buyers	**16**	23	12	51
Female buyers	22			
Total buyers	38			

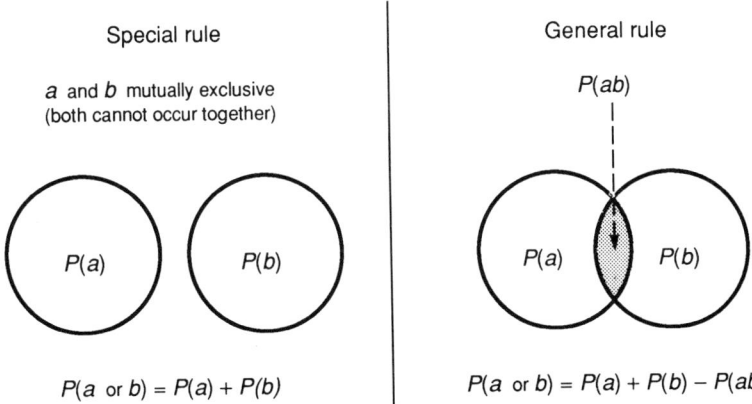

Figure 4.2 The addition rules

Examples of the addition rules

■ If a card is taken at random from a pack of 52 playing cards, what is the probability that it is an ace or a king?

Using the special addition rule:

P(ace or king) $= P$(ace) $+ P$(king)
$$= 4/52 + 4/52$$
$$= 8/52.$$

■ If a card is taken at random from a pack of 52 playing cards, what is the probability that it is an ace or a club?

Using the general addition rule:

P(ace or club) $= P$(ace) $+ P$(club) $- P$(ace of clubs)
$$= 4/52 + 13/52 - 1/52$$
$$= 16/52.$$

■ If, from experience, it is found that one or other of two cash dispensers is not working for some reason on 10% of days, while on 2% of days both are not working, what is the probability that at least one is not working on any one particular day?

Using the general addition rule:
P(cash 1 not working or cash 2 not working)
$= P$(cash 1 not working) $+ P$(cash 2 not working)
$\quad - P$(both not working)
$= 0.1 + 0.1 - 0.02 = 0.18$
i.e. this will occur on 18% of days over a period of time.

■ 4.6 The complement rule

There is certainty in the statement that an event E must either occur or not occur. This certainty is expressed mathematically as

$P(E$ occurs$) + P(E$ does not occur$) = 1$.

A rearrangement of this expression gives the probability of non-occurrence or the *complement rule*:

$P(E$ does not occur$) = 1 - P(E$ occurs$)$.

■ Using data from Table 4.1, P(female owner)
$= 1 - P$(male owner)$ = 1 - 51/83 = 32/83$, which can of course be found directly.
■ In taking a card from a pack, the probability it is not an ace is $1 - P(\text{ace}) = 1 - 4/52 = 48/52$.

This concept is important to description of the binomial distribution (Section 6.3) where the case (E occurs) is called a 'success' and (E does not occur) is called a 'failure'. Since one or other must occur

$P(\text{success}) + P(\text{failure}) = 1$.

■ 4.7 Conditional probability

Conditional probability describes the probability of an event occurring when some other event occurs. It is explained through the examples below and has the mathematical expression

$$P(b \mid a) = \frac{P(ab)}{P(a)}$$

where $P(b \mid a)$ is the probability that b occurs given that a occurs,
$P(a)$ is the probability that a occurs,
$P(ab)$ is the probability that a and b occur.

For example and referring to the data in Table 4.1, suppose that the required probability is that of selecting the buyer of a car in group 1 where the buyer is known to be male.

Let event a be (owner is male)
and event b (car is from group 1):

$$P(\text{group 1} \mid \text{male}) = \frac{P(\text{group 1 and male})}{P(\text{male})} = \frac{16/83}{51/83} = 16/51.$$

In this simple example, the probability could be calculated also by using the ratio rule because it is stated that the buyer is male, only males need be considered and 16 of the 51 men bought a group 1 car. Therefore

$$P(E) = m/n = 16/51.$$

However, it is not always possible to simplify problems in this way.

As illustration, consider the cash dispensers example in Section 4.5. One or the other was not working on 10% of days, and on 2% of days both were not working. Suppose somebody wanted to know the overlap of the two probabilities, i.e. on how many of the '10% days' situation did the '2% days' situation apply. In mathematical terms, the answer is the probability that both are not working given that at least one is not working.

Let event a be (at least one is not working)
and event b (both are not working).
Note in this particular example that if b occurs then a occurs as well, or

$$P(ab) = P(b) = P(\text{both not working}).$$

Also, $P(a)$ was calculated as 0.18 in Section 4.5.
Using the idea of conditional probability:

$P(\text{both not working/at least one not working})$

$$= \frac{P(\text{both not working})}{P(\text{at least one not working}}$$

$$= 0.02/0.18 = 0.111 \text{ to 3 decimal places.}$$

In other words, on 11.1% of days when one is not working, both are not working.

■ 4.8 Multiplication rule (for two dependent events)

The multiplication rule is used to calculate probabilities associated with events that occur together. For two events where one depends upon the other, its mathematical model is

$$P(ab) = P(a).P(b \mid a).$$

This model is a rearrangement of the conditional probability expression. Both sides of the expression have been multiplied by $P(a)$ and the symbols $P(a)$, $P(ab)$ and $P(b \mid a)$ retain the same meaning.

A use of the model is illustrated in Figure 4.3, which at the start shows 8 devices, two of which are defective. If two are taken, what is the probability that both are good?

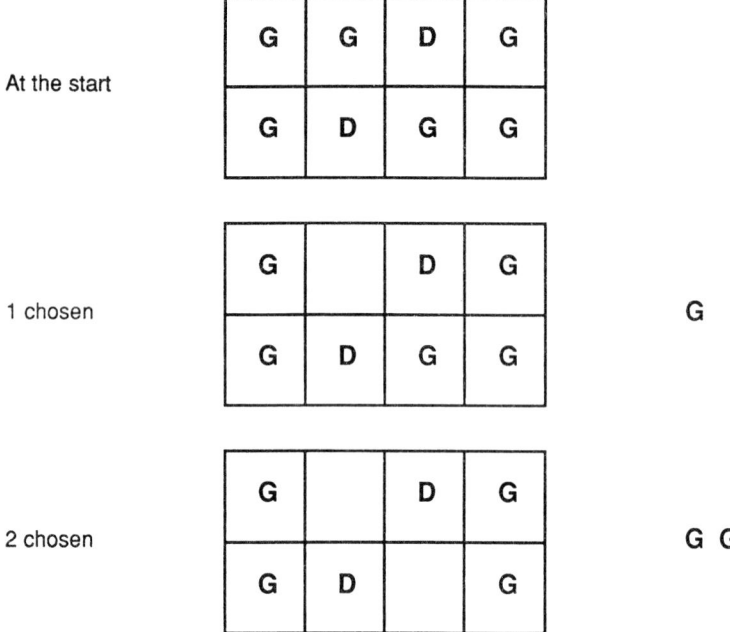

Figure 4.3 Selection of two devices

The probability is the same if both are taken together or if one device is taken and then the second is taken from the remaining seven. However, *use of the multiplication rule requires that an order is imposed.*

Therefore the first step in the calculation is to state the order, i.e.

Let event *a* be (the first device is good)
and *b* (the second device is good).

Using the multiplication rule for dependent events:

P(1st device is good and 2nd device is good)
$= P$(1st device is good).P(2nd is good *given* 1st is good).

From Figure 4.3 (at the start)

P(1st device is good) $= 6/8$.

Assuming a good device is chosen, see Figure 4.3 (1 chosen), there are 5 good ones in the 7 remaining and
P(2nd device is good) $= 5/7$.

Because of the assumption, this is the same as P(2nd is good *given* 1st is good).

Therefore, P(1st and 2nd are good) $= \dfrac{6}{8} \times \dfrac{5}{7} = \dfrac{15}{28} = 0.536$.

A similar calculation is involved to determine the probability that one of the two chosen devices is defective. First, impose an order by assuming they are taken one at a time, the first not being replaced before the second is taken. As before, this gives the same answer as two taken together.

P(one is good, one is defective)
$= P$(1st is good and 2nd is defective) $+ P$(1st is defective and 2nd is good)

$$= \frac{6}{8} \times \frac{2}{7} + \frac{2}{8} \times \frac{6}{7} = \frac{3}{7} = 0.429.$$

■ 4.9 Tree diagrams

A tree diagram is a pictorial representation of the multiplication rule. It illustrates the ordering of successive selections to results and can be helpful in sorting out some problems of calculation. Sometimes tree diagrams are drawn showing all possible results, but this can be unnecessarily complicated and it is really only necessary to include those results which give the event required.

For example, the tree diagram in Figure 4.4 represents the ways of obtaining one good and one defective device whose probability was calculated in the last example.

The tree is constructed from the starting point of 8 devices in the box. There are two possible results for the first device chosen: it might be good or it might be defective. Lines are drawn to each of the possibilities and the proportion good or defective is noted on each line as appropriate. The possibilities for the second device chosen depend upon the first. This is represented by extending each line to the next possibility and noting the proportion good or defective at this stage. Note that the totals good and defective do not change, but the number left in the box to choose from reduces.

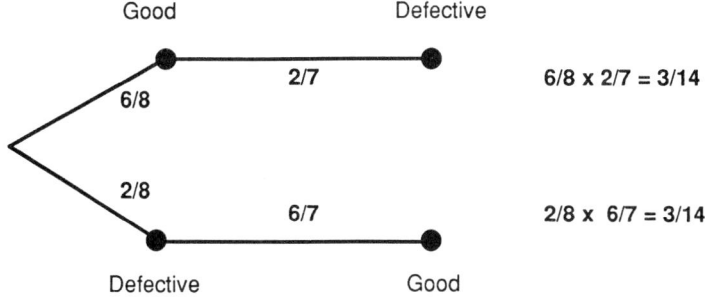

Figure 4.4 Tree diagram for P(one is good, one is defective)

The probability P(one is good, one is defective) is found by calculating the sum of the products of the individual probabilities along each branch. For example:

$$\frac{6}{8} \times \frac{2}{7} + \frac{2}{8} \times \frac{6}{7} = \frac{3}{14} + \frac{3}{14} = \frac{3}{7} \text{ (the same as calculated before).}$$

■ 4.10 Multiplication rule (for more than two dependent events)

The multiplication rule and tree diagrams can be applied to more than two events. The first example below is for three dependent events where the mathematical expression is

$$P(abc) = P(a)P(b \mid a)P(c \mid ab).$$

The probability of obtaining 2 good and 1 defective devices is required when 3 devices are chosen from the box of 6 good and 2 defective.

An order is imposed by assuming they are taken one at a time
P(1st is good, 2nd is good, 3rd is defective)
$= P$(1st good) $\times P$(2nd good *given* 1st good)
$\times P$(3rd defective *given* 1st and 2nd good)

$$= \frac{6}{8} \times \frac{5}{7} \times \frac{2}{6} = \frac{5}{28}.$$

Similarly
P(1st is good, 2nd is defective, 3rd is good)

$$= \frac{6}{8} \times \frac{2}{7} \times \frac{5}{6} = \frac{5}{28}.$$

Also
P(1st is defective, 2nd is good, 3rd is good)

$$= \frac{2}{8} \times \frac{6}{7} \times \frac{5}{6} = \frac{5}{28}$$

and the total probability is

$$3 \times \frac{5}{28} = \frac{15}{28} = 0.536.$$

It is not a coincidence that the three probabilities are all the same value (5/28). This always happens. To shorten the calculations, therefore, it is only

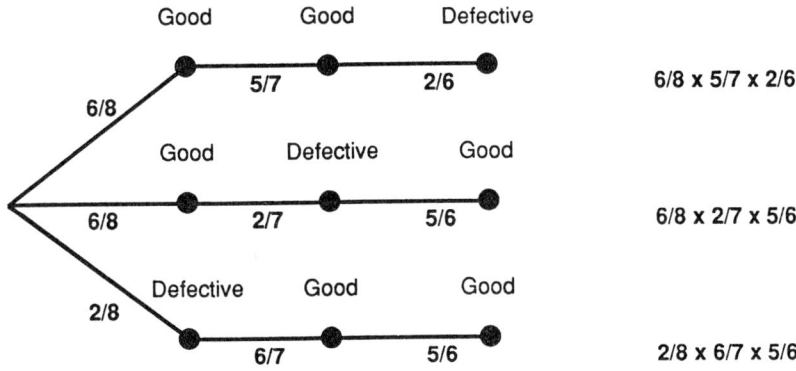

Figure 4.5 Tree diagram for P(two are good, one is defective)

necessary to calculate the probability in one particular order and then multiply by the number of rearrangements, i.e.

$$\begin{bmatrix}\text{Probability in} \\ \text{any order}\end{bmatrix} = \begin{bmatrix}\text{Probability in a} \\ \text{particular order}\end{bmatrix} \times \begin{bmatrix}\text{Number of} \\ \text{rearrangements}\end{bmatrix}.$$

A tree diagram for this example is shown in Figure 4.5, where multiplying along the branches and adding up the totals for each branch gives

$$\frac{6}{8} \times \frac{5}{7} \times \frac{2}{6} + \frac{6}{8} \times \frac{2}{7} \times \frac{5}{6} + \frac{2}{8} \times \frac{6}{7} \times \frac{5}{6} = \frac{15}{28}.$$

If four devices are taken from the box, the probability that just one is defective is given by

P(good,good,good,defective in order) × number of rearrangements

$$P(GGGD \text{ in order}) = \frac{6}{8} \times \frac{5}{7} \times \frac{4}{6} \times \frac{2}{5} = \frac{1}{7}.$$

There are four different rearrangements, i.e. *GGGD*, *GGDG*, *GDGG* and *DGGG*. Therefore, the probability just one of the four is defective is

$$4 \times \frac{1}{7} = \frac{4}{7} = 0.571.$$

■ 4.11 Multiplication rule (for independent events)

The discussion of the multiplication rule so far has involved dependent events. The probability in each case depends upon the previous results. For example, the probability that a second device from the box of 8 is defective depends upon whether the first one is. If the first is defective, the probability that the second is defective too is 1/7; but if the first is good, the probability that the second is defective is 2/7.

A more simple situation is that of independent events where probabilities do not change. This is often true of most manufacturing processes.

When applied to independent events, the mathematical expression of the multiplication rule is modified because $P(b \mid a) = P(b)$ since b does not now depend on a. The expression of the multiplication rule for independent events is

$$P(ab) = P(a).P(b).$$

For example, if it is known that 5% of components produced on a machine are defective and two are taken at random, the probability that both are defective is calculated as follows.

Using the multiplication rule for independent events:

P(1st is defective and 2nd is defective)
$= P$(1st is defective).P(2nd is defective)
$= 0.05 \times 0.05 = 0.0025.$

The probability of finding just one defective

If 3 are taken, the probability that just one is defective is

P(1st defective) $\times P$(2nd good) $\times P$(3rd good) $\times$ Re-arrangements
and there are 3 different arrangements, namely *DGG*, *GDG*, *GGD*.

Since the probability will be the same in any order and as
0.05 is the probability of a defective, and
0.95 is the probability of a good one,
the combined probability is $0.05 \times 0.95 \times 0.95 \times 3 = 0.135$.

This example is illustrated by the tree diagram in Figure 4.6.

The probability of finding at least one defective

When the probability of *a and b* is required, the probabilities are multiplied. When the probability of *a or b* is required, the probabilities are added.

Thus for the same machine, where 5% defective components are being produced, if 3 are taken, the probability that at least one defective is obtained can be found by calculating P(1 defective in 3, or 2 defectives in 3, or all 3 are defective).

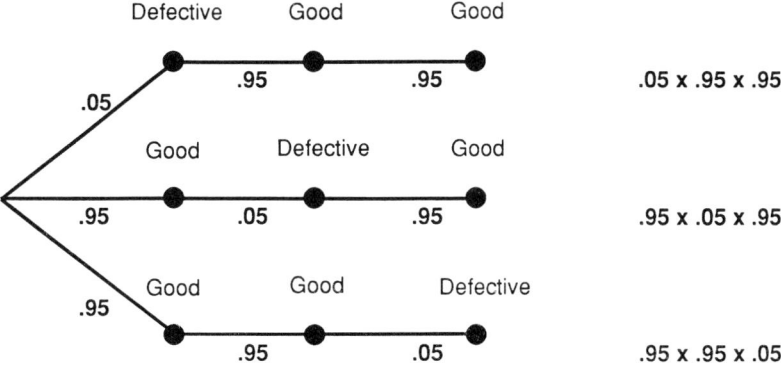

Figure 4.6   Tree diagram for one defective in three components

Because there are 3 ways of obtaining each of the first 2 results, but only one way of obtaining the last one, the required probability becomes

$P(DGG$ in order$) \times 3 + P(DDG$ in order$) \times 3 + P(DDD)$
$= P(D).P(G).P(G) \times 3 + P(D).P(D).P(G) \times 3 + P(D).P(D).P(D)$
$= 0.05 \times 0.95 \times 0.95 \times 3 + 0.05 \times 0.05 \times 0.95 \times 3$
$\quad + 0.05 \times 0.05 \times 0.05$
$= 0.1354 + 0.0071 + 0.0001$
$= 0.143.$

The probability of not finding a defective

If the probabilities of all possible results are calculated then they must sum to 1, since one of these results is certain to occur. Using this fact allows a more simple calculation than that in the previous example.

The probability required is $P(1$ defective in 3 or 2 defectives in 3, or all 3 are defective$)$
$= 1 - P($no defectives occur$)$
$= 1 - P(G$ and G and $G)$
$= 1 - P(G).P(G).P(G)$
$= 1 - 0.95 \times 0.95 \times 0.95$
$= 1 - 0.857$
$= 0.143.$

This technique, whereby the probability of not obtaining the event is calculated and subtracted from 1 in order to find the probability of the event occurring is a frequently used short-cut in some probability calculations. These calculations are based on particular distributions, and examples are given with the descriptions of the normal distribution (Section 5.7) and the binomial distribution (Section 6.3).

■ 4.12 System modelling

The multiplication rule is rarely recognized as such in commercial or manufacturing industry. However it is the basis of several widespread and frequently used techniques, the most common of which are availability, efficiency and reliability modelling of process and product systems. The numerical values used in these techniques are in fact probabilities of system components being available, efficient or reliable. For all practical purposes the techniques are the same, and the examples below are presented in terms of product system reliability.

The series model

The series model is the most simple and the most common representation of a system's reliability. It illustrates the situation where failure of any one component causes a total system failure. Failure of one component is assumed not to affect the reliability of any other components: in other words, failures are independent events.

> Note that the reliability of a component is the probability that the component operates successfully for a specified time period. Similarly for system reliability.

When a system has i components in series (see Figure 4.7) and R_1, R_2, R_3, ... R_i are the individual component reliabilities, the system reliability (R_S) is expressed mathematically as

$$R_S = R_1 \times R_2 \times R_3 \times \cdots R_i$$

An oil level sensor system is required to have a reliability of not more than 5% failures before 100 000 miles in service, i.e. $R_S = 0.95$. The system contains four components and the reliability of three is known. What must be the reliability of the fourth?

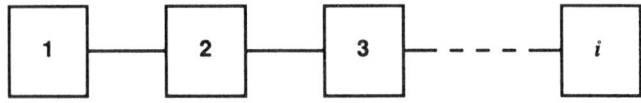

Figure 4.7 The series model

Sensor $R_1 = 0.985$.
Indicator $R_2 = 0.985$.
Terminals $R_3 = 0.990$.
Relay $R_4 = \text{unknown}$.
$R_S = 0.95 = 0.985 \times 0.985 \times 0.990 \times R_4$.
Hence $R_4 = 0.989$.

System reliability decreases as the number of components increases (see Figure 4.8). For example, 100% inspection by the human eye can result in between 1% and 2% faulty parts passing the control station. What reliability could be expected of a visual inspection system that screens 25 separate components of a motor vehicle for cosmetic faults?

Assuming that 1.5% of cosmetic faults are missed by the inspector, each component has $R = 1 - 0.015 = 0.985$ and a series model is assumed, $R_S = 0.985 \times \cdots \times 0.985 = 0.985^{25} = 0.685$.
In other words, about every third vehicle will have a cosmetic fault.

The parallel model

The reliability of a system can be increased by duplicating certain components, i.e. by including a mechanism that allows a component that fails to become redundant and not cause system failure – as, for example, in motor vehicle dual brake systems. The reliability of duplicated components is represented by a parallel model (Figure 4.9) whose reliability is expressed mathematically as

$$R_2 = 1 - (1 - R_{2a})(1 - R_{2b}).$$

On a window wiper/washer tank unit, two components contribute to the system reliability: a motor with $R_1 = 0.95$ and a pump with $R_2 = 0.90$.

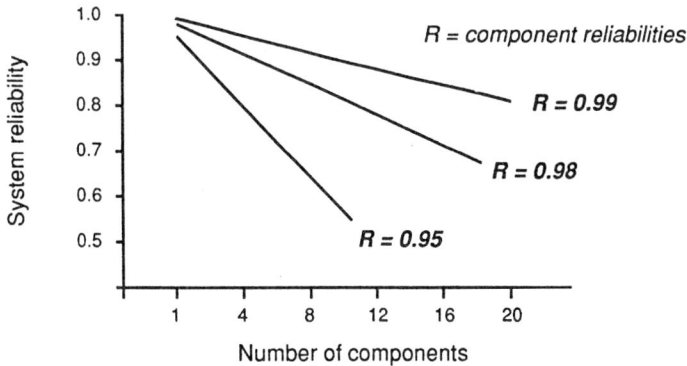

Figure 4.8 Effect on a system of components in series

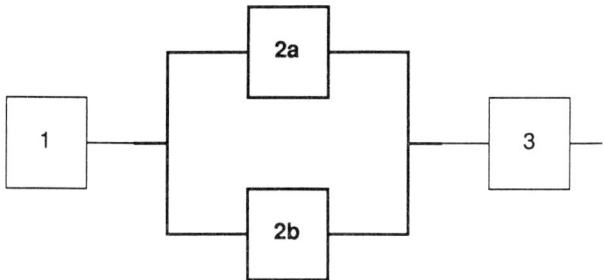

Figure 4.9 The parallel model

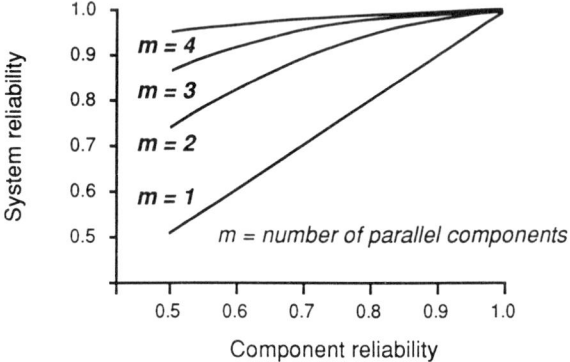

Figure 4.10 Effect on a system of parallel components

If there is one of each of the components,
$R_S = R_1 \times R_2 = 0.95 \times 0.90 = 0.855$
and the proportion of systems likely to fail is about 14.5%.

If the pump is duplicated,
$R_S = R_1\{(1 - (1 - R_{2a})(1 - R_{2b}))\}$
$= 0.95\{(1 - (1 - 0.90)(1 - 0.90))\} = 0.941$
and the proportion of systems likely to fail is about 5.9%.

Duplication in the example has reduced system failure probability to about one-third of its original value. More than one redundant component can be added in parallel, but beyond the third or fourth, the effect becomes small (see Figure 4.10).

■ 4.13 Expectations

The theory of expectations provides a model that is used to calculate mean values. These values are usually notional: that is, often they cannot occur in the real world (see the first example). However, they can provide information that is useful in wider contexts (see the second example). The mathematical expression of expectations is

$$E(x) = \Sigma \; x.P(x) = \mu = \text{mean of } x$$

where $E(x)$ is the expected value or expectation of x and $P(x)$ is the probability of x, the summation being over all possible values of x.

Example 1

Consider the six-sided cubical die used in most games. What is the expected number of spots showing if the die is thrown once?

Let $x =$ the number of spots on a side of the die. Then, since the probability of each number is $1/6$, $E(x) = \Sigma \; x.P(x)$

$$= \frac{1}{6} \times 1 + \frac{1}{6} \times 2 + \frac{1}{6} \times 3 + \frac{1}{6} \times 4 + \frac{1}{6} \times 5 + \frac{1}{6} \times 6$$

$$= 3.5.$$

In this example, the expected value of 3.5 never actually occurs, so what does it imply? In fact if the die were thrown a large number of times, the mean of the data would be about 3.5, this being the mean of the distribution of x.

Example 2

A particular component manufactured by a company is found to have no faults in 80% of cases, requires some repair work in 15% of cases and the remainder are so poor that they can only be sold as scrap. If the firm makes £20 profit on the good ones, £5 profit on those requiring further work and an overall loss of £8 on those sold as scrap, what is the expected profit on 100 components?

Consider just one component and, since the expected profit is required, let $x =$ profit. There are 3 possibilities for the component: near perfect, repair needed, sold for scrap. Therefore:

$$\begin{aligned} E(x) &= \Sigma \; x.P(x) \\ &= P(\text{near perfect}) \times £20 \text{ profit} + P(\text{repair needed}) \times £5 \text{ profit} \\ &\quad + P(\text{scrap}) \times \text{loss of } £8. \end{aligned}$$

Since a loss of £8 means a profit of $-£8$,

$$E(x) = 0.8 \times 20 + 0.15 \times 5 + 0.05 \times (-8)$$
$$= 16 + 0.75 - 0.4$$
$$= £16.35.$$

Although sometimes a component is satisfactory yielding a profit of £20, sometimes requires extra work with a profit of only £5 and sometimes is scrap with a resulting loss of £8, on the average a profit of £16.35 per item is made. Thus on 100 items, $100 \times 16.35 = £1635$ profit is expected.

Note that this mathematical expectation is the same expected profit as that referred to in everyday conversation. This is precisely how much the company expects to make on 100 items.

◼ 4.14 Probability distributions

Probability distributions were introduced in Section 3.12. Just as frequency distributions are based on discrete or continuous data, so probability distributions are discrete or continuous again depending on the values x can take.

Discrete probability distributions

The probability of the value x occurring is written as $f(x)$, where $f(x)$ is called the *probability function* for the distribution. The discrete distribution is represented by a histogram as shown in Figure 4.11.

Summing the probability terms for all values up to x_i gives the probability of a value less than or equal to x_i. This is denoted by $F(x_i)$ which is called the *cumulative probability function* or the *cumulative distribution function*.

$$F(x_i) = f(0) + f(1) + f(2) + \cdots f(x_i)$$

An important rule of probability distributions is that all probabilities sum to 1 or

$$\Sigma f(x) = 1.$$

This follows from the fact that one of the values x_1, x_2, x_3, x_4 or x_5 must occur, and for something that is certain the probability is 1.

For the components in Example 2 of the previous section, where x is the profit, the probability function is:

$$f(x) = 0.8 \quad \text{for } x = 20,$$
$$0.15 \quad \text{for } x = 5,$$
$$0.05 \quad \text{for } x = -8.$$

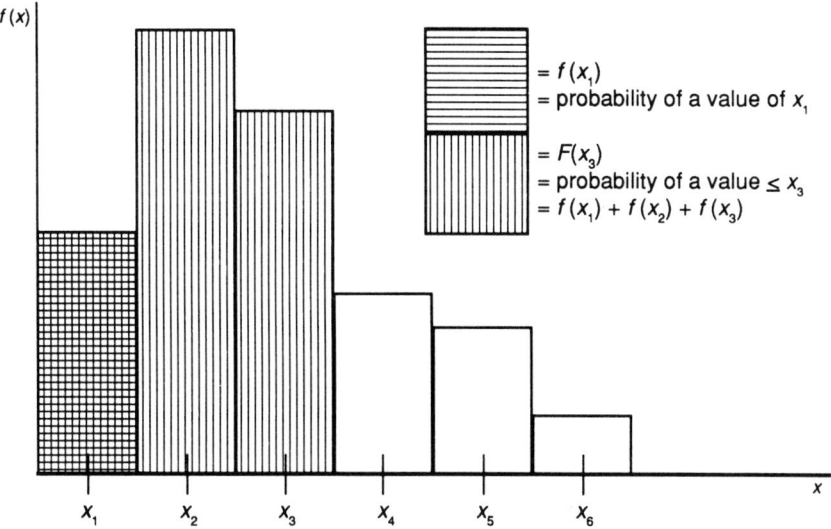

Figure 4.11 Histogram for a discrete probability distribution

Figure 4.12 is a histogram that pictures the probability function for the components example. Alternatively, this probability function could be written:

$$f(20) = 0.8,$$
$$f(5) = 0.15,$$
$$f(-8) = 0.05.$$

As another example, consider 2 tosses of a coin and the distribution of the number of heads. The possible results are 2 heads, 1 head and 1 tail, 2 tails. When x is the number of heads:
the probability of no heads $= f(0) = P(\text{tail 1st}) \times P(\text{tail 2nd})$

$$= \frac{1}{2} \times \frac{1}{2} = \frac{1}{4};$$

the probability of one head $= f(1)$
$= P(\text{tail 1st}) \times P(\text{head 2nd}) + P(\text{head 1st}) \times P(\text{tail 2nd})$

$$= \frac{1}{2} \times \frac{1}{2} + \frac{1}{2} \times \frac{1}{2} = \frac{1}{2};$$

the probability of two heads $= f(2) = P(\text{head 1st}) \times P(\text{head 2nd})$

$$= \frac{1}{2} \times \frac{1}{2} = \frac{1}{4};$$

i.e. $f(0) = 1/4$, $f(1) = 1/2$ and $f(2) = 1/4$.

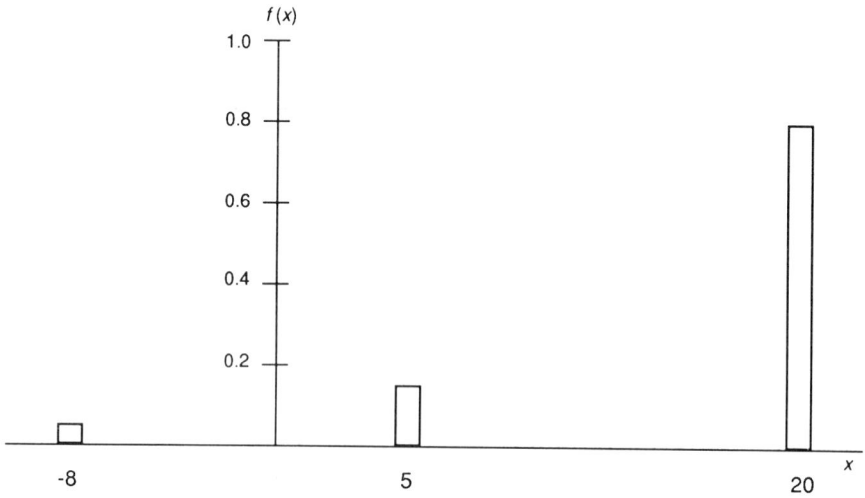

Figure 4.12   Histogram for components example

In many situations the probability function can be generalized and written in terms of x (see Chapter 6). In the coin example above, it can be written as

$$f(x) = \frac{1}{2[(1-x)^2 + 1]} \qquad \text{for } x = 0,1,2.$$

Or, even better, using $_2C_x$ which is explained in Section 4.15:

$$f(x) = \frac{_2C_x}{4} \qquad \text{for } x = 0,1,2.$$

Continuous probability distributions

In a continuous probability distribution, where x can take any value over a given range, $f(x)$ is called the *probability density function* (pdf) and defines the distribution.

The probability of a value between a and b is given by the area under the curve between a and b which is shaded in Figure 4.13. Alternatively, this area can be thought of as the proportion of the distribution between a and b. Examples of these distributions occur in the next two chapters.

The *cumulative distribution function* $F(x)$, gives the probability of a value less than x (alternatively less than or equal to x, which are the same probabilities in the continuous case). It is the area under the pdf curve lying to the left of x, as shown in Figure 4.14.

The total area under the curve is equal to 1 because x must take one of the values in the distribution.

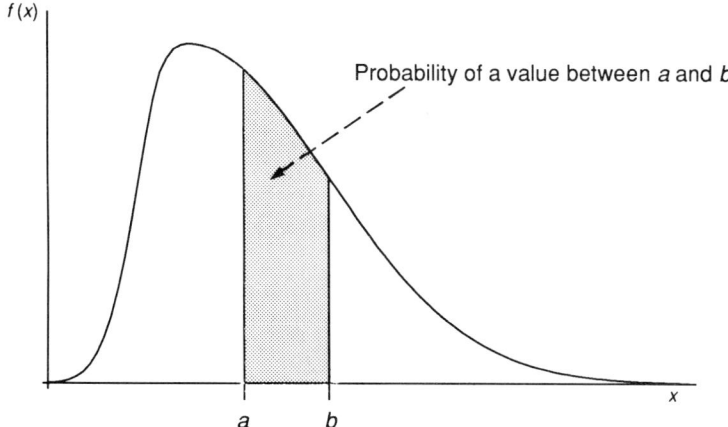

Figure 4.13 A continuous distribution

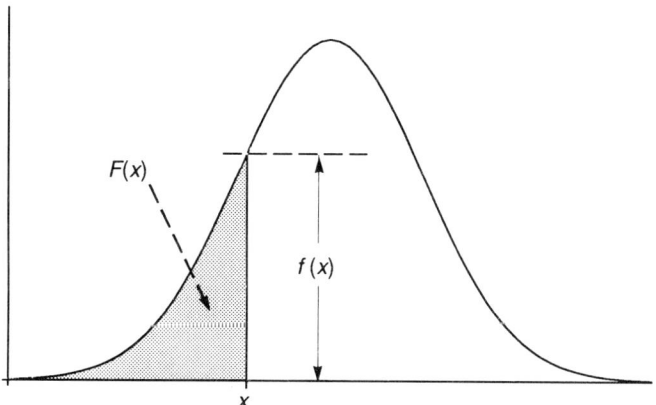

Figure 4.14 The cumulative distribution function

Since these probability distributions are completely defined, they are similar to the population described in Chapter 3. Therefore the mean is denoted by μ and the standard deviation by σ and the variance will be σ^2.

■ 4.15 Combinations

The *number of combinations* of r from n is the number of ways of choosing r from n.

For example, consider the number of ways of choosing 2 from 4; it does not matter whether it is 2 people from 4, 2 cars from 4, 2 routes from 4, so long as the 4 are different.

Represent the 4 by a, b, c and d, from which any 2 can be chosen in 6 ways, namely ab, ac, ad, bc, bd and cd. In other words, the number of combinations of 2 from 4 is 6.

In order to write down a general formula it is necessary to use $n!$ called *factorial n* (or n *factorial*), where

$$n! = n(n-1)(n-2)(n-3) \cdots (3)(2)(1).$$

For example: $5! = 5 \times 4 \times 3 \times 2 \times 1 = 120$

$0! = 1$ is defined as a special case.

The number of combinations of r from n is written as

$$_nC_r \text{ or } \binom{n}{r} \text{ where } _nC_r = \frac{n!}{r!(n-r)!}$$

For the example at the beginning of this paragraph where $n = 4$ and $r = 2$, substitution in the above formula gives

$$_4C_2 = \frac{4!}{2!2!} = \frac{4 \times 3 \times 2 \times 1}{(2 \times 1)(2 \times 1)} = 6$$

as obtained before by writing down all of the combinations.

With large n it would be extremely time consuming to write down all the possibilities and this is where the formula is useful. For example, the number of ways of choosing 18 from 20 is

$$_{20}C_{18} = \frac{20!}{18!2!} = \frac{20 \times 19 \times 18 \times 17 \times \cdots \times 3 \times 2 \times 1}{(18 \times 17 \times 16 \times \cdots \times 3 \times 2 \times 1)(2 \times 1)}$$

and cancelling $18 \times 17 \times 16 \times \cdots 3 \times 2 \times 1$ gives

$$\frac{20 \times 19}{2} = 190.$$

Note that $_{20}C_2 = \frac{20!}{2!18!} = 190$ as well.

In fact $_nC_{n-r} = {_nC_r}$ for all r, n

which is helpful if combinations are in regular use, as are the special cases below (which are valid for any n):

$$_nC_0 = {_nC_n} = 1,$$
$$_nC_1 = {_nC_{n-1}} = n.$$

■ References to further reading on the topics of Chapter 4

Cass (1973) and Chatfield (1983) give simple introductions to probability. Parzen (1960) and Walpole and Myers (1993) give more mathematics and more practical examples.
See Bibliography for details of titles and publishers.

■ Self assessment questions on Chapter 4

1. A box contains 9 components, 3 of which are defective.
    (a) If one is chosen at random, what is the probability it is defective?
    (b) If two are chosen at random, what is the probability that
        (i)   both are defective?
        (ii) just one is defective?
    (c) If three are chosen at random, what is the probability that
        (i)   all three are defective?
        (ii) just one is defective?
2. If in question 1, instead of choosing components from a box, they are taken from a process in which 20% are defective, what are the probabilities in this case?
3. What is the probability of obtaining
    (a) an even number when throwing a six-sided cubical die?
    (b) a total score of 10 with two dice?
    (c) a total score of 10 or more with two dice?
    (d) two fives with two dice?
    (e) one three and one four with two dice?
    (f) If three dice are thrown, what is the probability of failing to obtain a six?
4. On average, one in four sheets of plastic made by a certain process are near perfect, while one in three have only one small flaw. The remainder have two or more flaws. What is the probability that a sheet has fewer than two flaws?
5. A firm has two similar machines, one less reliable than the other. The better one has a probability of 0.9 of working throughout a week without the need of repair. The probability for the other is 0.7.
    (a) What is the probability of both machines working satisfactorily throughout the week?
    (b) What is the probability that at least one of them requires repair?
6. An experiment consists of tossing a coin and throwing a die. What is the probability of
    (a) head on the coin and 3 on the die?

   (b)   tail on the coin and anything on the die?

   (c)   head on the coin or 6 on the die?

7.   Near-miss collisions near an airport occur on average once every 20 days. In 5 consecutive days, what is the probability that no near-miss collisions occur?

8.   If the probability that any one person requires medical attention in one particular year is 0.25, what is the probability that

   (a)   four people live through the year without medical attention?

   (b)   just one of the four needs medical attention?

9.   Four types of substance are to be tested using 3 types of acid at 4 different temperatures. How many tests are required to cover every combination of substance, acid and temperature?

10. Find the values of $_5C_4$, $_7C_3$, $_{10}C_2$, $_{20}C_1$, $_{100}C_{98}$, $_nC_{n-2}$.

11. Three employees are to be chosen from 8 to go on a course. In how many ways can the 3 be chosen?

12. Four cards are taken at random from a pack of playing cards. What is the probability of choosing

   (a)   4 aces?

   (b)   no aces?

   (c)   1 ace and 3 kings?

13. Five per cent of items are defective and £5 profit is made on a good item, but there is a £1 loss on a defective item. What is the expected profit on

   (a)   1 item?

   (b)   10 items?

14. The owner of a garage where exhausts are fitted to cars estimates that £20 profit is made on each successful sale. However, 15% of replaced exhausts turn out to be faulty, 10% requiring a complete replacement with an overall loss to the garage of £10, the other 5% needing only an adjustment reducing the profit to £15. In one week sales of 25 exhausts should be achieved. What is the expected profit on these?

15. A unit consists of 8 components each with reliability 0.9 (i.e. the probability that the component works for the necessary time is 0.9). What is the reliability of the unit?

16. A manufacturing firm has 3 types of machine, A, B and C, each type used for a different manufacturing process. Defects can occur at any stage of the manufacturing process and a defect can be 'slight' or 'serious'. On any item only one type of defect can occur on each type of machine, i.e. at each stage the item has a slight defect caused by that machine, or a serious defect or no defect. The probabilities of defects at each stage are as follows:

Machine	Defect	
	Slight	Serious
A	0.1	0.05
B	0.2	0.04
C	0.1	0.02

What proportion of items have
(a)   no defects at all?
(b)   1 slight and 1 serious defect?
(c)   exactly 2 defects?
If £20 profit is made on an item with no defects, and £10 on an item with just one slight defect after the cost of repairs, but any others are sold as scrap with a loss of £2, what is the expected profit on 100 items?
17. Unknown to the fitter, a batch of 10 car tyres has 4 faulty. They are fitted one at a time at random and tested immediately afterwards. If x is the number of tyres fitted to find a good one, find the distribution of x.

<table>
<tr><td>

5

</td><td>

The Normal
Distribution

</td></tr>
</table>

■ 5.1 Introduction

In the technical language of statistics, the word *normal* always refers to the distribution described in this chapter. Although it is widely applicable, the normal distribution is not appropriate in some situations; this does not imply that such situations are 'abnormal' in the ordinary sense, merely that they are better described by some other distribution.

The mathematics of the normal distribution is the principal basis for interpretation of numerical data in manufacturing industry. There are several reasons for its widespread use:

■ Formulae, rules and techniques developed from its mathematics are well known and comparatively easy to use.

■ It is the distribution that occurs most frequently in measurements that are of practical interest.

■ Often, but under particular conditions, it approximates sufficiently to other distributions such as Poisson and binomial (Section 6.5), even though, strictly, it applies to continuous values and they might be of discrete values.

■ A mathematical concept known as the central limit theorem (Section 7.3) extends its use in the interpretation of other distributions, including the complex distributions that arise as a result of industrial processes.

Sometimes, the normal distribution is referred to as the Gaussian distribution. This is to honour the German mathematician K.F. Gauss (1777–1855), who made some of the most important contributions towards practical application of the theory, which had been published originally at the beginning of the eighteenth century.

This chapter describes the use of probability tables that have been drawn up to help calculations from data which are known to have a normal distribution. First, however, it might be necessary to check that the data does indeed have a normal distribution.

■ 5.2 Recognition of a normal distribution

Visual check

If there is sufficient data to construct a histogram – that is to say at least 30 and preferably more than 100 measurements – the shape of the histogram can be used as a rough check for normality. The normal distribution is bell shaped and is illustrated as the smooth outline of the 'expected distribution' in Figure 5.1.

Figure 5.1 also shows a histogram that has been constructed from data in Table 5.1, a frequency table that groups into 5-minute classes, measurements of the time taken to assemble the trim (seats, carpets, panel linings, decorative parts, etc.) in each of 100 motor vehicles during a training exercise.

The outline of the histogram is slightly different from the expected curve, but it has the same general bell shape. If fewer measurements had been used, the difference would probably have been greater; with more measurements the difference would probably have been less.

Numerical check

If the histogram outline is not easily identified as having a bell shape, there are some simple numerical checks that can be carried out. For these, it is necessary to calculate the mean, the median, the range and the standard deviation of the data.

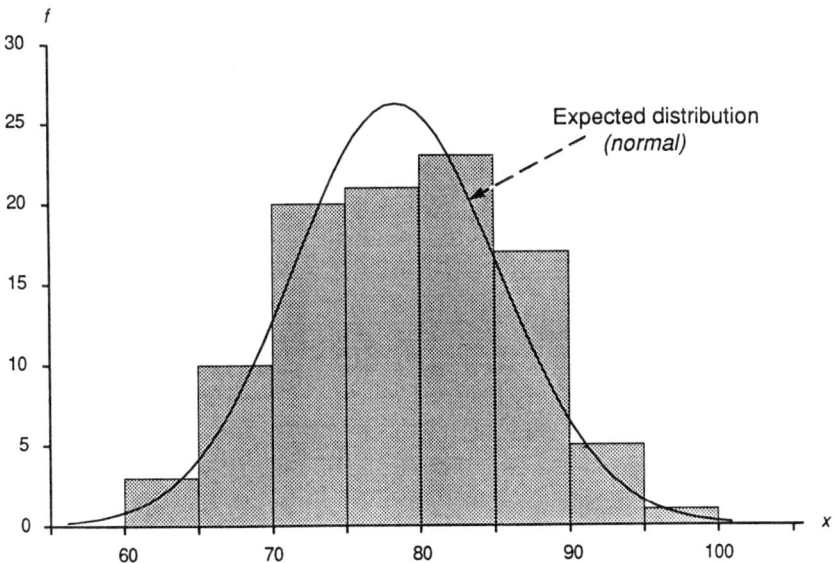

Figure 5.1 Histogram of frequencies in Table 5.1 and expected distribution

Table 5.1 Frequency table for times to assemble 100 sets of trim

Class (x)	Time (minutes)	Frequency (f)
1	60 to 65	3
2	65 to 70	10
3	70 to 75	20
4	75 to 80	21
5	80 to 85	23
6	85 to 90	17
7	90 to 95	5
8	95 to 100	1
	Total	100

For the data in Table 5.1, where all values are in minutes:

Mean $\bar{x} = 78.9$
Median $\tilde{x} = 79.0$
Range $R = 40$
Standard deviation $s = 7.65$ and hence $6s = 46$.

In most situations, data suggests a normal distribution if the following conditions are met:

- The histogram should rise fairly smoothly from both ends to a *single mode* (or peak).
- The mean and the median should be approximately equal. This indicates that the histogram is close to being *symmetrical*.
- The *range* should be approximately *six standard deviations*.
- Approximately two-thirds of the histogram area should lie within plus and minus one standard deviation from the mean. This requirement is slightly more difficult to check, especially when there is little data.

The histogram shown in Figure 5.1 meets all these requirements.

Other checks

If there is still uncertainty about a histogram meeting the requirements for normality, more elaborate tests can be used. For example:

- The chi-squared test (Section 11.5).
- The Kolmogorov–Smirnov test (Section 11.11).
- A graphical test using a probability paper (Section 8.3). Graphical tests are the most frequently used tests in manufacturing industry and are part of everyday routines for spot-checks or audits of process behaviour. They have the advantage that they can be applied to data sets which are too small for construction of a useful frequency table and histogram.

■ 5.3 Theoretical description of the normal distribution

The full mathematical description of a normal distribution is given by the formula for its probability density function (pdf), which is set out in Figure 5.2.

The formula for the pdf of the normal distribution is rarely used directly; much more important is the line that it describes, which is shown in Figure 5.3. The line is a smooth bell-shaped curve that is exactly symmetrical about its mean (μ) and whose curvature changes from concave to convex at one standard deviation on each side of the mean (that is, at $\mu + \sigma$ and at $\mu - \sigma$). Other characteristics are described in the next section.

Industrial use of the normal pdf depends upon knowing the values of μ and σ for a product or process feature. In practice, their values have to be

$$f(x) = \frac{1}{\sigma \sqrt{2\pi}} \exp\left[-\frac{1}{2}\left(\frac{x-\mu}{\sigma}\right)^2\right]$$

where x can have any value between $-\infty$ and $+\infty$,
           π is a constant (approximately 3.1416),
           $exp\,[....]$ is the same as $e^{(....)}$ where e is a constant (approximately 2.7183),
           μ is the theoretical or population mean and
           σ is the theoretical or population standard deviation.

Figure 5.2   The normal pdf formula

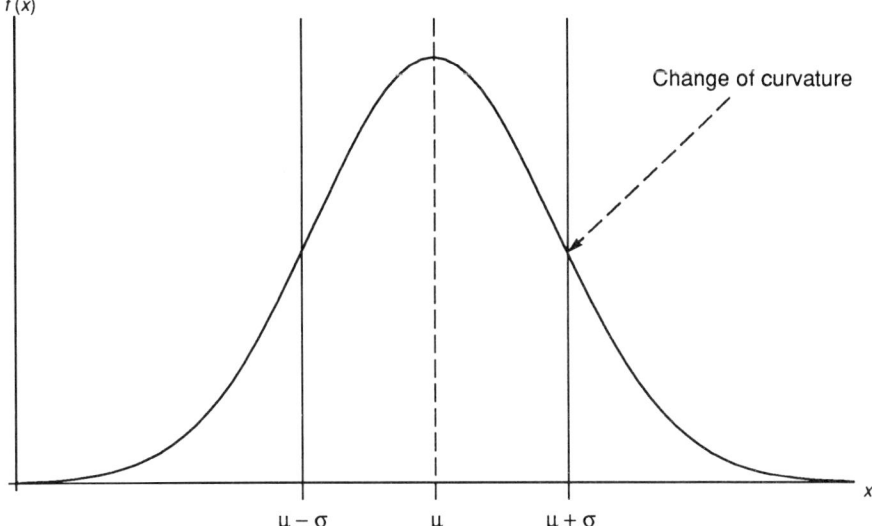

Figure 5.3   The pdf (probability density function) of a normal distribution

85

estimated from sample data. The accuracy of such estimates is discussed in Chapter 7.

- The population mean, μ, is estimated as $\bar{x}$ (the mean of a sample), and the population standard deviation, σ, as s (the standard deviation of a sample).
- Alternatively, when several small samples are available, such as in process control charting (Chapter 10), the population mean is estimated as $\bar{\bar{x}}$ (the mean of several sample means). The population standard deviation can be estimated from $\bar{s}$ (the mean of the standard deviations of several samples) or from $\bar{R}$ (the average sample range); both of these must be adjusted by correction factors that depend on the size of the samples (Section 7.7).

> For the assembly times in Table 5.1, $\bar{x}$ was 78.9 and s was 7.65.
> These values were used for μ and σ respectively to evaluate $f(x)$, from the normal pdf equation, at a sufficient number of x values to draw the smooth curve in Figure 5.1.
> The classes in this example have a width of 5 minutes and therefore the curve represents 5 times $f(x)$. This is because the vertical scale represents the numbers of measurements in each class rather than the numbers of each whole number value of $x(1, 2, 3, \ldots$ minutes).

Although it is easier to check the shape of a histogram by comparing it directly with the expected curve, the advantage must be offset against the effort of evaluating $f(x)$. In practice, therefore, it is usual to rely on recognizing the bell shaped outline, without calculating and drawing the theoretical expected curve.

■ 5.4 Characteristics of the normal pdf

Most practical applications of the pdf (normal or other) involve answering questions like the following:

- For purposes of process control: 'When does a significant change occur in the distribution of product or process measurements, and what is the size of such change?' For this example, the technique that uses the pdf is called a significance test (Chapter 11).
- For purposes of quality control: 'What proportion of product measurements are likely to be inside or outside some specified limit?' This type of question is answered using methods that employ tables derived from the pdf (Section 5.7).

The normal pdf curve is shown again in Figure 5.4. The area under the curve has been divided into bands that are one standard deviation wide and arranged

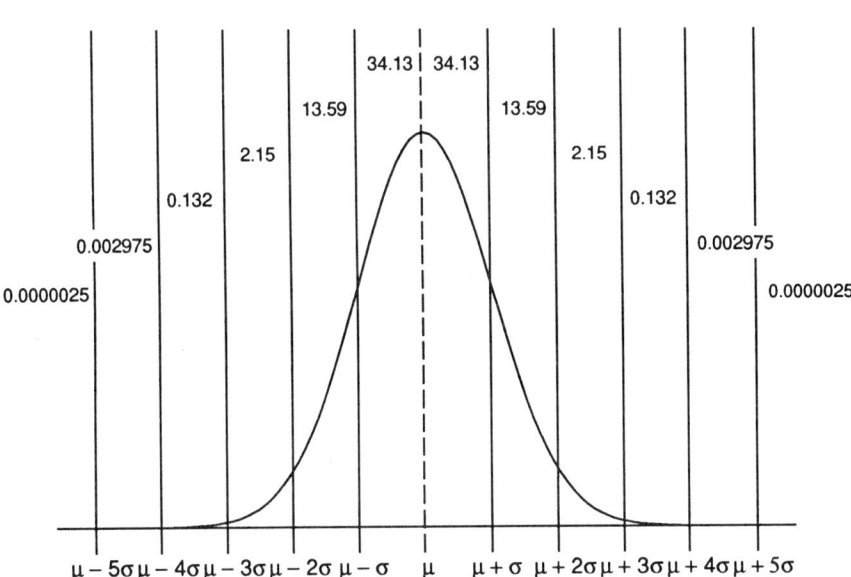

Area in band as % total area under curve

Figure 5.4 Percentage bands of the normal distribution

symmetrically about the mean. The value attached to each band is the percentage of the total area under the pdf curve that lies in that band; the total for all bands adds up to 100%. For practical purposes, it should be noted that values at the extremes (beyond $\mu \pm 3\sigma$) are notional and are included only to give a feel for the extremes. The values can be interpreted as either of the following:

■   The percentage of all measurements made that lie in a band.
■   The probability (multiplied by 100) that a single measurement, chosen at random, would lie in a band.

Figure 5.4 illustrates, for example, that 34.13% of measurements are expected to fall in each of its two central bands (i.e. between $\mu - \sigma$ and μ, also between μ and $\mu + \sigma$). In other words, 68.26% are expected to lie within $\mu + \sigma$ or within one standard deviation of the mean. Figure 5.5 is an alternative statement of this and other information derived from Figure 5.4. The way in which the percentages were calculated is described for *occurrences between values higher than mean* in Section 5.7.

The information in Figures 5.4 and 5.5 can be used to estimate the probable overall variation in products and processes from sample data.

87

IN A NORMAL DISTRIBUTION

about 2/3 of values lie within $\mu \pm \sigma$

about 95% of values lie within $\mu \pm 2\sigma$

nearly all (about 99.7%) of values lie within $\mu \pm 3\sigma$

approximately 3 in 1000 values are outside $\mu \pm 3\sigma$

approximately 6 in 100000 values are outside $\mu \pm 4\sigma$

approximately 6 in 10 million values are outside $\mu \pm 5\sigma$

The above characteristics are used in industrial process control
(Chapter 9)

The following characteristics are used in some control charts
(Section 10.19)

precisely 95% of values lie within $\mu \pm 1.96\sigma$

precisely 99.8% of values lie within $\mu \pm 3.09\sigma$

Figure 5.5   Characteristic proportions of normal distributions

For example, from the data in Figure 3.12 which records measurements (in units of ten-thousandths of an inch) of flange separations for 125 crankshafts:

■ $\bar{x}$ (the sample mean) is calculated as 54 and s (the sample standard deviation) as 16.
■ It is assumed that the sample of 125 crankshafts was randomly selected and represents total production (4000 units per week for 20 years – say, 3.68 million units allowing for holidays) and therefore that $\bar{x} = \mu = 54$ and $s = \sigma = 16$.
■ Using this information, Figure 5.4 has been redrawn as Figure 5.6, the 'quantities between particular separations' being the appropriate percentages of 3.68 million.
■ The two central bands in Figure 5.4 represent 68.26% of the total area under the curve. In Figure 5.6, these two bands (between $\mu - \sigma$ and $\mu + \sigma$) are rescaled and represent separations between $54 - 16$ and $54 + 16$, i.e. 68.26% of the total 3.68 million (about 2.51 million) lie between 38 and 70. The quantities in the four middle bands represent $2(13.59 + 34.13)\% = 95.44\%$ of 3.68 million $= 3.51$ million; and so on.

Figure 5.6 illustrates the probable variability of crankshaft flange separation *provided the manufacturing process and hence the distribution remain unchanged* throughout the 20 years of production. The information obtained can be used to make 'snap-shot' statements about process capability (is the

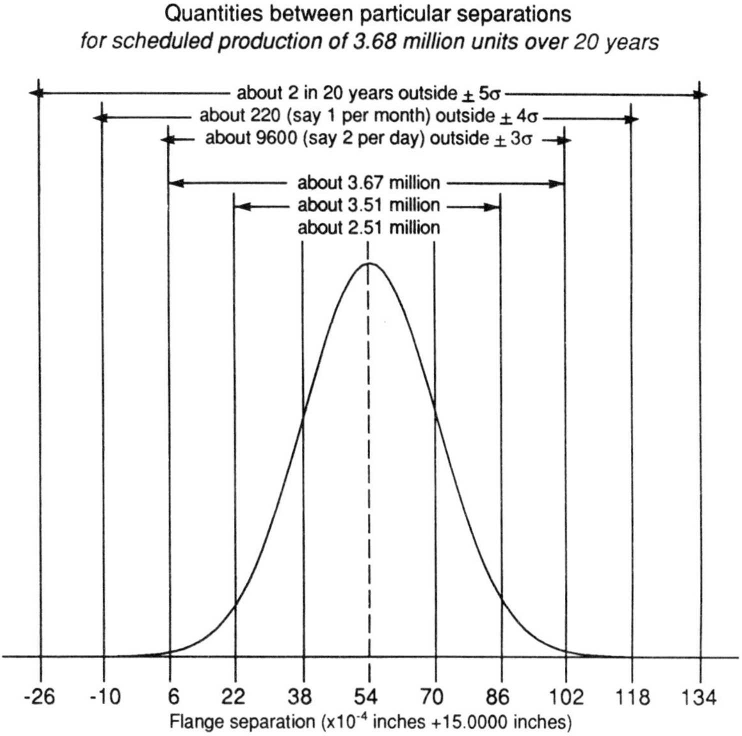

Quantities between particular separations
for scheduled production of 3.68 million units over 20 years

Figure 5.6   An estimation of the distribution of crankshaft flange
separations

variability, at a point in time, acceptable to customers?); also, the information
can be used as a 'stake in the ground' to monitor future production (will
results from later samples be the same?). The theory of the normal distribution
is the principal basis for process capability estimations and process monitoring
(Chapter 9).

■ 5.5 Standardization

The information of probable variability of crankshaft flange separation
(shown in Figure 5.6) is of only academic interest until it is used to help answer
important questions about customer satisfaction and economic process design.
For example, how many crankshafts will rub against their casing? Does
machinery need to be reset or replaced? Is selective assembly a possibility and,
if so, what ranges need to be accommodated?

Answers to these questions can be found from knowledge of areas under
the pdf curve of relevant measurements. To calculate areas under a normal

curve, it is necessary to use *tables* based upon a particular normal distribution, called the *standard normal distribution*.

The standardized deviate

In statistics, standardization is the simple process of converting measurements (*x* values) into units of standard deviation that are called *standardized deviates*, and which are referred to as *z*-values. The population mean (μ) and standard deviation (σ) must be known so that measurements can be converted into standardized deviates by using the formula

$$\text{standardized deviate, } z = \frac{x - \mu}{\sigma}.$$

Expressed in words, the standardized deviate (*z*) is the number of standard deviations (σ) by which *x* exceeds the mean (μ). This can be seen more clearly if the formula is rearranged:

$$x = \mu + z\sigma$$

The standard normal distribution

Standardization transforms all normal distributions of *x* values (whatever their individual mean and standard deviation values) into a single normal distribution of *z* values that always has $\mu = 0$ and $\sigma = 1$. This is the *standard normal distribution*.

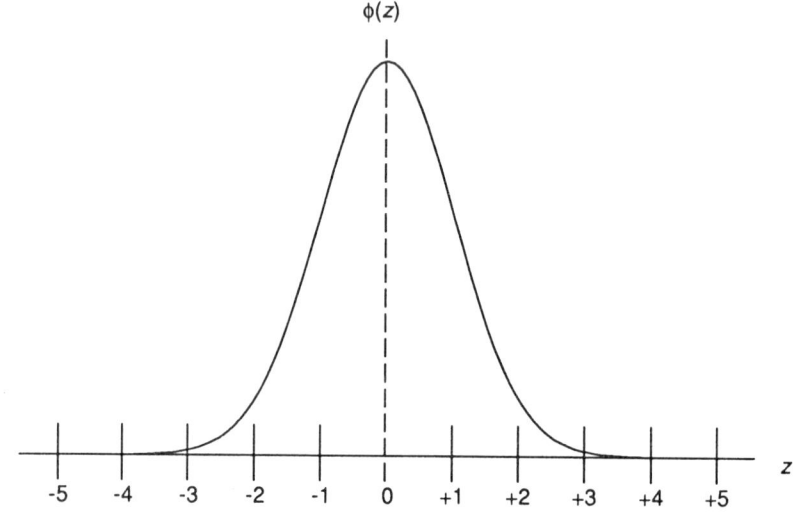

Figure 5.7 The standard normal distribution

Normal distribution theory and the normal pdf characteristics apply to the standard normal distribution. When the values $\sigma = 1$ and $(x - \mu)/\sigma = z$ are put into the normal pdf equation (Figure 5.2), the result is the mathematical description of the standard normal distribution. It is the formula for the pdf of z, designated $\phi(z)$ where ϕ is the Greek letter 'phi'.

$$\phi(z) = \frac{1}{\sqrt{2\pi}} \, exp \left[-\frac{1}{2} z^2 \right]$$

This pdf of z is illustrated in Figure 5.7. It is centred on zero and, although it is shown as extending from -5 to $+5$, virtually the whole distribution lies between -3 and $+3$.

■ 5.6 Standard normal probability tables

The *standard normal probability table* in Appendix B gives the proportions of the area under the $\phi(z)$ curve that lie to the *right* of selected positive values of z and are sometimes called the *areas in the tail* of the standard normal distribution.

For example, as illustrated in Figure 5.8, the value shown in the table against $z = 2.00$ is 0.0228. This value is the probability (P) that a single measurement (chosen at random) would be greater than $z = 2.00$ and is referred to as an α value. Alternatively, when multiplied by 100, it represents the percentage of all x values that have $z > 2.00$ (i.e. 2.28%).

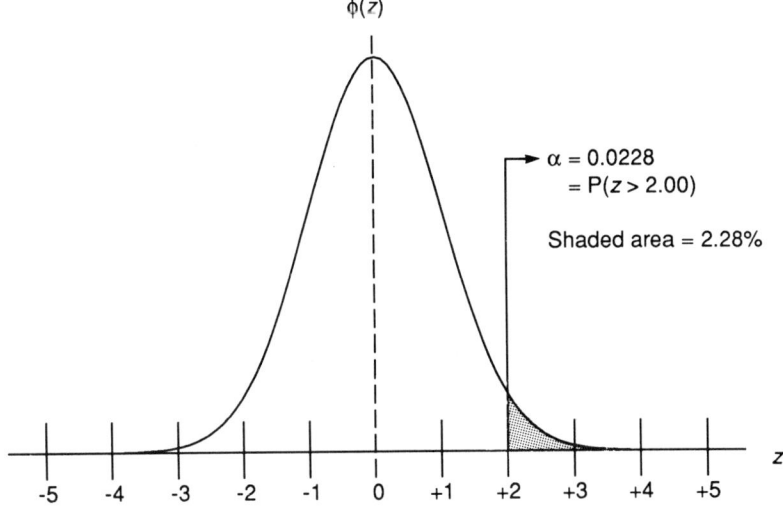

Figure 5.8 Illustration of α values in standard normal tables

The table in Appendix B sets out the probabilities of exceeding positive values of z. However, the table can also be used to calculate other probabilities, such as probabilities of being less than positive values of z, less or more than negative values of z, between two values of z, etc. Calculations associated with these other applications of the table utilize two properties of the standard normal pdf:

■ The curve is symmetrical about its mean, which is zero.
■ The total area under the curve is 1.

■ 5.7 Normal probability determinations

General procedure

The procedure for determining probabilities, in situations of occurrences above, below or between specified values, is as follows:

■ Transform the x values into standard normal deviates (z values).
■ Determine an α value for each z value and calculate the required probability, using a method to suit the situation in hand. When choosing a method, it is helpful to sketch the distributions of measurements (x values) or derived information (z values) and the required probabilities. Rough sketches are an invaluable aid to logic in any problem-solving activity.

The remainder of this section describes the principal methods:

■ Direct use of the standard normal probability table.
■ Use of the symmetry property.
■ Use of the total area property.
■ Various combined applications of the above.

Direct use of the standard normal probability table

For occurrences above a value higher than mean

For purposes of scheduling, it was important to know how often the time taken for a certain maintenance task might exceed one hour. In other words, what is the probability that the time (x), on any particular occasion, is greater than 60 minutes? Actual times were observed on 100 occasions; their histogram looked bell shaped, their mean was calculated as 50 minutes and their standard deviation as 5 minutes.

The required probability is determined by the following method which uses the standard normal probability table. When there is familiarity with the method, some of the detail may be left out − it is given here for completeness.

1. *Sketch the distribution for x.* See Figure 5.9(a) where the curve is symmetrical and bell shaped and the distribution has a mean (μ) of $x = 50$ minutes (at the peak of the curve) and standard deviation (σ) of 5 minutes (the curve approaches the horizontal axis at $\mu \pm 3\sigma$, i.e. at about $x = 35$ and 65 minutes). The required probability is represented by the shaded area, which lies to the right of $x = 60$ minutes.

    Note that a rough sketch will do – details do not need to be accurate.

2. *Standardize* the *x* value of 60 minutes to a *z* value: in other words, convert 60 to its corresponding standardized deviate:

$$z = \frac{x - \mu}{\sigma} = \frac{60 - 50}{5} = 2.00.$$

3. *Sketch the standardized distribution.* See Figure 5.9(b), where the curve is identical to the curve in Figure 5.9(a), but *x* values have become their corresponding *z* values. The probability is represented still by a shaded area now to the right of $z = 2.00$.

    Note that the shaded areas of Figures 5.9(a) and 5.9(b) are exactly equal.

4. *Use the table* in Appendix B. The shaded area is the α-value (0.0228) against $z = 2.00$. Since the two shaded areas are equal, the probability that the task takes more than 60 minutes on any one occasion is 0.0228.

5. *Report* the finding in familiar language. For example, if the task is done on many occasions, it would be expected to take more than 60 minutes on just over 2% of the occasions, or about one time in fifty.

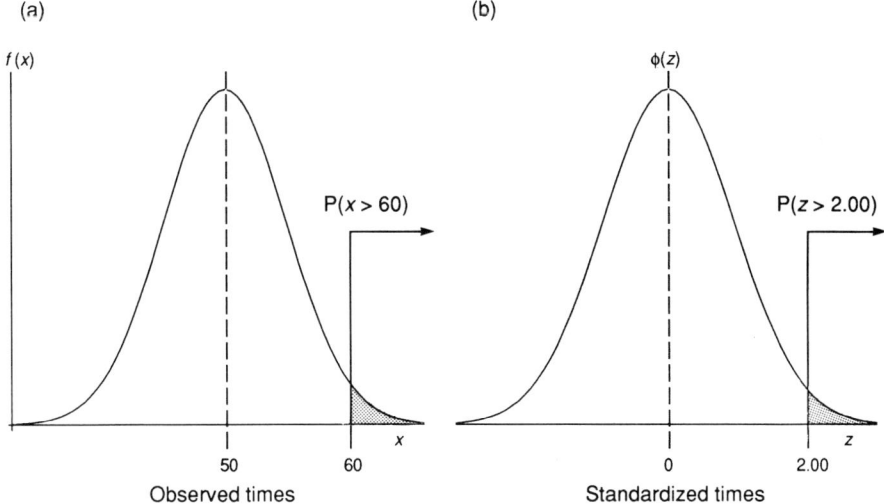

Figure 5.9 Distribution of times for a maintenance task

Using the symmetry of the standard normal distribution

For occurrences below a value lower than mean

In the situation described above, it was also necessary to know the probability that the time taken for the task might be less than 42.3 minutes.

1. *Note* that only one diagram needs to be sketched.
2. *Find the relevant z value.* In this example, the required probability is for occurrences less than $x = 42.3$, i.e.

$$z = \frac{42.3 - 50}{5} = -1.54.$$

The probability is expressed mathematically as $P(z < -1.54)$.

3. *Use the table* in Appendix B. It gives an α value of 0.0618 against $z = +1.54$, which is the shaded area to the right of $z = 1.54$ in Figure 5.10.
4. *Because of the symmetry* of the distribution, this area is equal to the shaded area to the left of $z = -1.54$. Hence $P(z < -1.54) = 0.0618$.
5. *Alternatively*: the percentage of tasks which are accomplished in less than 42.3 minutes is expected to be 6.18%.

Using the area under the standard normal curve is 1

For occurrences below a value higher than mean

In the same situation as the previous example, suppose that the requirement was to find the probability that the task might take less than 57.7 minutes.

1. *Find the relevant z value.* In this example, the required probability is for occurrences less than $x = 57.7$, so

$$z = \frac{57.7 - 50}{5} = 1.54.$$

The probability is expressed mathematically as $P(z < 1.54)$.

2. *Use the table* in Appendix B. It gives an α value of 0.0618 against $z = 1.54$.
3. *The total area under the distribution curve is 1*. Therefore, the shaded area in Figure 5.11 to the left of $z = 1.54$ is $1 - 0.0618 = 0.9382$, which is $P(z < 1.54)$.
4. In other words, the probability is 0.9382 that any one specific task is completed within 57.7 minutes. Alternatively, 93.8% of all tasks should be completed within this time.

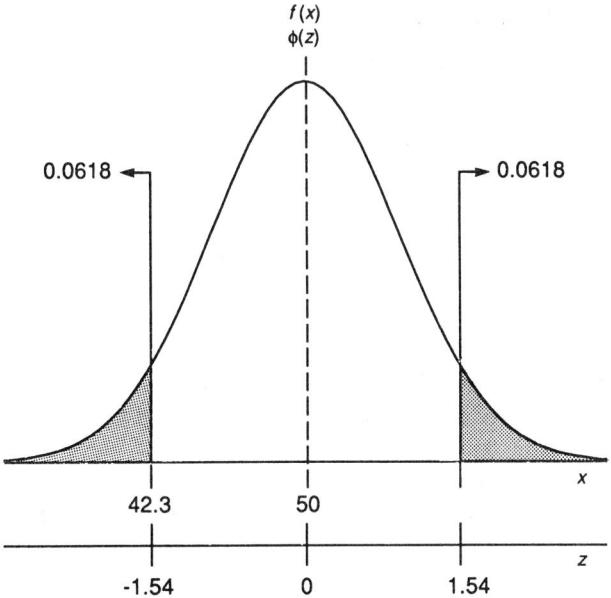

Figure 5.10 Probabilities below values lower than mean

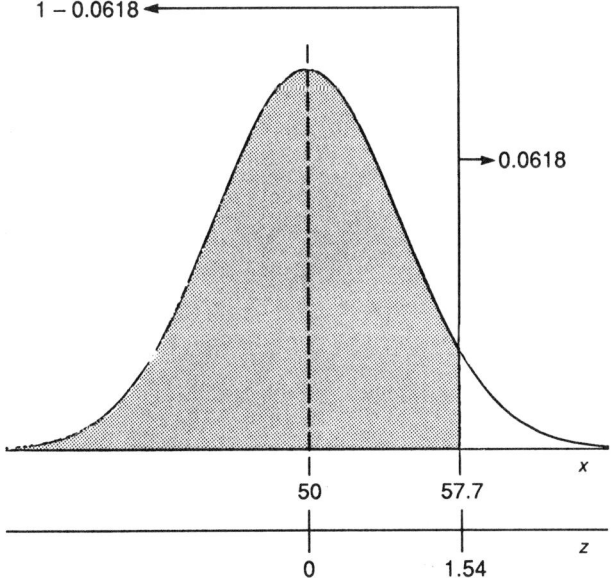

Figure 5.11 Probabilities below values higher than mean

Combined applications of methods

For occurrences between two values, both higher than mean

Again in the situation described for previous examples, suppose that the requirement was to find the probability that tasks might take between 55 and 60 minutes.

1. *Find the relevant z values.* In this example, the required probability is for occurrences between $x = 55$ and $x = 60$, i.e.

$$z = \frac{55 - 50}{5} = 1 \quad \text{and} \quad z = \frac{60 - 50}{5} = 2.$$

The probability is expressed mathematically as $P(1 < z < 2)$.

2. *Use the table* in Appendix B. It gives values of $\alpha_1 = 0.1587$ against $z = 1$ and $\alpha_2 = 0.0228$ against $z = 2$. The α values represent the areas lying to the right of their respective z values.

3. *The required probability* is the shaded area in Figure 5.12, i.e.

$$\alpha_1 - \alpha_2 = 0.1587 - 0.0228 = 0.1359.$$

4. *Alternatively*: 13.6% of tasks are expected to take between 55 and 60 minutes.

5. *Also*, this example illustrates that in any normal distribution, 13.59% of values are likely to fall between 1 and 2 standard deviations above the mean. This is the value given in Figure 5.4, where the other percentages were calculated in a similar way.

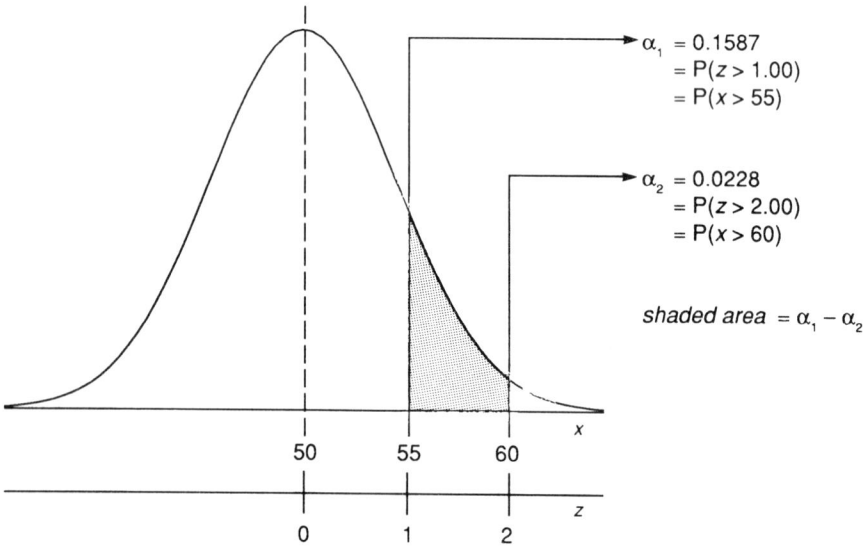

Figure 5.12 Probabilities between two values higher than mean

For occurrences above a value lower than mean

This situation is illustrated in Figure 5.13(a). The procedure is the same as for occurrences above a value higher than mean, but bearing in mind the symmetry property: in other words, where α_1 is the value given in the probability table against $+z_1$:

$$P(z > -z_1) = 1 - \alpha_1.$$

For occurrences between two values, both lower than mean

This situation is illustrated in Figure 5.13(b). The procedure is the same as for occurrences between two values higher than mean, but bearing in mind the symmetry property. So, where α_2 and α_3 are the values given in the table against $+z_2$ and $+z_3$ respectively:

$$P(-z_3 < z < -z_2) = \alpha_2 - \alpha_3.$$

For occurrences between two values, one higher and one lower than mean

This situation is illustrated in Figure 5.14 and the procedure is described through an example that uses the crankshaft flange separation data shown in Figure 5.6. The distribution of separations (x values) has a mean (μ) of 54 and a standard deviation (σ) of 16, and the units are ten-thousandths of an inch (as explained previously, x values are plus 15.0000 inches).

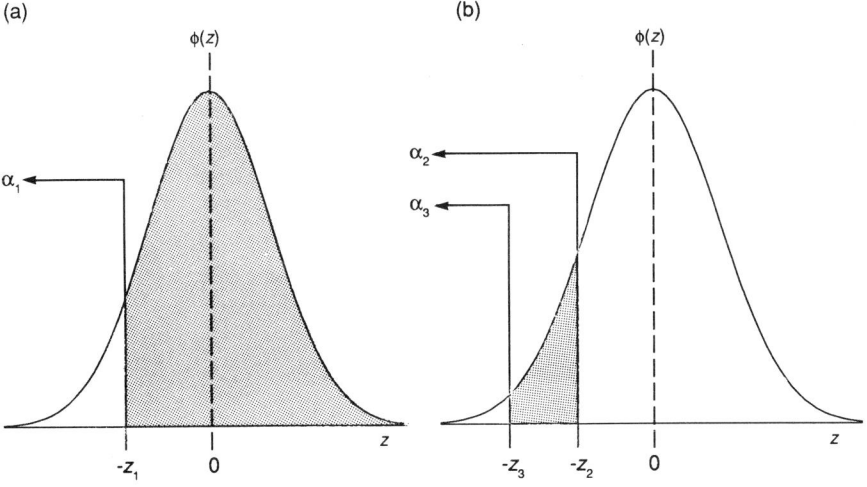

Required probabilities = Shaded areas

Figure 5.13 Use of symmetry

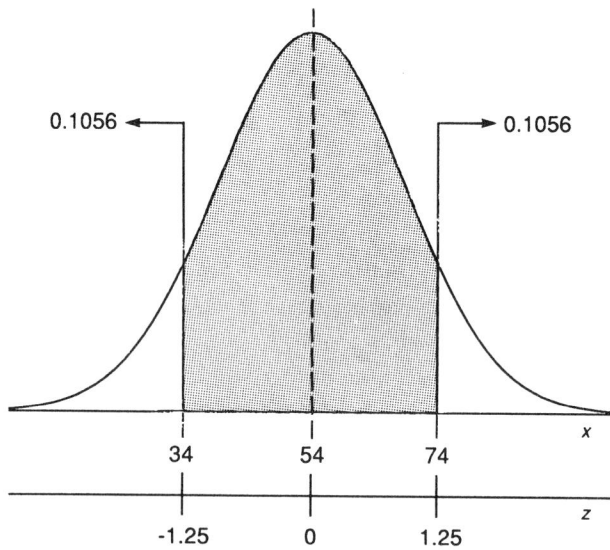

Required probability = Shaded area = 1 − 2(0.1056) = 0.7888

Figure 5.14 The probability of separations between $\mu \pm 20$

The equipment that machines the flange separation is modern and, although its setting can be adjusted (to achieve the nominal mean of 50), it is unlikely to produce less variation in separations. Figure 5.6 indicates that between $\pm 4\sigma$ about the mean, the likely range is 128 and about one crankshaft per month will be outside this range. The machine is able automatically to measure, sort and identify crankshafts in groups according to size (such as small, medium and large).

At the next stage in the manufacturing process, the equipment will only accept a range of 40. This equipment can be reset on a batch basis to accommodate several ranges of 40. The production manager has to plan an economic batch and resetting schedule. Part of the information needed is the relative occurrence of small, medium and large separations.

For convenience, small-sized separations were specified as between 60 and 20 units below mean, medium as mean ± 20, and large as between 20 and 60 above mean. It was accepted that there could be a few units outside these ranges; any loss would be measured against the alternative of more crankshaft groups.

The procedure adopted to obtain information for the production manager about the proportions of small, medium and large separations was as follows:

1. *Obtain data.* Measurements were made on 125 components.
2. *Examine the data.* The measurements were normally distributed with a mean of 54 and a standard deviation of 16.

3. *Sketch the data.* Figure 5.6 showed a normal distribution (it is symmetrical and bell shaped) with mean (μ) at $x = 54$ (at the peak of the curve) and standard deviation (σ) of 16 (the curve approaches the horizontal axis at $\mu \pm 3\sigma$, i.e. at about $x = 6$ and 102). The sketch is repeated in Figure 5.14, where the probability of medium-sized components occurring is represented by the shaded area, which lies within $x = 34$ and $x = 74$.

4. *Standardize* the x-values of 34 and 74 to z values. In other words, convert 34 and 74 to their corresponding standardized deviates:

$$z = \frac{x_1 - \mu}{\sigma} = \frac{34 - 54}{16} = -1.25,$$

$$z = \frac{x_2 - \mu}{\sigma} = \frac{74 - 54}{16} = 1.25.$$

5. *Sketch the standardized data.* The distribution of z values is identical to the distribution of x values. Hence, in Figure 5.14, z values have been shown below their corresponding x values. The required probability, represented by the shaded area, also lies within $z = -1.25$ and $z = 1.25$.

6. *Use the table* in Appendix B. The unshaded area to the right of $z = 1.25$ is the α value of 0.1056. Because of symmetry, the unshaded area to the left of $z = -1.25$ has the same α value. Therefore (bearing in mind that the total area under the curve is 1), the probability that any crankshaft's flange separation lies within $\mu \pm 20$ is

$$1 - 2(0.1056) = 0.7888.$$

7. *The report* is that nearly 80% of components are likely to occur in the middle-size group and about 10% are likely to occur in each of the small and large groups.

8. *In addition*, a rough estimate was made of the likely number of components that might be larger than the large-size group.

$$z = \frac{x - \mu}{\sigma} = \frac{114 - 54}{16} = 3.75.$$

For $z = 3.75$, α is approximately 0.000089 (the value has been interpolated from the table). Alternatively, 0.0089% of about 3.68 million in 20 years (say, 1 or 2 per month if there is a constant production rate) would probably be larger than the large-size group and, because of symmetry, a similar quantity would probably be smaller than the small-size group.

■ References to further reading on the topics of Chapter 5

Cass (1973) and Chatfield (1983) give clear introductions to the use of the normal distribution. More details are given by Walpole and Myers (1993).
See Bibliography for details of titles and publishers.

■ Self-assessment questions on Chapter 5

1. What is the proportion of the standard normal distribution lying to the right of
    (a) $z = 1.24$?
    (b) $z = 2.32$?
    (c) $z = -1.27$?
    (d) $z = 1.135$?
2. A normal distribution has a mean of 60 and a standard deviation of 5. If an item is taken at random, what is the probability that its value is
    (a) > 70?
    (b) < 54.8?
    (c) > 58.8?
    (d) $> 53.6, < 63.6$?
    Also, between what values does the middle 50% of the distribution lie?
3. The average life of a particular headlamp bulb is 2200 hours with a standard deviation of 250 hours. Assuming that bulb life has a normal distribution, how many out of a batch of 1000 bulbs would be expected to need replacement
    (a) before 1200 hours?
    (b) before 1800 hours?
    (c) before 2500 hours?
    (d) before 3000 hours?
4. The inside diameter measurements of a batch of piston rings are normally distributed with a mean of 3.00 cm and a standard deviation of 0.01 cm. What percentage of the rings will have an inside diameter
    (a) exceeding 3.02 cm?
    (b) under 3.0105 cm?
    (c) between 2.99 cm and 3.01 cm?
5. 'Official' information states that the petrol consumption for a particular motor vehicle is 30 miles per gallon. Calculation from actual measurements, of a valid sample of these vehicles, shows that the standard deviation is 0.8 miles per gallon. If the manufacturer wants to make sure that less than 1% of these vehicles have petrol consumption worse than the 'official' statement, what should be the nominal or optimum petrol consumption specification?

6. Packets of a manufactured product have weights that are normally distributed and the packets are advertised as containing an average of 250 g. The purchaser will not bother if a packet is overweight, but the manufacturer has to guard against packets being underweight! From experience, the manufacturer has found that no more than 1 in 50 packets should contain less than 250 g (in other words, the risk of a packet being underweight is to be less than 0.02). If the manufacturer wants to aim at a nominal weight of 255 g, what is the maximum allowable value of the standard deviation, and what is likely to be the variation in weight among the middle 99% of packets?

6 | Other Probability Distributions

■ 6.1 Introduction

Although the normal distribution provides an appropriate mathematical model for many practical situations, there are some situations in which the probability distribution is not adequately represented by the symmetrical bell-shaped probability density curve that was described in Chapter 5. Also, as described in Chapter 4, only continuous variables have probability densities and the curves representing these may be of many different shapes.

A full description of a continuous probability distribution is given by quoting the formula for its probability density function, $f(x)$, and including the values of any parameters (such as the constants μ and σ in the formula for the normal probability density function).

Similarly, for a discrete distribution, a full description is provided by the formula for its probability function, which is also designated $f(x)$.

However, for many purposes, the full descriptions are not necessary and an adequate description of any discrete or continuous distribution can be provided by quoting the following:

- The family to which the distribution belongs.
- A measure of middle value: for example, the value of the mean.
- A measure of spread: for example, the value of variance or standard deviation.
- Possibly, some other measure of shape: for example, skewness.

These measures can be expressed, usually quite simply, in terms of the parameters of the distribution, so that if the values of the measures are known, the values of the parameters can be found and vice versa. The accuracy of the relationship between measure and parameter values is discussed in Chapter 7.

■ 6.2 Uniform distributions

A uniform distribution describes a situation where all possibilities in a situation have an equal probability of occurring. The possibilities might have discrete or continuous values.

Discrete uniform distributions

In tossing six-sided dice, the score from one toss of a single die can have possible values 1, 2, 3, 4, 5 and 6. These can be considered as values of a discrete variable, x, each having the same probability of 1/6 if the tossing is unbiased.

The last digit of the registration number (or chassis or engine number) of a vehicle could be any one of the ten digits 0 to 9. If the vehicle were selected at random, it would be reasonable to consider the last digit as a discrete variable, x, for which each possible value had probability 1/10. If there were good reason to suppose that certain digits were favoured or avoided, the probabilities should not be equal.

The probability distributions for x in both these examples belong to the same family, where x can take k possible discrete values, each value having a probability of $1/k$ so that the probability function may be illustrated as in Figure 6.1 and written as

$$f(x) = 1/k \text{ for } x = a, a+1, \ldots b$$

where a denotes the smallest possible value and b the largest, so that $b = a + k - 1$.

This formula describes what is called the *discrete uniform family of distributions*. Once the smallest value has been specified (usually as 0 or 1), the only parameter that needs to be known is k, which is the number of possible values of x.

Continuous uniform distributions

If, in the above formula, k becomes very large, x becomes effectively a continuous variable and the distribution should then be described by a probability density function which is constant over the range a to b. In other words, the description becomes

$$f(x) = 1/(b-a) \quad \text{for any } x \text{ between } a \text{ and } b.$$

This expression describes the typical probability density function for the continuous uniform family of distributions. These are sometimes called rectangular distributions because of the shape of their probability density function, which is illustrated in Figure 6.2.

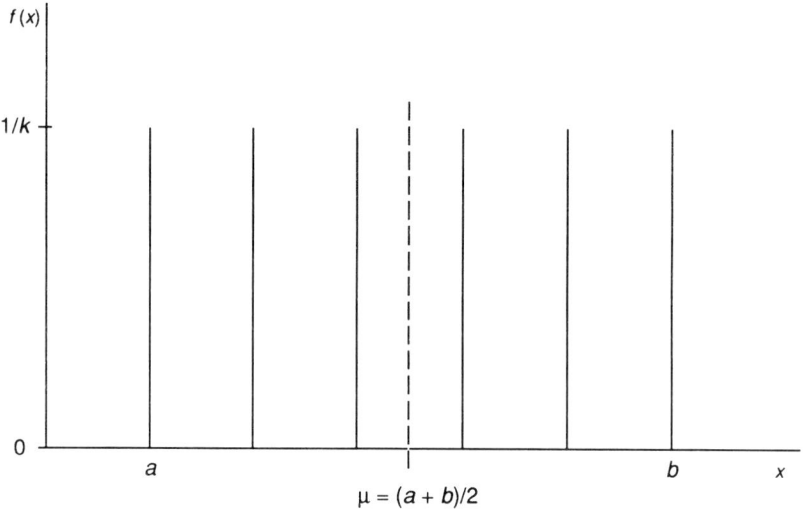

Figure 6.1 Probability function for a discrete uniform distribution

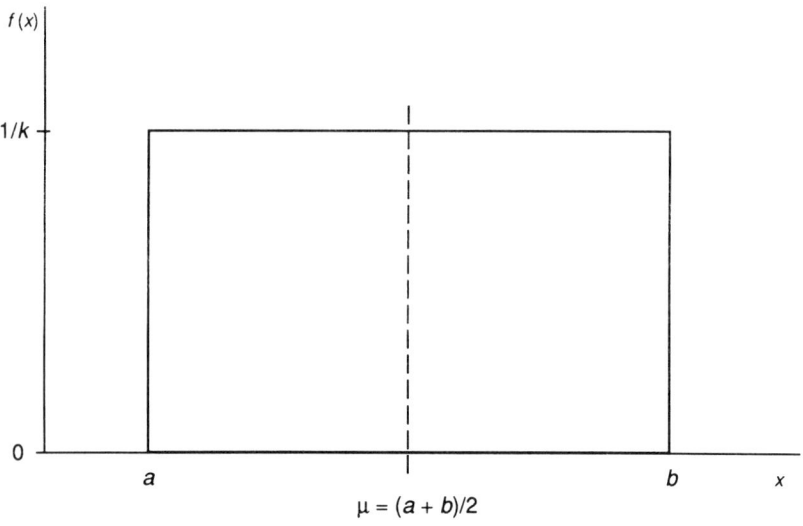

Figure 6.2 Probability function for a continuous uniform (or rectangular) distribution

Formulae for uniform distributions

Table 6.1 summarizes formulae that give $f(x)$, the mean and the variance, for the basic discrete and continuous distributions considered in this book.

Table 6.1 Basic probability distributions

Discrete	Probability function $f(x)$	Range	Mean	Variance
Uniform	$1/(b - a + 1) = 1/k$	a to b	$(a + b)/2$	$(b - a)(b - a + 2)/12$ $= (k^2 - 1)/12$
Binomial	$\dfrac{n!}{x!(n - x)!}\, p^x q^{n-x}$	0 to n	np	npq
Poisson	$\dfrac{\lambda^x \exp(-\lambda)}{x!}$	0 to ∞	λ	λ
Geometric	pq^{x-1}	1 to ∞	$1/p$	q/p^2
Continuous	Probability density $f(x)$	Range	Mean	Variance
Uniform or rectangular	$1/(b - a) = 1/k$	a to b	$(a + b)/2$	$(b - a)^2/12 = k^2/12$
Normal	$\dfrac{1}{\sigma \sqrt{2\pi}} \exp\left[-\dfrac{1}{2}\left(\dfrac{x - \mu}{\sigma}\right)^2 \right]$	$-\infty$ to ∞	μ	σ^2
Exponential	$r \exp(-rx)$	0 to ∞	$1/r$	$(1/r)^2$
Weibull	$\dfrac{\beta}{\eta}\left(\dfrac{x - \gamma}{\eta}\right)^{\beta - 1} \exp\left[-\left(\dfrac{x - \gamma}{\eta}\right)^{\beta} \right]$	0 to ∞		(see below)

The formulae for the mean and variance of a Weibull distribution involve the *gamma function*, which can be found in mathematical tables. However, when $1/\beta$ is a whole number, the definitions can be given in terms of factorials:

$$\text{mean} = \gamma + \eta(1/\beta)! \qquad \text{variance} = \eta^2\{(2/\beta)! - [(1/\beta)!]^2\}$$

When $\beta = 1$ and $\gamma = 0$, the Weibull is the same as an exponential distribution with $r = 1/\eta$.

Note that, for both the discrete and the continuous versions of the uniform distribution, the mean is simply the average of the largest and smallest possible values, $(a + b)/2$. Applying this to a six-sided die, a mean score of $(1 + 6)/2 = 3.5$ is obtained. As pointed out in Section 4.13, this score does not correspond to any observable value of x, since the only possible scores are whole numbers between 1 and 6. As a general rule, for any discrete distribution, the mean need not be an observable value.

A useful formula for $F(x)$, the cumulative probability or cumulative distribution function, of a continuous uniform distribution is

$$F(x) = (x - a)/(b - a).$$

In other words, the probability of a value less than or equal to x is the ratio of the difference between x and the lowest possible value to the difference between the highest and lowest possible values (the range).

Random numbers

There are various electronic and mathematical methods of generating numbers that behave as if they were observations from a uniform distribution; such numbers are called *random numbers*. Some statistical uses of random numbers are described in Section 12.8. Among the non-statistical uses is the selection of premium bond prize winners in the United Kingdom.

■ 6.3 Binomial distributions

The *binomial family of distributions* is applicable to any process that can be considered as a series of independent trials with only two possible outcomes at each trial (for example, a screening activity where each item checked is only either go or no-go).

Success and failure

In this situation, it is usual to label one type of outcome as *success*, and to say that this has probability p at each trial. The other type of outcome is labelled *failure* and has probability $q = 1 - p$ at each trial. The number of trials being considered is denoted by n and the number of successes in n trials is denoted by x.

The choice of which type of outcome is to be called success is more or less arbitrary. However, it is common practice to use this label for the type of outcome which is considered to be the most important even though, sometimes, this may seem to be slightly contradictory.

> For example, suppose that a study were being carried out into damage to the paintwork of new cars in transit from factory to dealer. Inspection of each car would reveal 'damage' or 'no damage': in other words, there are two possible outcomes, so the first condition for a binomial distribution is satisfied. Since the study is directed towards detecting damaged cars, it would be usual to consider finding a damaged car as a success, even though the occurrence of the damage would represent some sort of failure in the arrangements for safe delivery.

Independent trials

Having sorted out the meaning of 'success' and 'failure' in the context of a particular problem, the next step in seeing whether a binomial distribution is applicable is to check that both the following conditions are satisfied:

■ That the trials are conducted independently.
■ That the probability of success remains the same at each trial.

Provided that these conditions are satisfied, the probability of observing x successes in n trials can be calculated by using the probability function of the binomial distribution.

Binomial probability function

The expression for the probability function of the binomial distribution includes the notation for combinations that was introduced in Section 4.15.

$$f(x) = {_n}C_x p^x q^{n-x} \text{ for } x = 0, 1, 2, \ldots n.$$

The formula has two parts:

${_n}C_x$ = the number of patterns of success and failure that contain exactly x successes and $n - x$ failures in n observations;

$p^x q^{n-x}$ = the probability of each individual pattern that contains exactly x successes and $n - x$ failures in n observations.

For example, suppose that 3% of cars suffer damage to paintwork in transit to dealers. Find the probability that a dealer receiving 50 cars will find that none of them is damaged. This is treated as a binomial problem and finding a damaged car is a 'success'.

Since 3% of cars are damaged, the probability of selecting a single car at random and finding it to be damaged is $3/100 = 0.03$, which is the value of p for this problem. The corresponding value of q is $1 - 0.03 = 0.97$.

The number of cars received by the dealer represents the number of trials, so $n = 50$. The number of damaged cars in this 50 will be x, so $x = 0$ if there are no damaged cars. Substitute these values into the formula for $f(x)$:

$$f(0) = {_{50}}C_0 p^0 q^{50-0}.$$

There will only be one pattern corresponding to no successes and 50 failures in 50 trials, so ${_{50}}C_0$ will equal 1. This can be checked by evaluating

$${_{50}}C_0 = 50!/[0!\ 50!].$$

Note that $0! = 1$ (Section 4.15), and because $p^0 = 1$, the required probability reduces to

$$f(0) = q^{50} = (0.97)^{50} = 0.218 \text{ to 3 decimal places.}$$

Exactly the same numerical answer would have been obtained by considering defining an undamaged car as a 'success', so that $p = 0.97$ and $q = 0.03$. The problem then becomes that of finding the probability of 50 successes in 50 trials, so n is still 50 but x is now also 50.

$$f(50) = {_{50}}C_{50} p^{50} q^0 = (0.97)^{50} = 0.218 \text{ as before.}$$

As long as p and x both refer to the type of outcome that has been defined as 'success' the formula will always give the correct answer. If there is any doubt about labelling the outcomes, 'success' should be defined as the outcome with the smaller probability (in other words, always make p smaller than q); usually, this will make it easier to use tables, as described below.

Binomial probability tables

When there are several values of x, direct application of the formula will involve an inconvenient amount of calculation. This can be avoided by using tables that have been calculated from the formula. These tables may show the probabilities of individual values, $f(x)$, cumulative probabilities, $F(x)$, or the complements of the cumulative probabilities, $1 - F(x)$. A wide range is readily available in books devoted to mathematical tables.

The following two examples illustrate the advantages of using tables. They use the table in Appendix C, which shows cumulative probabilities, $F(x)$, for a selection of values of n and p.

Example 1: Determination of cumulative probability

Suppose that a dealer receives 50 cars through a delivery process which, over an extended period, results in 3% of cars receiving paintwork damage. Find the probability that more than 2 cars in the batch have damaged paintwork.

Taking 'damage' as 'success', $p = 0.03$ and $n = 50$, as before. The required probability is $P(x > 2)$ which could be found by writing

$$P(x > 2) = f(3) + f(4) + \cdots + f(50)$$

and using the formula to evaluate each $f(x)$ term for $x = 3$ to 50.

A more efficient method is to use the complement rule and write

$$P(x > 2) = 1 - P(x \leqslant 2) = 1 - F(2)$$

where $F(2) = f(0) + f(1) + f(2)$ and only involves evaluating three $f(x)$ terms. However, using the table in Appendix C, it is only necessary to look up $F(2)$ and subtract this from 1.

For $n = 50$, $p = 0.03$ and $x = 2$, the value $F(2) = 0.8108$. Hence the probability of more than 2 damaged cars in the batch of 50 is $1 - 0.8108 = 0.1892$. In other words, just under 19% of batches of 50 cars would be expected to include more than 2 damaged cars per batch.

Example 2: Determination of individual probability

Tables showing values of the cumulative distribution function, $F(x)$, can be used also to find values of the probability function, $f(x)$. For example, it was noted in Example 1 above that the cumulative probability for $x = 2$ is $F(2) = f(0) + f(1) + f(2)$.

From the definition of $F(x)$, the corresponding cumulative probability for $x = 1$ is

$$F(1) = f(0) + f(1)$$

which differs from $F(2)$ by not including the term $f(2)$. Hence $f(2)$ can be evaluated by finding the difference between $F(2)$ and $F(1)$:

$$f(2) = F(2) - F(1).$$

In general, $f(x) = F(x) - F(x - 1)$ for $x > 0$ and $f(0) = F(0)$.

> Use the tables in Appendix C to find the probability that the car dealer will find exactly one car with damaged paintwork in a batch of 50, assuming that $p = 0.03$ as before.
>
> This involves evaluating $f(x)$ for $x = 1$, so look up $F(1)$ and $F(0)$, then find the difference:
>
> $$f(1) = F(1) - F(0)$$
> $$= 0.5553 - 0.2181$$
> $$= 0.3372 \text{ or } 0.337 \text{ to 3 decimal places.}$$
>
> In other words, approximately one-third of all batches of 50 cars would be expected to include exactly one car with damaged paintwork. The same result can be obtained by using the formula for $f(x)$ with $x = 1$, but that involves much more calculation.

Approximations to binomial probabilities

Although binomial distributions provide a basic model for many industrial situations, it is not always convenient to use this exact model. The binomial probability formula is awkward to use, and tables of binomial $f(x)$ or $F(x)$ need to cover a wide range of values of n and p, so they tend to be rather cumbersome. Fortunately, in many practical situations it is possible to get good approximations to the exact binomial probabilities by using some other distribution. These approximations are explained in Section 6.5.

■ 6.4 Poisson distributions

The *Poisson* family of distributions is used to describe situations that are concerned with counting the number of times that a certain type of event occurs, within some specified *opportunity frame*. The opportunity frame will depend upon what type of event is being counted. It might be a *space frame*, representing some physical region such as the surface of a car door, if defects in paintwork are being counted. It might be a *time frame*, say a week, if something like the numbers of issues from a tool store are being counted.

Poisson probability function

The calculation of probabilities, that events occur, requires knowledge of the mean number of events expected in the appropriate opportunity frame. This mean number is usually denoted by the Greek letter λ (lambda). The probability of observing exactly x events in the specified frame is then given by the Poisson probability function

$$f(x) = \lambda^x e^{-\lambda}/x! \text{ for } x = 0, 1, 2, \ldots$$

Theoretically, this function has non-zero values for all whole-number values of x up to infinity, but in practice the values of $f(x)$ become so small for large values of x that they can be ignored.

When control charts are used to monitor the number of events (Section 10.12), the letter c is generally used instead of x to denote the *count*: that is, the number recorded in the specified opportunity frame.

Poisson process

When events occur in such a way that the probability of observing x events in the specified frame is given by this formula, the events are said to be following a *Poisson process*. There are strict mathematical conditions for such a process, but in ordinary language the main requirements are as follows:

- Events should occur at a constant underlying average rate.
- They should occur rather rarely within the opportunity frame (so there are many possible opportunities when they do not occur).
- They should occur without affecting each other, i.e. they occur 'at random'.

Counts of events such as 'number of pot-holes per mile of highway', 'weekly number of accidents at a particular crossroad', 'daily number of demands for components from a dealer' or 'number of goals in a football match' behave very nearly as if they follow a Poisson process, and so this can be used as an appropriate model in theoretical studies.

When a time frame is being considered, the process is defined by the *average rate of occurrence per unit time*, denoted by r, and the *time period* being considered, denoted by t. The mean number of events expected in time t is then $\lambda = rt$. This value of λ is then used in the formula for $f(x)$. Notice that, as shown in Table 6.1, this parameter λ gives the value of both the mean and the variance of a Poisson distribution.

Example using the Poisson distribution

The Poisson distribution formulae can be used, in many counting situations, to help management planning: for example, in determining the need for additional maintenance resource to cope with breakdowns in a machine shop, where it has been established that the breakdowns meet the conditions of a Poisson process and that the mean rate is 2 per day.

Bearing in mind the existing resource, the information required was the probabilities of (a) exactly 10 and (b) more than 15 breakdowns in a working week of 5 days.

The average rate is $r = 2$ breakdowns per day, the period is $t = 5$ days, so $\lambda = rt = 10$ breakdowns in 5 days, on average.

(a) The required probability is $f(10)$, which can be calculated either from the formula for the Poisson $f(x)$ or from the tables of the Poisson $F(x)$ in Appendix D.

Using the formula

$$f(10) = \lambda^{10}e^{-\lambda}/10! = 10^{10}e^{-10}/10!$$
$$= 0.125 \text{ to 3 decimal places.}$$

Alternatively, using the tables with $\lambda = 10$:

$$f(10) = F(10) - F(9) = 0.5830 - 0.4579$$
$$= 0.125 \text{ to 3 decimal places.}$$

In other words, since $0.125 = 1/8$, exactly 10 breakdowns should be expected in about one week out of eight over an extended period.

(b) To find the probability of more than 15 breakdowns in any given week, the only sensible method is to use tables of cumulative probabilities and the complement rule.

$$P(x > 15) = 1 - P(x \leqslant 15) = 1 - F(15)$$
$$= 1 - 0.9513 = 0.049 \text{ to 3 decimal places.}$$

Hence, if the maintenance department had sufficient capacity to deal with up to 15 breakdowns per week (50% above the mean level), it should be able to cope on all but about 5% of occasions: that is, except during 5 weeks in every 2 years.

■ 6.5 Approximations

Poisson approximation to the binomial

The Poisson distribution can be used as an approximation to the binomial when n is large (say 50 or greater) and p is small (less than about 0.1).

If these conditions are met, the means of the two distributions can be equated by putting $\lambda = np$. This ensures that both distributions are centred on the same value.

The variance of the Poisson is also λ (Table 6.1) and now equals np; the variance of the binomial is npq. Provided p is small, q will be close to 1 and so the variance of the binomial will be almost equal to np as well.

Hence, the two distributions will be almost the same shape as far as middle and spread are concerned, so they should have similar probabilities at each value of x.

> Use of the Poisson approximation is illustrated by the following example, to find the probability of exactly one car with paintwork damage in a batch of 50 cars when the underlying process produces 3% damaged cars.
>
> For the binomial, $n = 50$ and $p = 0.03$. Hence n is sufficiently large and p is sufficiently small for the approximation to be reasonable.
>
> Therefore put $\lambda = np = 1.5$ and use the Poisson tables of $F(x)$ in Appendix D:
>
> $$f(1) = F(1) - F(0) = 0.5578 - 0.2231 = 0.3347.$$

This compares with 0.3372 obtained from an appropriate binomial table. Similarly, $f(2)$ is 0.2555 from the binomial and 0.2510 from the Poisson approximation. The values of $f(x)$ for other values of x also differ slightly in the third decimal place, but are effectively the same for most practical purposes.

Normal approximation to discrete distributions

The binomial and Poisson distributions are for discrete data. However, their probabilities for individual values of x may be represented as areas or blocks that are one unit wide and centred on the values of x to which they relate (Figure 4.11). Taken together, these blocks make up a probability histogram for the discrete distribution.

The probability density function of a normal distribution can be super-imposed on the probability histogram of a discrete distribution. If both distri-butions have the same mean and variance, as shown in Figure 6.3, the areas of the blocks of the histogram may be compared with areas under the normal curve between the same values of x.

In some circumstances, the areas under the normal curve will provide good approximations to the histogram areas. Under these conditions, the theory of the normal distribution can be usefully applied to discrete distributions.

Normal approximation to the binomial

For a good normal approximation to the binomial distribution, p should be close to 0.5 and n greater than about 10, so that np is greater than about 5.

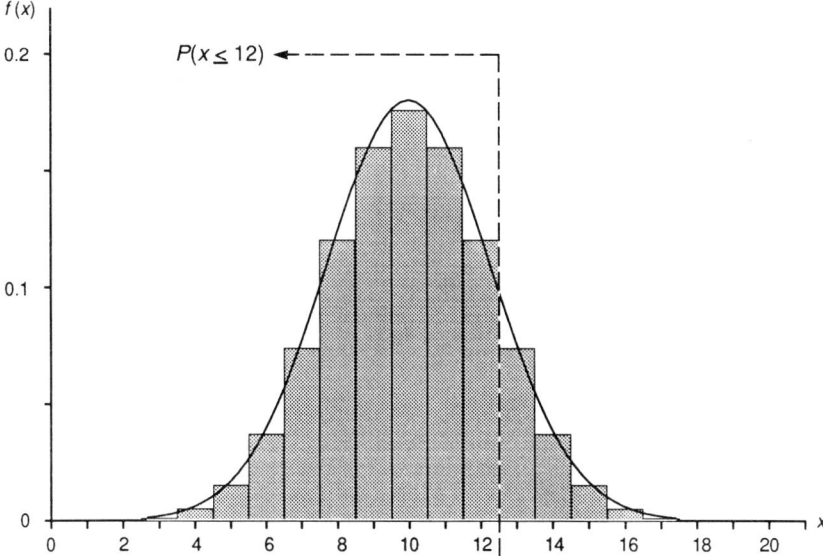

Figure 6.3 Normal approximation to a binomial distribution with
$n = 20$ and $p = 0.5$

As n increases, p can depart further and further from 0.5, but np should still be greater than about 5 (this is to ensure that the normal approximation does not allocate any appreciable probability to negative values of x, which are impossible for the true binomial).

If the values of n and p satisfy these conditions, the mean of the normal distribution can be put equal to np and its variance equal to npq.

For example, compare probabilities that are shown for the binomial distribution in Appendix C where $n = 20$ and $p = 0.5$ with the corresponding probabilities obtained from the normal approximation.

■ Equating means and variances gives $\mu = np = 10$ and $\sigma^2 = npq = 5$, so $\sigma = \sqrt{5}$.

■ Consider first the probability that x equals 10, represented by the histogram block lying between $x = 9.5$ and $x = 10.5$.

■ The methods described in Section 5.7 are used to find the area under the normal curve between these two values. The z value corresponding to $x = 10.5$ is $0.5/\sqrt{5} = 0.22$ and that corresponding to $x = 9.5$ is $-0.5/\sqrt{5} = -0.22$. The α value corresponding to $z = 0.22$ is 0.4129, so the required probability is $P(-0.22 < z < 0.22) = 1 - 2(0.4129) = 0.1742$.

From Appendix C, the true binomial probability is $f(10) = F(10) - F(9) = 0.5881 - 0.4119 = 0.1762$, so *the normal approximation is slightly out in the third decimal place.*

■ Next consider the probability that x is less than or equal to 12. From Appendix C, the true binomial value is $F(12) = 0.8684$, which is represented in Figure 6.3 by the total histogram area to the left of $x = 12.5$, which is the upper limit of the block centred on $x = 12$.

For the normal approximation, the probability will be
$$P(x < 12.5) \quad = P[z < (12.5 - 10)\sqrt{5} = 1.12] = 1 - 0.1314 = 0.8686,$$
where the fourth decimal place is slightly out.

■ However, notice that the area lying to the left of $x = 12$ under the normal curve equals $P[z < (12 - 10)\sqrt{5} = 0.89] = 1 - 0.1867 = 0.8133$, which is *a much worse approximation* because it leaves out approximately half the area of the histogram block centred on $x = 12$.

Unless n is very large, it is most important to remember this *half-unit correction* when using a normal approximation to any discrete distribution. When n is very large, there are many possible values of x over which the total probability is distributed, and correspondingly smaller probabilities are associated with each x.

Normal approximation to Poisson

For the normal distribution to provide a good approximation to the Poisson, the mean (λ) should be greater than about 20. If this condition is satisfied, means and variances can be equated and the calculation method is the same as for the binomial.

For example, usage of a certain type of cutting tool follows a Poisson process with a mean of 25 tools per week. The normal approximation can be used to find the probability that a stock of 35 of these tools will be sufficient to cover all demands in any particular week.

In this case, the weekly tool usage is x. Its mean is $\lambda = 25$, which is greater than 20, so the normal approximation should be reasonable. Put $\mu = 25$ and $\sigma^2 = 25$, so that $\sigma = 5$. The stock of 35 will be sufficient if x is less than or equal to 35.

Applying the half-unit correction, the required probability is represented by all the area to the left of $x = 35.5$, i.e.

$$P(x < 35.5) = P[z < (35.5 - 25)/5 = 2.1]$$
$$= 1 - 0.0179 = 0.9821.$$

The true Poisson probability in this case is 0.9775. Whichever is used, it can be seen that the stock of 35 should be sufficient during approximately 98 weeks out of 100: in other words, it will be insufficient about once a year.

■ 6.6 Geometric distributions

In Section 6.2 it was shown that the binomial distribution governed the probability of particular numbers of successes in a fixed number of independent trials. Sometimes, it is necessary to look at such a series of trials from a different point of view: in particular, when the required information is the probability of a particular number of trials being required in order to observe the first success. Instead of being fixed, the number of trials is now considered as the variable, x.

As before, if the probability of successes at each trial is p, the corresponding probability of failure is $q = 1 - p$. Requiring x trials to get the first success implies getting a pattern of $x - 1$ failures followed by the one success. The multiplication rule gives the probability of this pattern as the product of $x - 1$ failure probabilities, q, and one success probability, p. The order in which the terms are multiplied does not affect the numerical value of the result, and the usual convention is to write the p term first so that the probability function for the required number of trials is

$$f(x) = pq^{x-1}.$$

Obviously, at least one trial will be needed to get the first success, but there is no fixed upper limit, so this formula for $f(x)$ is valid for all whole-number values of x from 1 to infinity. Putting successive values of x into the formula gives

$$f(1) = p, \ f(2) = pq, \ f(3) = pq^2, \ f(4) = pq^3, \text{ etc.}$$

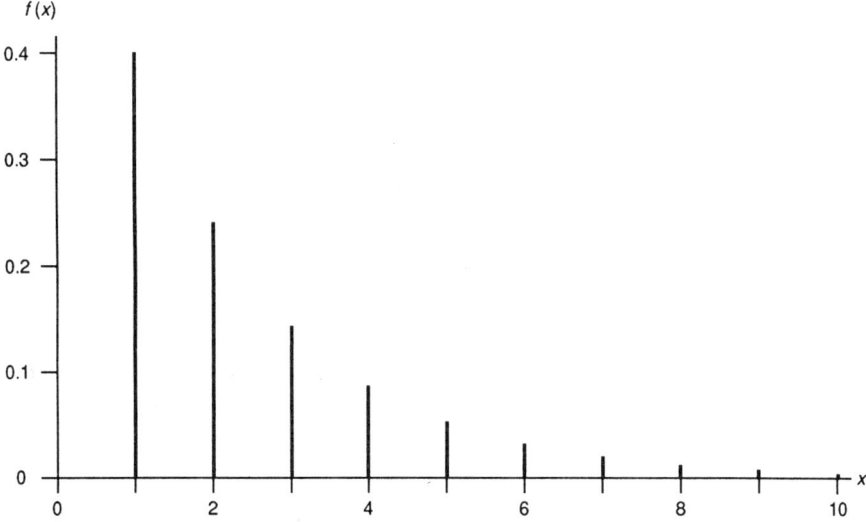

Figure 6.4 Geometric distribution with $p = 0.4$ and $q = 0.6$

which shows that the probabilities form what is called a *geometric progression*, a sequence in which each term is a fixed proportion (*q* in this case) of the previous one. This is why the name *geometric distribution* is used when probabilities are given by this formula.

Notice that the underlying probability of success at each trial remains constant (equal to *p*), but, as shown in Figure 6.4, the values of $f(x)$ get smaller as *x* increases.

Table 6.1 shows that the mean of a geometric distribution is $1/p$; this agrees with intuition, as shown in the following simple example.

It has been found that, on average, one in 20 telephone calls made by a sales office result in an order for a particular product. Supposing that the calls are independent of each other, find the mean number of calls needed to get the first order on a particular day.

Making the calls can be regarded as a series of independent trials with 'getting an order' as 'success', so $p = 1/20$. The mean number of trials required is then $1/p = 1/(1/20) = 20$.

In other words, the original information can be interpreted as saying that, on average, 20 calls will be needed to secure one order.

■ 6.7 Exponential distributions

The *exponential family of continuous distributions* provides models which are used to study many industrial phenomena such as time between machine breakdowns, length of queues at repair or processing facilities and the reliability of electronic systems. The typical probability density function for this family is of the form

$f(x) = r \exp(-rx)$ for any *x* between 0 and infinity.

The corresponding cumulative distribution function is

$F(x) = 1 - \exp(-rx)$

Graphs of this $f(x)$ and $F(x)$ are shown in Figure 6.5; expressions for the mean and variance are shown in Table 6.1. Note that the median is 0.693 times the mean and approximately 63% of possible values lie below the mean for any exponential distribution.

Exponential distributions can be thought of as the continuous equivalent of geometric distributions – compare the shape of the graphs for $f(x)$ in Figures 6.4 and 6.5.

The exponential distribution is related to the Poisson in much the same way as the geometric is to the binomial. In a Poisson process, the time between events (or the time to the next event if observation starts when the process is already running) has an exponential distribution. If the average rate of the

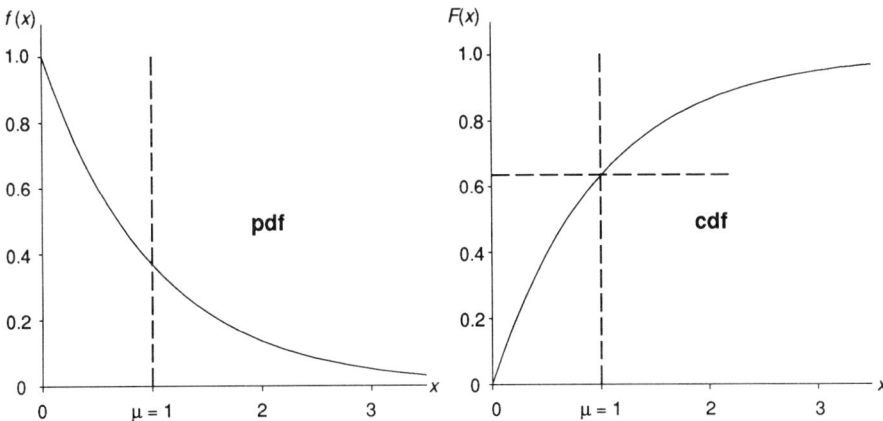

Figure 6.5 The pdf and cdf for an exponential distribution with $r = 1$

Poisson process is r, and t denotes the time between events (or time until the next event), the probability density function of t is

$$f(t) = r \exp(-rt).$$

The mean time between events is $1/r$ or the reciprocal of the rate at which the events occur.

Reliability

When the variable t represents time between breakdowns or failures of some kind, it is usual to introduce the notation $R(t)$ to denote the *reliability* of the equipment being considered. This reliability represents the probability that the equipment is still operating at some specified time, say t_1, hence its time to failure is greater than the specified value t_1. It follows that $R(t_1)$ is simply the complement of the cumulative distribution function, $F(t_1)$. In other words, as illustrated in Figure 6.6

$$R(t_1) = P(t > t_1)$$
$$= 1 - P(t < t_1) = 1 - F(t_1).$$

Hence, for an exponential distribution with mean $1/r$, the reliability function at any time t is

$$R(t) = \exp(-rt).$$

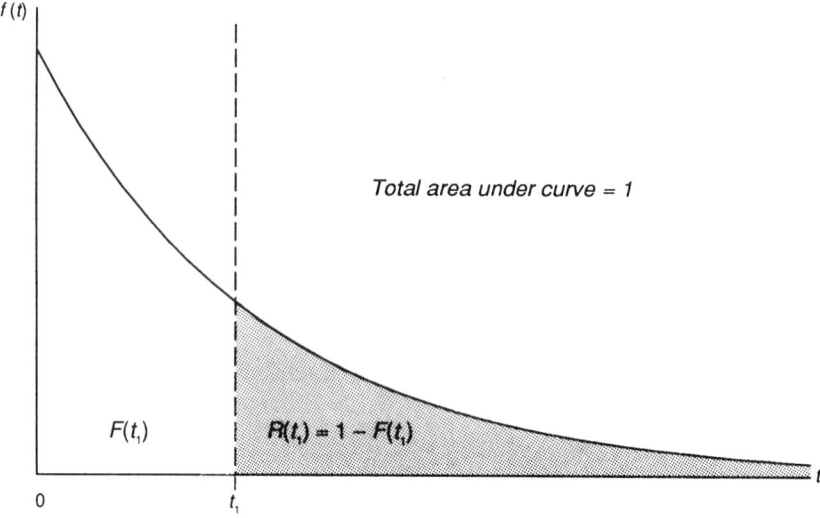

Figure 6.6 The reliability as the complement of the cdf for an exponential distribution

Hazard rate

Another function that is used in studying equipment failure is the *hazard function* or *hazard rate*, which is denoted by $h(t)$. It is related to the probability density function of time to failure and the reliability by

$$h(t) = f(t)/R(t).$$

The precise theoretical definition of the hazard rate is rather complicated. However, broadly speaking, it measures the intensity of the risk that equipment which has survived to age t will fail very shortly after reaching that age. The age at failure is often called the *lifetime*.

When the lifetime of the equipment is exponentially distributed

$$h(t) = r \exp(-rt)/\exp(-rt) = r$$

so the hazard rate is constant: in other words, the intensity of the risk of instant failure is independent of the age of the equipment in this case. Compare this result with the constant probability of success in the trials considered above in the discussion of the geometric distribution.

The practical meaning of the hazard rate is considered below and other models are introduced to represent situations where the hazard rate may vary with the age of the equipment.

118

Estimation of hazard rates

Many specialized methods have been developed to deal with complicated situations that arise in studying the observed lifetimes of various types of equipment. The underlying mathematical reasoning also applies to certain problems of human survival that are of interest to medical researchers and actuaries concerned with life insurance. The methods considered here can only be applied to certain types of data, but they do provide an insight into what hazard rates actually measure.

> Table 6.2 shows a record of failures of a particular sub-assembly in a batch of 100 air-conditioning systems, covering the first four years after the systems had been installed. All 100 systems were under observation during that period, but failure of the sub-assembly was found in some cases only as a result of routine maintenance, so lifetimes are known only within six-month intervals.

The hazard rate for each interval is estimated by dividing the number of failures in the interval by the average number at risk in the interval, then dividing the result by the width of the interval. The result therefore represents an estimate of the proportion of survivors of a particular age which will fail per unit of time (per month in this case).

The method assumes that failures may take place at any time during the interval and that the hazard rate is constant over the interval.

The average number at risk is found by subtracting half the number of failures from the number of survivors at the start of the interval concerned, which is equivalent to taking an average of the number of survivors at the start of this interval and the corresponding number at the start of the next.

> For the first six-month interval, the average number at risk $100 - 10/2$ or $(100 + 90)/2 = 95$
>
> and so hazard rate $= \dfrac{\text{number of failures}}{\text{average number at risk}} \times \dfrac{1}{\text{width}}$
>
> $\qquad = (10/95) \times (1/6) = 0.0175.$
>
> Similarly, for the second interval,
>
> $\qquad$ hazard rate $= (3/88.5) \times (1/6) = 0.0056$
>
> and so on, as shown in Table 6.3.

The results are shown in graphical form in Figure 6.7, which represents a kind of histogram of hazard. The broken line superimposed on the histogram indicates the approximate shape of the underlying theoretical hazard function $h(t)$.

The area under the histogram to the left of a specified time is called the *cumulative hazard* up to that time. Its value at the end of an interval is the

Table 6.2 Air-conditioning system failures

Interval (months	Number of survivors (at start of interval)	Number of failures (in interval)
0 to 6	100	10
6 to 12	90	3
12 to 18	87	4
18 to 24	83	7
24 to 30	76	7
30 to 36	69	10
36 to 42	59	9
42 to 48	50	8

Table 6.3 Air-conditioning system hazard rates

Months	Hazard rate	Cumulative hazard (at end of interval)
0 to 6	$(10/95)/6 = 0.0175$	0.1053
6 to 12	$(3/88.5)/6 = 0.0056$	0.1392
12 to 18	$(4/85)/6 = 0.0078$	0.1862
18 to 24	$(7/79.5)/6 = 0.0147$	0.2743
24 to 30	$(7/72.5)/6 = 0.0161$	0.3708
30 to 36	$(10/64)/6 = 0.0260$	0.5271
36 to 42	$(9/54.5)/6 = 0.0275$	0.6922
42 to 48	$(8/46)/6 = 0.0290$	0.8661

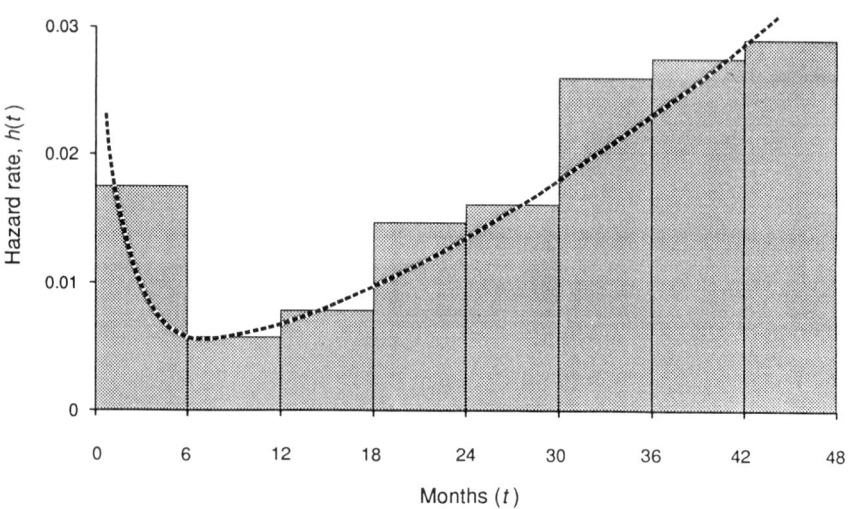

Figure 6.7 Histogram of air-conditioning system hazard rates

cumulative total of the ratios of numbers of failures to average numbers at risk that are used to calculate the hazard rates. Thus, for the air-conditioning system, the cumulative hazard at the end of the first interval (at 6 months) is $10/95 = 0.1053$; at the end of the second interval (at 12 months) it is $10/95 + 3/88.5 = 0.1392$ and so on.

Like the frequency histograms considered in Chapter 3, the hazard histogram should be considered as providing a rough indication of the shape of a smooth curve that represents the underlying distribution. If the number of observations were increased and more smaller intervals were considered, the histogram would come closer and closer to the smooth curve representing $h(t)$.

The cumulative hazard calculated by summing areas under the histogram would come closer and closer to the area under $h(t)$, which is the theoretical cumulative hazard and is denoted by $H(t)$. From the relationships between the various functions that describe any distribution of lifetimes, it can be shown that the cumulative hazard is the natural logarithm of the reciprocal of the reliability,

$$H(t) = \ln[1/R(t)].$$

so an estimate of $H(t)$ can be used to obtain an estimate of $R(t)$ and vice versa.

Figure 6.7 seems to indicate a hazard rate curve that is initially fairly high, but which has dropped to a very low level after about 6 months. From about 12 months onwards it rises, passing the original level at about 24 months.

This kind of variation in hazard rate is often called a *bath-tub distribution* – imagine the taps and the plug-hole at the left! It arises when equipment is exposed to at least two competing kinds of risk, one of which is very intense in the early stages and the others which are appreciable only when the equipment has already been in service for a considerable time.

A similar pattern applies to human mortality: babies are exposed to high risks during the first few days of life, older children and young adults are exposed to fairly low risks, and the intensity of risks starts to rise with the onset of middle age.

The method of estimating the hazard rate described above is based on that used by actuaries. Another method, which is not recommended, is sometimes used because it is felt that there is too much arithmetic involved in calculating the average number at risk (to use in the denominator). In this short-cut method, the number of survivors at the start of the interval is used, without any allowance for the decrease in the number at risk as failures occur during the interval. In effect, this method assumes that all failures take place at the start of the intervals. For example, in the first interval this would give $(10/100)/6 = 0.0167$ instead of the 0.0175 calculated above. This method is not recommended because it will always underestimate the true hazard rate.

An alternative method enables the hazard rates to be estimated very easily using a calculator that can take natural logarithms. It allows for failures being distributed over each interval in a slightly different way. This *logarithmic method* estimates hazard rates as follows:

1. Divide the number of survivors at the start of the current interval by the number of survivors at the start of the next interval.
2. Take the natural logarithm, using the 'ln' key on the calculator.
3. Divide by the width of the interval.

Thus, for the first six months of the air-conditioning system,

hazard rate = [ln(100/90)]/6 = 0.0176.

The actuarial method gave 0.0175 to four decimal places; the difference between the two results is actually less than 2 in the fifth decimal place.

In this example, the two methods give effectively the same results for all intervals, only showing a difference of 1 in the fourth decimal place for the last three intervals. The two methods will give appreciably different results only when the hazard rate is changing very rapidly, in which case slightly different methods should be used anyhow.

The main reason for estimating the shape of the hazard distribution, as described in this section, is to find an appropriate model for the distribution of lifetimes. If the air-conditioning data had shown the hazard function to be approximately constant over all intervals, an exponential model would have been appropriate, as explained at the beginning of this section. The next section introduces a generalization of the exponential model, which allows for hazard rates that may increase or decrease with time.

■ 6.8 Weibull distributions

It was shown previously that the reliability function corresponding to exponentially distributed lifetimes is $R(t) = \exp(-rt)$. If t in the expression on the right is replaced by some power of t, the resulting equation $R(t) = \exp(-rt^\beta)$, defines what is called a *Weibull distribution* with shape parameter β (beta).

The exponential distribution is a special case of the Weibull, in which $\beta = 1$.

This family of distributions is named after the Swedish engineer W. Weibull, who first used them in studies of metal fatigue. Since they can describe quite complicated variations of reliability over time, they provide a very useful class of models for life testing of mechanical, electrical or electronic components. Although other suitable families of distributions are available, Weibull distributions are often used in cases of what is called *censored sampling* or *time truncated tests* (Section 12.9), where some components may be withdrawn from testing before they have failed. Difficulties then arise concerning how to combine the information about those which were observed until they failed with the information about those which were withdrawn (censored) but were still working when last seen.

> For example, a model was required to represent the lifetime of an engine part. A sample of only 15 units was available for test. The sample was tested and, for whatever reason, it was only possible to carry out the test for 2015 hours, by which time 9 of the 15 parts had failed and the rest were still in working order. The time (in hours) that each failed part lasted was recorded as below, where C denotes a test that was censored at 2015 hours.

Unit number	1	2	3	4	5	6	7	8	9	10	11	12	13	14	15
Time to failure (t_i)	1281	1768	C	C	468	1806	C	2015	C	950	C	1177	C	1427	1313

Even if the estimated mean lifetime of the parts is sufficient information from the test, some assumption needs to be made about the complete distribution lifetimes. It is easy to see the flaws in some approaches which try to avoid this necessity. For example:

> If the 6 censored parts are ignored, an average of the remaining 9 observed times to failure will underestimate the overall mean.
> If the censored parts are assumed to have failed at 2015 hours, the estimate would still be too low because the actual times to failure would be various values greater than 2015 hours.

This is a situation where Weibull might help. In practice, a graphical method of fitting a Weibull distribution to the data would be used (Section 8.8). The method provides a check on whether it is reasonable to assume that the Weibull is a suitable model. Also, it provides estimates of all the relevant parameters, including the mean.

Because of the general relationships between $f(t)$, $F(t)$ and $R(t)$ for any distribution of lifetimes, it can be shown that the probability density $f(t)$ represents the rate at which the reliability $R(t)$ decreases as time t increases. Another way of expressing this is to say that $f(t)$ represents the derivative $dR(t)/dt$ with the sign changed.

Thus for the Weibull distribution, the probability density function is

$$f(t) = -dR(t)/dt = r\beta t^{\beta-1}\exp(-rt^\beta).$$

The corresponding hazard function is

$$h(t) = f(t)/R(t) = r\beta t^{\beta-1}$$

and the cumulative probability function is

$$F(t) = 1 - \exp^{(-rt)\beta}.$$

When $\beta = 1$, the hazard function reduces to r; that is $h(t) = r$. The parameter β is the most important for determining the shape of the distribution of times to failure. Its effect on the pdf is shown in Figure 6.8.

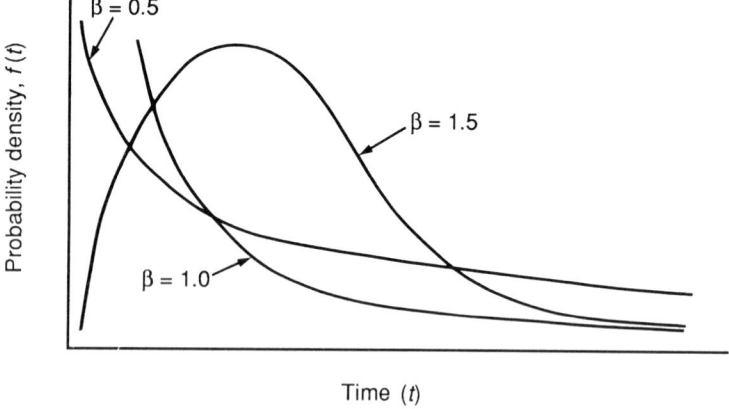

Figure 6.8   Weibull β and the probability density

The way in which changing β affects the hazard rate is shown in Figure 6.9.

$\beta < 1$ Hazard rate decreases with time; β must be greater than 0 and is usually at least 0.5.

$\beta = 1$ Characterizes the exponential model with a constant hazard rate that represents random failures.

$\beta > 1$ Hazard rate increases with time. Distribution approaches normal when $b \simeq 3.5$. Usually β is less than 4. A large β, determined from reliability test data (Section 12.9), often indicates the existence of some particular flaw that 'kills' virtually all the parts by the time they reach a particular age.

Sometimes, the Weibull distribution is described in a slightly different way so that the reliability might be expressed in the form

$$R(t) = \exp\left[-\left(\frac{t-\gamma}{\eta}\right)^{\beta}\right].$$

In this formula

β (beta)   is called the *shape parameter* or the form parameter.

γ (gamma) is called the *location parameter*. It represents the shortest possible lifetime and is sometimes designated t_0.

η (eta)   is the *scale parameter* or *characteristic life*.

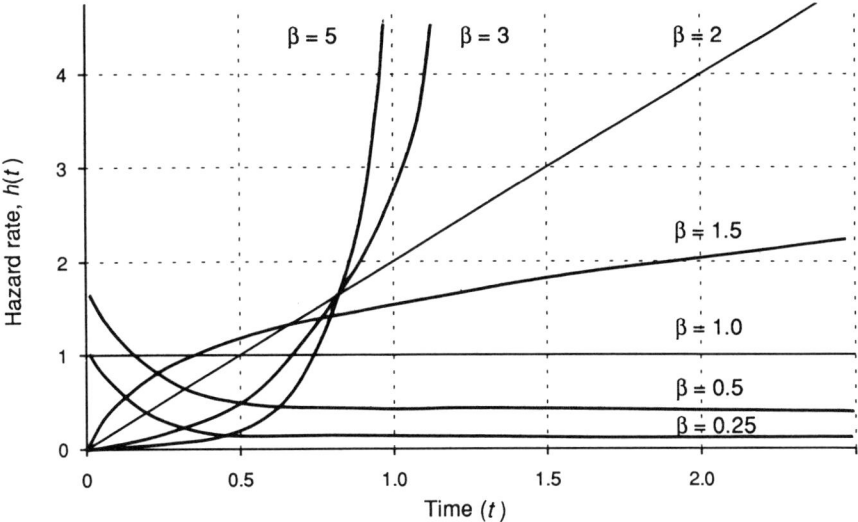

Figure 6.9 Weibull β and the hazard rate

Using this parameterization, with the location parameter (γ) and the scale parameter (η) instead of the average rate (r), the formulae for the probability density function and the hazard function become

$$f(t) = \left(\frac{\beta}{\eta}\right) \left(\frac{t-\gamma}{\eta}\right)^{\beta-1} \exp\left[-\left(\frac{t-\gamma}{\eta}\right)^{\beta}\right],$$

$$h(t) = \left(\frac{\beta}{\eta}\right) \left(\frac{t-\gamma}{\eta}\right)^{\beta-1}.$$

The names for the parameters β, γ and η are best explained by considering their effect upon the graph of the probability density function.

Shape parameter

The β in the new formula is the same as that described previously. Changing it alters the shape of the curve as shown in Figure 6.8.

Location parameter

Changing γ moves the whole curve to the left or to the right: in other words, it alters the location of the curve, as shown in Figure 6.10(a).

(a) Varying γ when β = 2 and η = 1

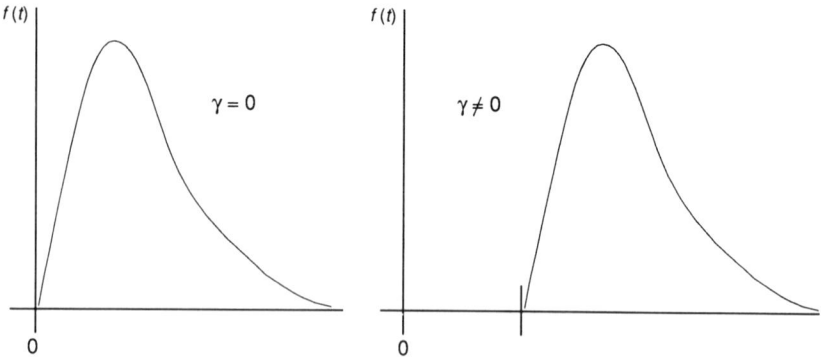

(b) Varying η when β = 2 and γ = 0

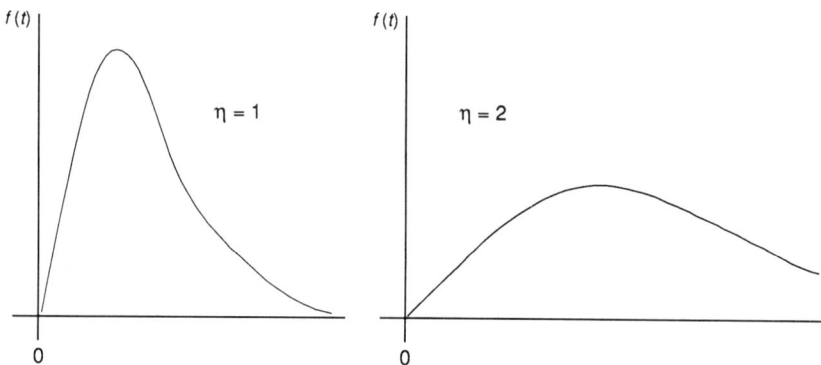

Figure 6.10 Effect of changing Weibull γ and η

Characteristic life

Changing η alters both the horizontal and the vertical scales of the curve, an effect shown in Figure 6.10(b). However, the term deserves more explanation.

Consider a population of components whose lifetimes have a Weibull distribution with $\gamma = 0$. This means that some components may fail as soon as they are brought into use. Putting $t = \eta$ (the characteristic life) into the equation for reliability gives

$$R(\eta) = \exp(-1) = 0.368.$$

hence *the characteristic life is the time at which only 36.8% of the population still survives.*

In general, it is the time elapsed since $t = \gamma$ at which this percentage survives. This may not seem to be a particularly useful concept, but in practice it provides a simple basis for estimation of the mean.

The expression for the mean of a Weibull distribution is a rather complicated function and it is difficult to estimate directly from censored data. However, β and η can be estimated quite easily using a graphical method (Section 8.5) and the mean is estimated by adjusting the estimate of η by a factor that depends on β.

For the exponential distribution (Weibull with $\beta = 1$), no adjustment is required and *the mean is equal to the characteristic life.*

In the original notation used above, r represents the *average rate* at which failures occur, so that $1/r$ is the *mean time between failures (MTBF)*, the same as for the special case of the exponential distribution. The general relationship between *MTBF* and characteristic life for any Weibull distribution is

$$MTBF = 1/r = \eta^\beta.$$

Special methods have been developed to find appropriate values for the parameters in the various versions of the Weibull formulae, using grouped data or individual observations, with or without censoring. The graphical method, mentioned above, is frequently used in engineering but also has applications in biology and medicine, such as studying the survival of patients after major surgery.

In summary, to fit a distribution to data and hence allow estimation of the population mean, it is necessary to find the values of η and β. In addition, the value of β indicates if the parts are failing in early life, are wearing out or have no predominant failure mode.

■ 6.9 Mixed and composite Weibull distributions

Although a single Weibull distribution cannot be used to model the bath-tub shape of the type of hazard distribution discussed in Section 6.7, there are various ways of combining several Weibull distributions to produce reasonable approximations.

Mixed Weibull distribution

In the mixed Weibull distribution, the probability density function of lifetimes, $f(t)$, is a weighted average of one Weibull probability density function, $f_1(t)$, which has shape parameter $\beta_1 < 1$ (decreasing hazard rate), and

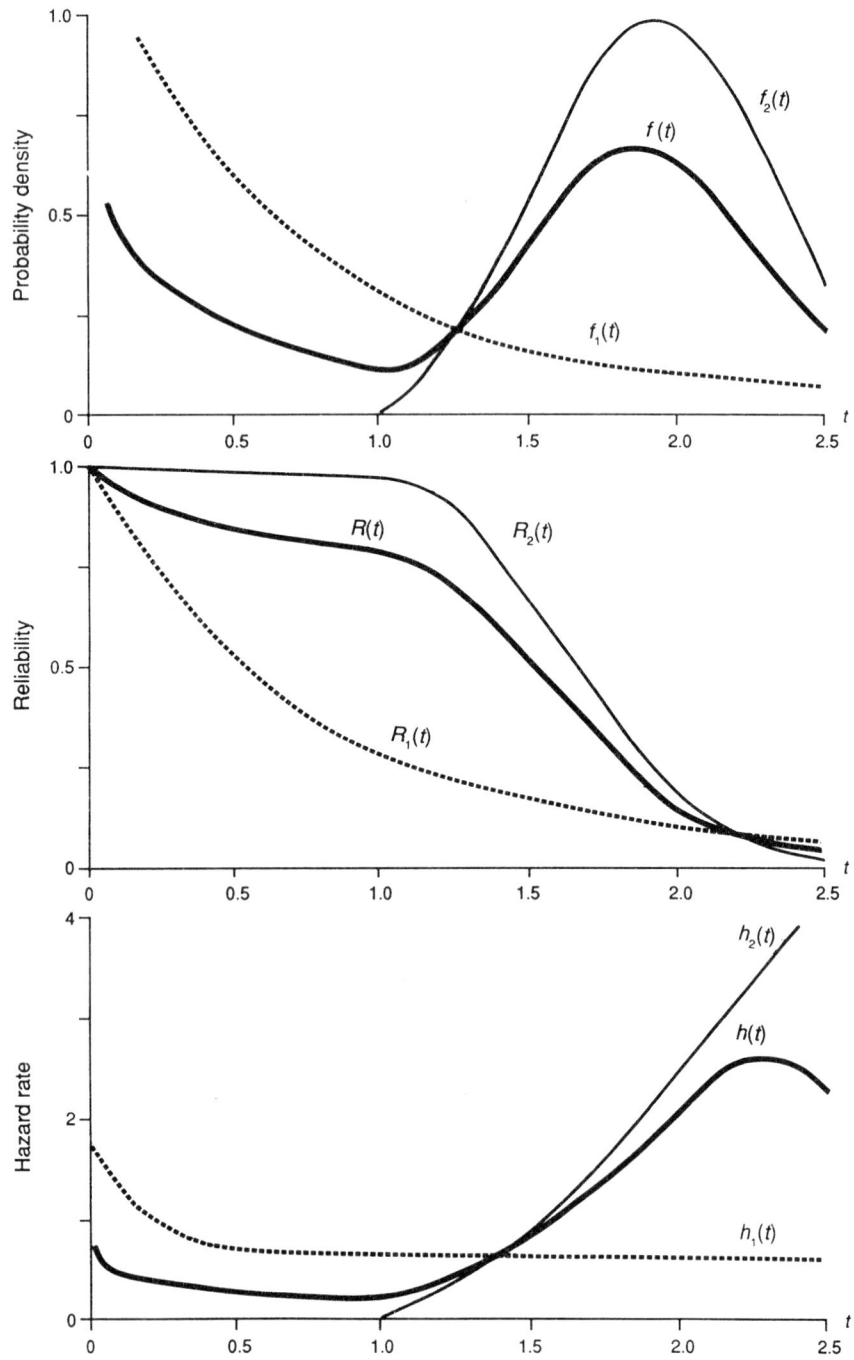

Figure 6.11 Mixed Weibull distribution with $w_1 = 1$, $w_2 = 0.6$, $\beta_1 = 0.8$, $\beta_2 = 2.5$, $\gamma_1 = 0$, $\gamma_2 = 1$, $\eta_1 = \eta_2 = 1$

another, $f_2(t)$, which has shape parameter $\beta_2 > 1$ (increasing hazard rate), so that

$$f(t) = w_1 f_1(t) + w_2 f_2(t)$$

where w_1 and w_2 are the weightings; they are positive and are chosen so that $w_1 + w_2 = 1$. The location parameter for distribution 1 is $\gamma_1 = 0$, but that for distribution 2 is $\gamma_2 > 0$. In other words, the latter is a positive value that represents a delay before the distribution with the increasing failure rate becomes effective. Figure 6.11 shows the probability density function, reliability and hazard function for a typical mixed Weibull distribution and the two distributions from which it is built up.

The hazard function, $h(t)$, of the mixed distribution, has the required bath-tub shape. However, the parameters in this mixed model can be estimated only if there is sufficient information about causes of failure, so that individual failures can be allocated either to distribution 1 or to distribution 2.

Composite Weibull distribution

If cause information is not available, an alternative approach is the composite Weibull distribution (a description coined by John Kao at Cornell University). In principle, the composite model assumes that distribution 1 (with decreasing hazard rate) initially operates on its own; it is then switched off and distribution 2 (with increasing hazard rate) takes over. This model has limitations which are discussed below.

In the model, the location parameters (γ) of both distributions are taken as zero. The other parameters are β_1 and η_1 for distribution 1 and β_2 and η_2 for distribution 2. The transition from distribution 1 to distribution 2 is taken as taking place at time T, so, for distribution 1

$$f(t) = f_1(t), \ R(t) = R_1(t) \text{ and } h(t) = h_1(t) \text{ for } t < T$$

and for distribution 2

$$f(t) = f_2(t), \ R(t) = R_2(t) \text{ and } h(t) = h_2(t) \text{ for } t > T.$$

The two parts of the composite distribution are joined by equating their reliability functions at the transition time. In other words, equating and rearranging

$$R_1(T) = \exp\left[-\left(\frac{T}{\eta_1}\right)^{\beta_1} \right] \text{ and } R_2(T) = \exp\left[-\left(\frac{T}{\eta_2}\right)^{\beta_2} \right]$$

gives an equation that allows the transition time (T) to be calculated once the other parameters are known

$$T = \exp\left[\frac{\beta_2 \ln \eta_2 - \beta_1 \ln \eta_1}{\beta_2 - \beta_1} \right]$$

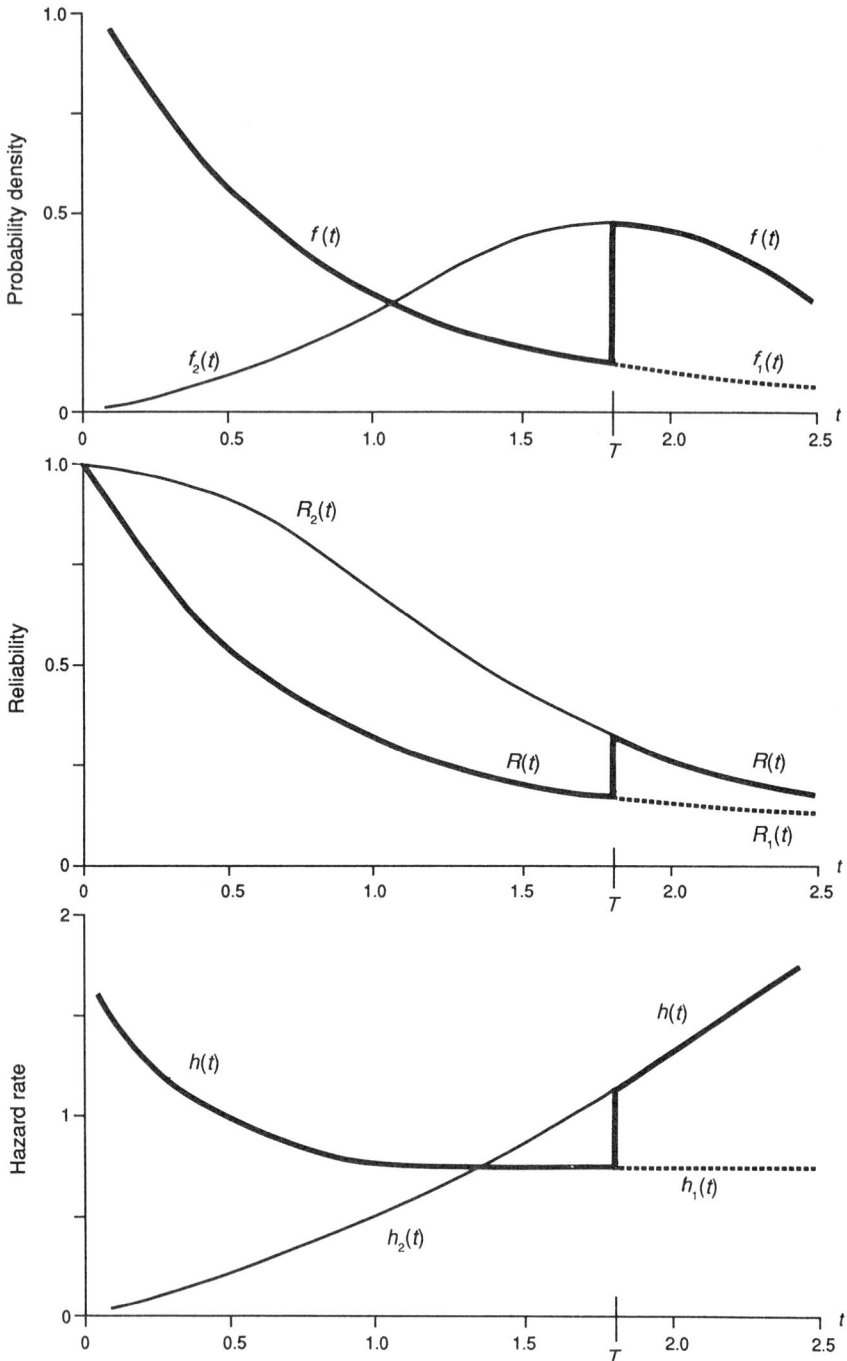

Figure 6.12 Composite Weibull distribution with $\beta_1 = 0.8$, $\beta_2 = 2.5$, $\gamma_1 = \gamma_2 = 0$, $\eta_1 = 1$, $\eta_2 = 2$ and $T = 1.8$

The behaviour of the probability densities, reliabilities and hazard functions for a typical composite Weibull distribution is shown in Figure 6.12.

The limitation of this composite model is the discontinuity in $f(t)$ and $h(t)$ at the transition time, $t = T$. However, so long as it is understood that the model is not very realistic near the transition time, the composite Weibull can provide a useful alternative to the more complicated mixed Weibull model. The parameters of the composite Weibull are quite easily estimated by a simple extension of the graphical method used for a single Weibull distribution. This method is described in Section 8.7, where it is used to fit a composite Weibull to the air-conditioning system data in Tables 6.2 and 6.3.

■ References to further reading on the topics of Chapter 6

Chatfield (1983), Hahn and Shapiro (1968), Parzen (1960) and Walpole and Myers (1993) give good general accounts of the main types of probability distribution that are likely to be used in industry. Lawless (1982) and Mann, Schafer and Singpurwalla (1974) are useful for the Weibull and other distributions used in reliability work. Kao (1959) and David and Moeschberger (1978) describe ways of modelling situations where there may be several potential causes of failure.

See Bibliography for details of titles and publishers.

■ Self-assessment questions on Chapter 6

1. For planning purposes, a maintenance department uses a 'standard job time' which is 'mean job time plus 1.5 standard deviations'. It is thought that the time taken for a particular routine maintenance task may vary between 60 and 75 minutes, and that all times between these are equally likely so that the time for this task can be considered as having a continuous uniform distribution.

   Find the probability that the task will be completed within the standard job time if
   (a) the assumption about the uniform distribution is correct;
   (b) the true distribution is normal, with the same mean and standard deviation as the uniform distribution.

2. A high-speed process produces fastening pins whose diameters are normally distributed with a mean of 4 mm and a standard deviation of 0.1 mm. A pin is judged to be 'oversize' if its diameter is greater than 4.2 mm.

(a) Use the tables of the standard normal distribution to find the probability of a pin being oversize.

(b) Using the result of (a) as the value of p in the binomial formula, calculate the probability of less than 2 oversize pins in a random sample of 20.

3. Vehicles arrive at a loading bay according to a Poisson process at an average rate of one vehicle every 80 minutes.

Find the probability of the following numbers of vehicles arriving in a period of 4 hours.

(a) Fewer than three.

(b) Three or more.

(c) Fewer than twice the expected number.

4. Weekly sales of headlight bulbs are thought to be Poisson distributed with a mean of 25. Use the normal approximation to the Poisson to find the probability of selling more than 30 bulbs in a week.

5. An engineer trying to order spare parts has found that, on average, the suppliers' telephone is engaged 3 times out of 4 when contact is attempted. Use the geometric distribution to find the probability that the engineer will get through to the suppliers by making fewer than 4 calls.

6. The lifetime of a particular system has a Weibull distribution with $\beta = 2.2$, $\gamma = 0$ and $\eta = 1$. Calculate and plot the hazard rate, $h(t)$, and the reliability, $R(t)$, for $t = 0$, 0.5, 1.0, 1.5, 2.0 and 2.5.

7 | Estimation

■ 7.1 Introduction

To be able to use the theoretical probability distributions, described in Chapters 5 and 6, it is necessary to have values for the parameters (or constants) which occur in the various formulae that define the distributions.

If a sample $x_1, x_2, \ldots x_n$ is drawn from a distribution that is described by some parameter, it is very unusual for the exact true value of that parameter to be known, but various *estimates* of the parameter can be made from the observed x values.

There are two main kinds of estimate:

■ A *point estimate*, which represents a 'best' single value for the unknown parameter. The value of a point estimate will depend upon how the word 'best' is interpreted.

■ An *interval estimate*, which represents an interval or range of values within which the true value will almost certainly lie. The size of the interval will depend upon the required degree of certainty which is called the *confidence level*.

Both types of estimate can usually be expressed fairly simply in terms of certain *statistics*. In this context, the word 'statistic' is just the general name for a quantity such as the sample mean ($\bar{x}$) which is calculated from the observed values.

Since the observed values will usually differ from sample to sample, the corresponding values of any statistics will also vary, even though the same basic formula is used to calculate them each time. This variation is governed by what is called the *sampling distribution* of the statistic. This sampling distribution is, in turn, governed by the underlying distribution of the individual x values.

■ 7.2 Point estimates based on the sample mean

In Chapter 5, it was explained that the theoretical mean of a normal distribution is equal to the parameter μ, whose value may not be known. When a sample is drawn from the normal distribution and its mean, $\bar{x}$, is calculated, that value can be used as an estimate of μ (mathematically, this is described as equating the sample mean and the theoretical mean). It is usual to distinguish an estimate of a parameter from its true (but unknown) value by writing a circumflex or hat (^) on top of the symbol for the parameter. In other words, for the theoretical mean of a normal distribution, μ:

estimate of $\mu = \hat{\mu} = \bar{x}$.

The same procedure, equating theoretical and sample means, can be used for estimating parameters in several of the distributions described in Table 6.1. For example, Table 6.1 showed that the theoretical mean of a binomial distribution with parameters n and p is np.

The basic variable x represents the number of successes in n trials, so the parameter n is usually known; but p, the probability of success in any single trial, is usually unknown. Suppose that m sets of n trials are carried out and that the number of successes is found to be x_1 in the first set, x_2 in the second set, and so on. The observed mean number of successes in m sets of n trials will then be

$$\bar{x} = \frac{1}{m} \sum_{j=1}^{m} x_j = \frac{\text{total number of successes}}{\text{number of sets of trials}}.$$

If this expression of the value of $\bar{x}$ is equated to the expression for the theoretical mean with p replaced by $\hat{p}$ (in other words, $n\hat{p} = \bar{x}$) and then each side is divided by n, an estimate of p is given by:

$$\text{estimate of binomial } p = \hat{p} = \frac{\bar{x}}{n} = \frac{\text{total number of successes}}{\text{total number of trials}}.$$

If $m = 1$ (in other words, only one set of n trials is carried out) and x successes are observed, this formula simplifies to

$$\hat{p} = x/n.$$

Similar results for some other distributions are shown in Table 7.1. Beside being easy to calculate, the parameter estimates in the table can be shown to be 'best' when judged by various theoretical criteria.

The effect of sample size on estimates

When the underlying distribution of individual values of x is exactly normal, the distribution of the values of $\bar{x}$ is also exactly normal.

Table 7.1 Parameter estimates for basic distributions

Distribution	Theoretical mean	Sample mean	Parameter estimate
Normal	μ	$\bar{x}$	$\hat{\mu} = \bar{x}$
Binomial	np	$\bar{x}$	$\hat{p} = \bar{x}/n$
Poisson	λ	$\bar{x}$	$\hat{\lambda} = \bar{x}$
Exponential	$1/r$	$\bar{x}$	$\hat{r} = 1/\bar{x}$

In this case the implications of varying the sample size can be seen in a diagram such as Figure 7.1, where the curve (a) shows the distribution of individual values of x (corresponding to sample size $n = 1$). Curves (b) and (c) show the corresponding distributions for $\bar{x}$ when $n = 4$ and $n = 16$ respectively.

The shaded regions contained between lines drawn at one standard deviation on either side of μ represent the probability of observing values that fall within one standard deviation of the mean; since all the distributions are normal they represent the middle 68% of each distribution.

When $n = 1$ the width of the shaded region is simply 2σ; when $n = 4$ it is σ, and when $n = 16$ it is $\sigma/2$. Thus the width is halved every time the sample size is increased by a factor of 4. When n gets very large the distribution of $\bar{x}$ becomes concentrated in a very narrow band on either side of μ, so there is a very high probability that $\bar{x}$ will take a value very close to μ and hence provide a good estimate of μ.

■ 7.3 The central limit theorem

In many practical situations it is not reasonable to assume that the individual x values are normally distributed. However, an extremely important rule called the *central limit theorem* says that, depending on sample size, *the distribution of the $\bar{x}$ values from a sample of size n will be (at least approximately) normal with mean μ and standard deviation $\sigma/\sqrt{n}$.*

This statement is true for any form of underlying distribution of the individual x values, provided only that it has mean μ and standard deviation σ. The quantity $\sigma/\sqrt{n}$ is often referred to as the *standard error of the mean*.

The distribution of $\bar{x}$ is exactly normal for any value of n when the underlying distribution is exactly normal.

For any other form of underlying distribution, the approximation gets better as n increases. The value of n at which it can be assumed that the distribution of $\bar{x}$ is effectively normal will depend upon the shape of the underlying distribution, particularly upon how symmetrical this distribution is.

A symmetric distribution, such as the continuous uniform distribution, will give an effectively normal $\bar{x}$ when n is as small as 4. On the other hand,

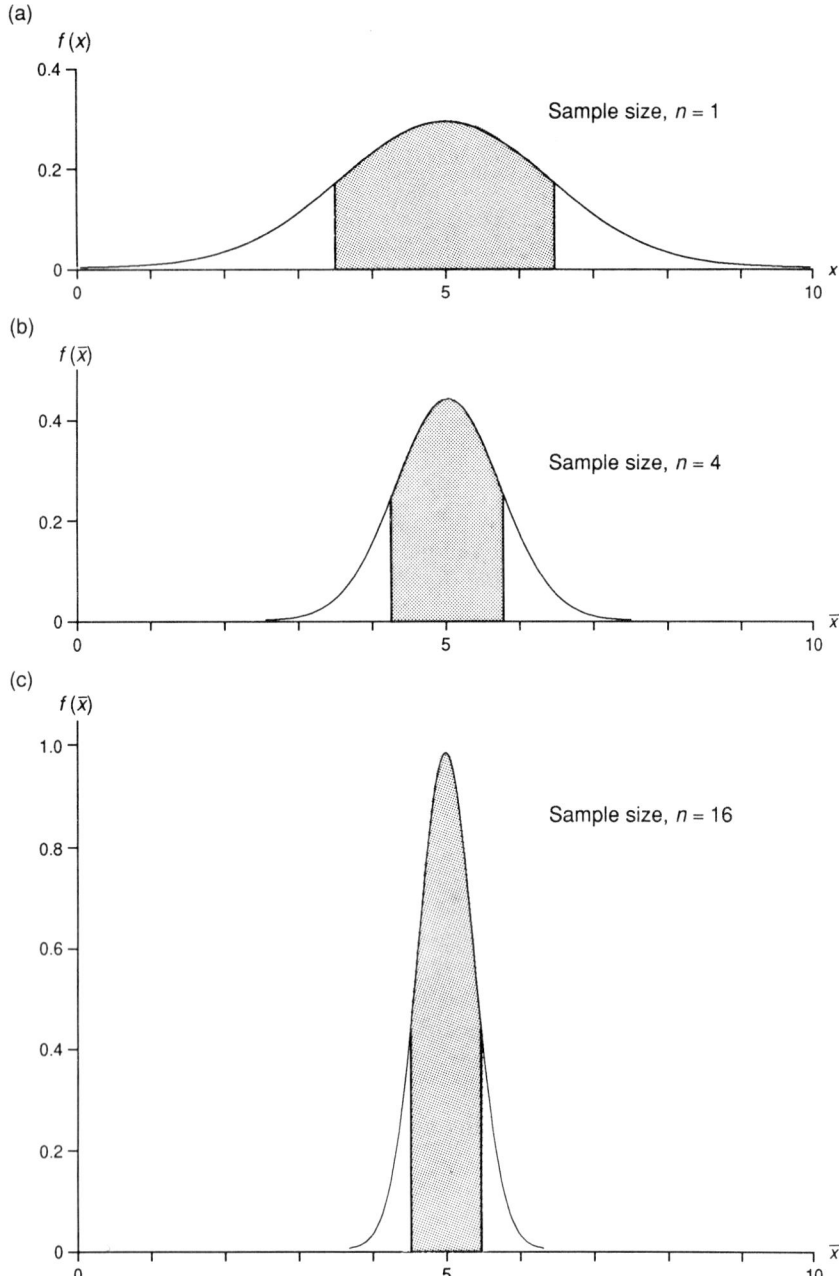

Figure 7.1 Distributions of sample means from a normal distribution with $\mu = 5$ and $\sigma = 1.5$

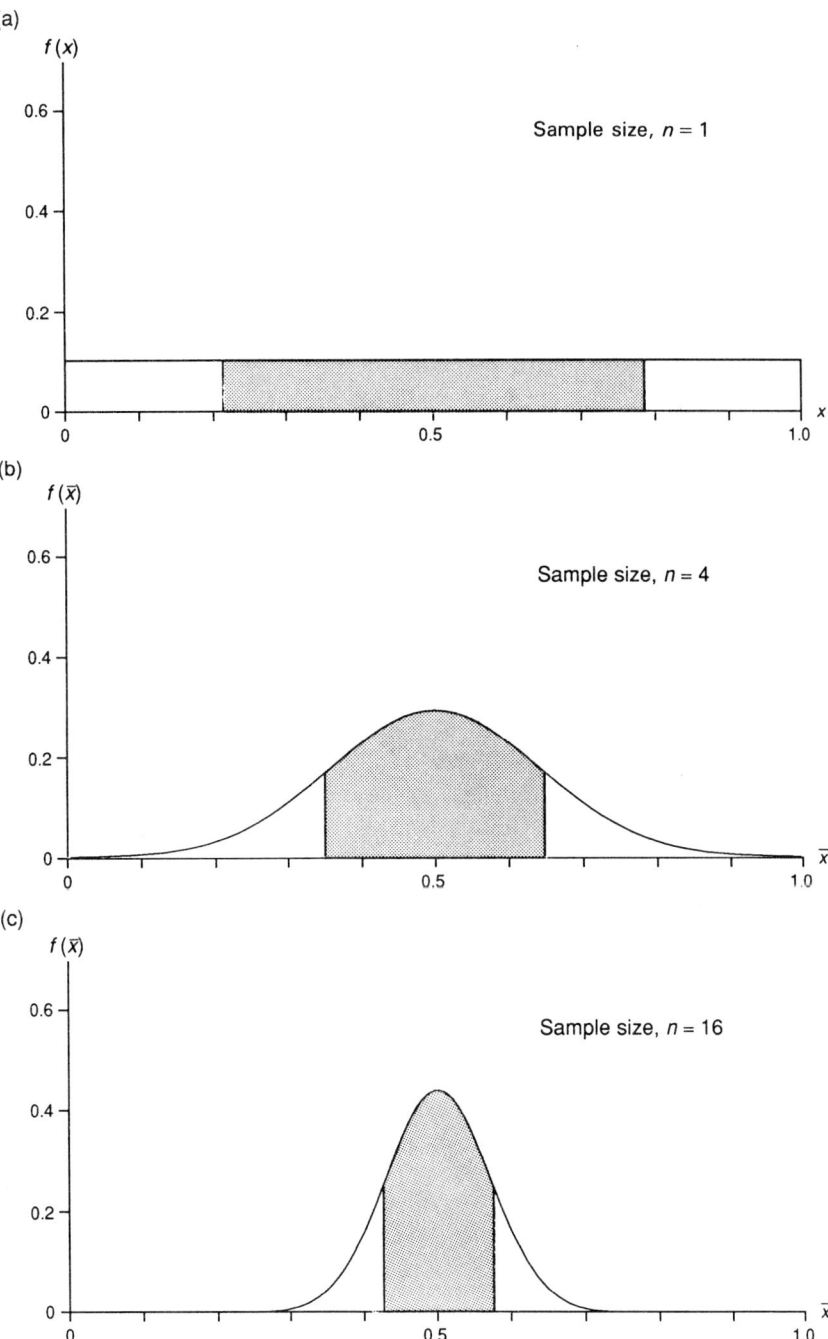

Figure 7.2 Distributions of sample means from a uniform distribution
on the interval $x = 0$ to $x = 1$

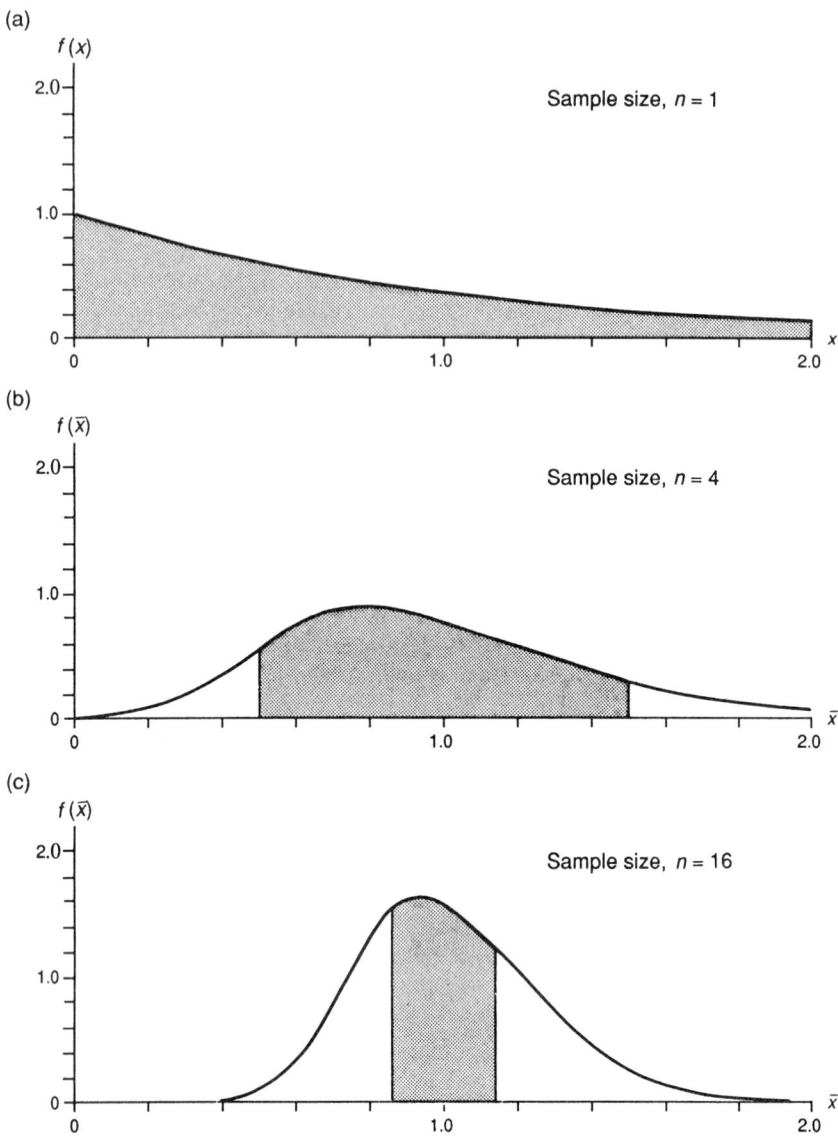

Figure 7.3 Distributions of sample means from an exponential distribution with mean and variance equal to 1

a skewed distribution, such as the exponential, would require n to be considerably larger, say 16 or more. This is illustrated in Figures 7.2, 7.3 and 7.4, where the shaded areas again represent probabilities of values falling within one standard deviation of the mean.

Note on skewness when sampling from exponential distributions

A rough idea of the amount of skewness in the distribution of $\bar{x}$ when sampling from an exponential distribution can be obtained by noting that the ratio of mode to mean in this distribution of $\bar{x}$ is $(n-1)/n$. When $n = 16$ this gives the mode as $15/16$ of the mean. Figure 7.3 shows that the distribution of $\bar{x}$ is almost symmetrical about the mode but is obviously not symmetrical about the mean.

■ 7.4 Confidence limits for normal μ when σ is known

From the previous section, if x is normally distributed with mean μ and standard deviation σ, then $\bar{x}$ will be normally distributed with mean μ and standard deviation $\sigma/\sqrt{n}$. This distribution of $\bar{x}$ can be transformed to a standard normal distribution (Section 5.5) and the standardized deviate corresponding to a particular value of $\bar{x}$ will be

$$z = \frac{\bar{x} - \mu}{\sigma/\sqrt{n}} \ .$$

If z_α denotes the value of z which corresponds to an area α, under the right tail of the standard normal probability density function, then $P(z > z_\alpha) = \alpha$. Also, because of the symmetry of the probability density function, $P(z < -z_\alpha) = \alpha$. It follows from these two probability statements that (as shown in Figure 7.4) because the total area under the pdf curve is 1,

$$P(z_\alpha > z > -z_\alpha) = 1 - 2\alpha.$$

When $(\bar{x} - \mu)/(\sigma/\sqrt{n})$ is substituted for z, this equation becomes a probability statement about the difference between $\bar{x}$ and μ:

$$P(z_a > \frac{\bar{x} - \mu}{\sigma/\sqrt{n}} > -z_\alpha) = 1 - 2\alpha.$$

The terms inside the bracket on the left-hand side can then be rearranged to give a probability statement about μ, which is expressed as

$$P(\bar{x} - z_\alpha\sigma/\sqrt{n} < \mu < \bar{x} + z_\alpha\sigma/\sqrt{n}) = 1 - 2\alpha.$$

In other words, μ lies between $x - z_\alpha\sigma/\sqrt{n}$ and $x + z_\alpha\sigma/\sqrt{n}$ with probability $1 - 2\alpha$. It is usual to express this probability as a percentage and call it the *confidence level*.

confidence level $= 100(1 - 2\alpha)\%$.

The values on either side of $\bar{x}$, which contain μ with the specified confidence level, are called *confidence limits*. The interval between them is called the *confidence interval*.

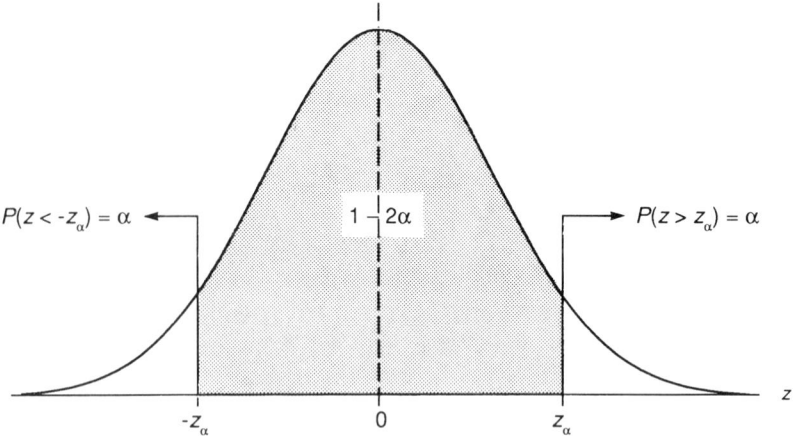

Figure 7.4 Values of standard normal *z* used to find confidence limits

If the confidence level is increased (to be more certain that μ will lie between the confidence limits) then z_α becomes larger and the confidence limits lie further apart; the result is that a more vague statement is made about the relative values of μ and $\bar{x}$.

To combine reasonably high confidence and close confidence limits, it is most usual to choose a confidence level of 95%. This corresponds to $1 - 2\alpha = 0.95$ and $\alpha = 0.025$, hence $z_\alpha = z_{0.025}$ which cuts off a tail area of 0.025 (or 2.5%) at the right of the standard normal probability density function. From the standard normal table (Appendix B) it can be seen that $\alpha = 0.025$ when $z = 1.96$ and so

95% confidence limits for $\mu = \bar{x} \pm 1.96\sigma/\sqrt{n}$.

The 95% confidence limits for μ are often interpreted as 'the limits within which one is 95% confident the true mean lies'. Alternatively, from a practical point of view, the true mean will lie within these limits on 95% of the occasions when the formula is used.

> For example, the wear of 4 tyres (all of same specification) is measured on a new model of car. The mean distance covered before the tyres wear to the legal minimum tread depth is calculated as 45 000 miles. Previous data suggests that similar tyres have a standard deviation of 8000 miles.
>
> The 95% confidence limits on the calculated mean are
>
> $$45\,000 \pm \frac{1.96 \times 8000}{\sqrt{4}} = 37\,160 \text{ miles and } 52\,840 \text{ miles.}$$
>
> In other words, although the best estimate of the average mileage of this tyre on the new car is 45 000 miles, the true average could easily be any value between 37 160 miles and 52 840 miles.

Whereas $\bar{x}$ is an estimate of the true mean, the confidence interval gives an indication of the *margin of error* and so provides more information from the data.

■ 7.5 Confidence limits for normal μ when σ is not known

In most practical situations the formulae in Section 7.4 cannot be used to find confidence limits for μ because σ, the population standard deviation, is not known. The best estimate of σ will be the sample standard deviation, s, but simply substituting s for σ in the formulae will not give the correct limits unless a very large sample (more than about 50 values of x) has been used to calculate s.

The reason for this is the statistical behaviour of the quantity

$$t = \frac{\bar{x} - \mu}{s/\sqrt{n}}$$

which is often called *Student's t*, after the statistician W.S. Gosset who wrote under the pen-name 'A Student of Statistics'.

Superficially, the formula for t looks like that for z at the start of Section 7.4, but the presence in the denominator of s (which will vary from sample to sample) rather than the constant σ makes t more variable than z. This is reflected in the shape of the pdf curve for t, which is centred on zero and symmetrical, like that for the standard normal z, but wider and flatter. The amount of widening and flattening depends upon a quantity called the *degrees of freedom*, which varies according to the sample size (Section 11.3). 'Degrees of freedom' is sometimes abbreviated to DF but is more usually designated by the Greek lower-case letters ϕ (phi) or ν (nu). The last of these is used in this chapter, but DF is also used in Chapter 12.

Figure 7.5 shows how the shape of the t-distribution varies with ν. It is very wide and flat for small ν (or small sample sizes), but as ν increases it gets closer and closer to the shape of the standard normal distribution. Because of this, the value of t which cuts off the probability α in the right tail (denoted by t_α) is very different from z_α for small ν but approaches z_α as ν gets large.

The formula for $100(1 - 2\alpha)\%$ confidence limits on μ when σ is estimated by s from a sample of size n is

$$\bar{x} \pm t_\alpha s\sqrt{n}$$

where t_α refers to the t-distribution with $\nu = n - 1$.

Values of t_α are given in Appendix E, where it can be seen that when ν equals infinity the 2.5% point for t is 1.96, exactly the same as that for the standard normal deviate, z. However, for a small sample size such as $n = 5$ which

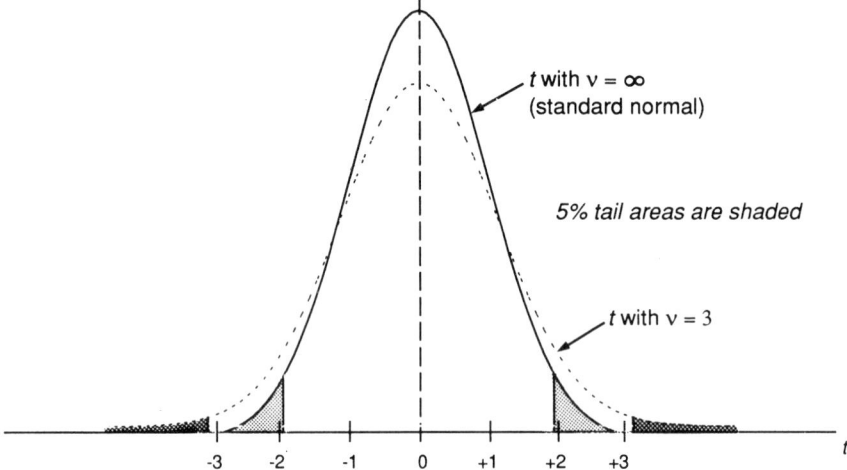

Figure 7.5 Variation of the *t*-distribution with degrees of freedom (*v*)

corresponds to $v = 4$, $t_{0.025}$ equals 2.78, which is considerably larger than the 1.96 for $z_{0.025}$. Simply substituting s for σ in the formula for 95% confidence limits when σ is known would therefore give the limits as

$$\bar{x} \pm 1.96 s / \sqrt{5}$$

instead of the correct values,

$$\bar{x} \pm 2.78 s / \sqrt{5}.$$

Using z instead of t when σ is replaced by s will always give a false impression of accuracy, suggesting that the confidence limits are closer together than they really are. Only when the sample size is sufficiently large for the appropriate *t*-distribution to be effectively the same as the standard normal can z be used as an approximation to the true value of t; in practice this approximation is reasonable if the sample size is greater than about 50.

> For small samples, always use the true *t* value for finding confidence limits on μ if these are to be based on a sample standard deviation.

■ 7.6 Confidence limits on binomial *p*

In Section 7.2 it was shown that the point estimate of p, the expected proportion of successes in any set of n binomial trials, is

$$\hat{p} = x/n$$

where x is the number of successes observed in a particular set of n trials.

If several sets of trials are carried out, $\hat{p}$ will vary from one set to another even if the underlying binomial distribution is the same in each case. If n is reasonably large, the central limit theorem can be used to show that the distribution of the values of $\hat{p}$ will be effectively normal. The mean of this distribution is the true value of p for the underlying distribution, and the standard deviation is $\sqrt{[p(1-p)/n]}$, which can be estimated by replacing p with $\hat{p}$.

Reasoning similar to that used to derive the $100(1-2\alpha)\%$ confidence limits for normal μ can then be used to show that the approximate $100(1-2\alpha)\%$ confidence limits for p are

$$\hat{p} \pm z_{\alpha}\sqrt{[\hat{p}(1-\hat{p})/n]}.$$

For 95% confidence, the approximate limits are

$$\hat{p} \pm 1.96\sqrt{[\hat{p}(1-\hat{p})/n]}.$$

Application of this last expression is illustrated in the following example. Information is required within approximate 95% confidence limits on the proportion of all motorists who would choose a red car. A random sample of 100 motorists were asked what colour they would choose for their next car. Twenty-eight replied that they would choose a red car.

In this case, choosing a red car will be counted as a 'success', so $n = 100$ and $x = 28$; hence $\hat{p} = 28/100 = 0.28$ and $1 - \hat{p} = 1 - 0.28 = 0.72$. The approximate 95% confidence limits on p will therefore be

$$0.28 \pm 1.96\sqrt{[0.28(0.72)/100]} = 0.28 \pm 0.088.$$

Expressing this in terms of percentages rather than proportions, it is reasonable to assume (on the basis of the sample and with 95% confidence) that the proportion of all motorists who would choose a red car lies between 36.8% $(= 28\% + 8.8\%)$ and 19.2% $(= 28\% - 8.8\%)$.

Although the sample size was 100, which would be considered large if one were estimating normal μ, the confidence interval is rather wide in this case. *In general, very large sample sizes are needed in order to estimate p accurately.*

Because large samples are expensive, the final choice of sample size is usually determined by balancing the *margin of error* (the width of the confidence interval) against the cost of sampling. Suppose that the final estimate is required in the form

$$\hat{p} \pm e$$

where e is an acceptable error. Comparison with the formula for the 95% confidence limits shows that

$$e = 1.96\sqrt{[\hat{p}(1-\hat{p})/n]}.$$

Squaring both sides of this expression and rearranging the terms then shows that

$$n = [1.96/e]^2 \hat{p}(1 - \hat{p}).$$

If e is specified and a preliminary or 'pilot' sample is taken to get a rough estimate of $\hat{p}$, this formula can be used to find the approximate value of n that would be required in order to achieve the specified accuracy.

In the case of the red cars, suppose the final percentage was required to within 2%, so that $e = 0.02$. The 100 drivers already questioned can be treated as the pilot sample so that $p = 0.28$ can be used in the formula, which then gives

$$n = [1.96/0.02]^2 \; 0.28(0.72) = 1936 \text{ (approximately)}.$$

This indicates that a total sample of about 1900 drivers (1800 in addition to the 100 already questioned) would be needed to achieve the specified accuracy. If e were set at 0.03 the corresponding value of n would be 861 (only about 750 more drivers needed), so this wider margin of error might be chosen on second thoughts if the sample of 1900 drivers was found to be too expensive.

■ 7.7 Using several small samples to estimate normal μ and σ

Small samples can give only vague estimates of parameters in an underlying distribution; this will be reflected in wide confidence intervals. However, information from several small samples can be combined to give more accurate estimates. This approach is used frequently in process control when charted data is available (Section 9.11). In this type of work, sample sizes of about 5 are commonly used. As an example, consider the data in Table 7.2, which represents 25 samples, each of size 5, from an underlying distribution that is believed to be normal. The problem is to find good estimates of μ and σ for that underlying distribution.

Estimating μ is quite straightforward. First calculate $\bar{x}$ for each sample. Then find the average of these $\bar{x}$ values and denote this by $\bar{\bar{x}}$ (x double bar, which is sometimes called the *grand mean*). If $\bar{x}_j$ denotes the mean for sample j and there are m samples, each of size n, then

$$\bar{\bar{x}} = \frac{1}{m} \sum_{j-1}^{m} \bar{x}_j.$$

Table 7.2 Estimation from small samples

Sample	Measurements					Mean	Std dev.	Range
j	x_1	x_2	x_3	x_4	x_5	$\bar{x}_j$	s_j	R_j
1	81.5	66.6	84.2	70.1	72.2	74.9	7.57	17.6
2	72.3	83.6	76.3	84.5	76.0	78.5	5.28	12.2
3	78.3	99.8	88.9	83.0	87.9	87.6	8.04	21.5
4	75.9	66.7	94.2	93.2	85.7	83.1	11.76	27.5
5	69.0	71.8	81.6	86.8	94.0	80.6	10.38	25.0
6	66.4	81.4	63.3	87.3	70.6	73.8	10.19	24.0
7	75.6	85.9	84.4	89.8	89.7	85.1	5.80	14.2
8	73.8	83.6	77.5	75.4	78.1	77.7	3.73	9.8
9	82.3	77.3	84.9	80.2	73.7	79.7	4.35	11.2
10	76.9	74.6	88.2	83.7	67.4	78.2	8.09	20.8
11	83.1	81.8	72.4	85.8	75.6	79.7	5.55	13.4
12	71.0	79.5	81.9	69.6	83.7	77.1	6.44	14.1
13	71.3	86.4	70.5	77.3	76.2	76.3	6.36	15.9
14	80.0	72.4	53.9	75.6	85.8	73.5	12.08	31.9
15	65.1	68.5	75.0	80.0	80.3	73.8	6.82	15.2
16	78.6	72.1	85.2	80.8	68.4	77.0	6.75	16.8
17	80.8	86.8	79.5	78.9	80.0	81.2	3.21	7.9
18	76.8	85.6	65.0	70.2	81.5	75.8	8.33	20.6
19	88.8	64.6	74.5	72.6	70.5	74.2	8.97	24.2
20	74.9	91.4	91.2	72.3	85.6	83.1	9.01	19.1
21	77.5	78.2	81.9	70.1	76.8	76.9	4.28	11.8
22	74.8	88.7	76.7	80.7	78.6	79.9	5.39	13.9
23	69.1	76.0	78.3	72.3	81.5	75.4	4.88	12.4
24	72.3	76.1	87.8	85.7	94.0	83.2	8.85	21.7
25	85.5	70.4	81.4	79.9	68.3	77.1	7.40	17.2
	Totals					1963.4	179.51	439.9
	Averages					78.5	7.18	17.6

Since each $\bar{x}_j$ is the sum of all $\bar{x}$ values in sample j divided by the number of values in that sample, the formula could alternatively be written

$$\bar{\bar{x}} = \frac{\text{grand total}}{\text{number of samples} \times \text{number in each sample}}.$$

Although this version of the formula could be used, with the grand total calculated by simply adding up all the x values, it is preferable to calculate $\bar{\bar{x}}$ by averaging the $\bar{x}$ values for two reasons:

■ It reduces the risk of mistakes in hand calculations.
■ The $\bar{x}$ values themselves provide information about the underlying distribution (as explained in Chapter 10).

Using the results from Table 7.2, where $m = 25$,

$$\bar{\bar{x}} = 1963.4/25 = 78.5.$$

Estimating σ is slightly more complicated. There are two commonly used ways of approaching the problem. The first method is based on the *average sample standard deviation*. The sample standard deviation, s, is calculated for each sample and the results are averaged over all m samples to give

$$\bar{s} = \frac{1}{m} \sum_{j-1}^{m} s_j.$$

The idea of bias, which was introduced in Section 3.19, needs to be considered again here. To produce an unbiased estimate of σ, this $\bar{s}$ must be divided by a correction factor c_4, which depends upon the sample size n. A table that includes values of c_4 for various sample sizes is given in Appendix F.

This table shows that $c_4 = 0.940$ when $n = 5$, so for the data in Table 7.2, where $s = 179.51/25 = 7.18$, the corresponding unbiased estimate of σ is

$$\hat{\sigma} = \bar{s}/c_4 = 7.18/0.940 = 7.6 \text{ (to 1 decimal place).}$$

The second method is based on the *average sample range*. The sample range R is calculated for each sample and the results are averaged over all m samples to give

$$\bar{R} = \frac{1}{m} \sum_{j-1}^{m} R_j.$$

A different correction factor, d_2, converts $\bar{R}$ into another unbiased estimate of σ. Values of d_2 are also tabulated in Appendix F, which shows that $d_2 = 2.326$ when $n = 5$, so that for the data in Table 7.2, where $\bar{R} = 439.9/25 = 17.6$, the corresponding unbiased estimate of σ is

$$\hat{\sigma} = \bar{R}/d_2 = 17.6/2.326 = 7.6 \text{ (to 1 decimal place).}$$

In this case, both methods give the same value of σ when this is rounded to the same number of decimal places as the original data. The two methods usually give very similar results, but the second method is generally preferred for hand calculation because it is much easier to calculate R than s.

It might be thought that it would be better to avoid any method that involved using correction factors and simply estimate σ by treating all the observed values of x as a single large sample and calculating s for this large sample. However, if the small samples have been collected at different times and there is a possibility that the underlying mean may have shifted when some of them were collected, then the resulting s, considered as an estimate of σ, will be inflated by such a shift. The methods based on the average spread in a small sample ($\bar{s}$ or $\bar{R}$) avoid this risk.

■ 7.8 Estimating binomial *p* from several samples of different sizes

In Section 7.2 it was shown that, if the total number of successes is recorded over *m* sets of binomial trials, each set consisting of *n* trials, the estimate of *p* could be expressed as the total number of successes divided by the total number of trials.

This result is still valid if the number of trials differs from one set to another. Suppose that there are n_1 trials in the first set, resulting in x_1 successes, n_2 trials in the second set, resulting in x_2 successes, and so on up to n_m trials and x_m successes in the last set. The estimate of *p* obtained by combining these results is usually denoted by $\bar{p}$ (*p* bar).

$$\bar{p} = \frac{x_1 + x_2 + \cdots + x_m}{n_1 + n_2 + \cdots + n_m}.$$

The assumption behind this formula is that the underlying value of *p* was the same when each set of trials was being carried out. This can be checked using control chart techniques (Section 10.10).

■ References to further reading on the topics of Chapter 7

Chatfield (1983) and Montgomery (1985) give good outlines of the ideas behind estimation. Walpole and Myers (1993) give a more detailed treatment.
See Bibliography for details of titles and publishers.

■ Self-assessment questions on Chapter 7

1. Use the appropriate formula from Table 7.1 to estimate the probability of success in a single binomial trial, given that 3 sets of 30 trials resulted in 3, 2 and 4 successes respectively.
2. Use the appropriate formula from Table 7.1 to estimate the mean number of blemishes in paintwork on a refrigerator, assuming that the number of blemishes has a Poisson distribution and given that 5 refrigerators were found to have 5, 9, 8, 6 and 5 blemishes respectively.
3. The time between breakdowns in a machine shop, *x* hours, is exponentially distributed with probability density $f(x) = r \exp(-rx)$. Obtain a point estimate of *r*, given that times between the last 6 breakdowns have been 7, 12, 4, 15, 17 and 8 hours respectively.

4. Components are manufactured at factory A and then delivered by road to factory B. The journey time is thought to be normally distributed. Drivers' logs show that for the last 20 journeys the mean time was 3 hours 47 minutes and the standard deviation was 15 minutes. Use tables of the t distribution to find 95% confidence limits for μ, the mean of the underlying distribution of journey times.

5. A survey of 100 cars, randomly selected from those first registered in a particular month, showed that 87 had been driven more than ten thousand miles in the first 12 months since registration. Use this result to find approximate 95% confidence limits for the percentage of all such cars which had been driven more than ten thousand miles in the first 12 months. Also find approximately how many more cars should be sampled so that the margin of error in the estimate is no more than 5% ($e = 0.05$).

6. At the start of a study of process capability, four measurements of the feature under investigation were made in five successive shifts. The results are shown in the table below.

Shift	Measurements			
1	25.6	10.5	21.5	20.0
2	12.3	12.6	10.2	18.9
3	21.2	14.5	15.0	19.5
4	11.4	15.8	25.6	26.8
5	14.9	28.7	16.8	20.3

Calculate the standard deviation (s) and range (R) for each shift and average these to find $\bar{s}$ and $\bar{R}$. Then divide by the appropriate values of c_4 and d_2 to obtain two separate estimates of the process standard deviation.

8 | Probability Plotting

■ 8.1 Introduction

Before using observed data to estimate parameters of a particular type of theoretical distribution, it is often necessary to check whether that distribution provides a suitable model for the data. For some important distributions there are simple graphical methods for doing this; the general name for them is *probability plotting*.

The technique is widely used in industry, and specially printed graph papers, known as *probability papers*, are available from a number of suppliers. These are designed to provide rough parameter estimates; at the same time they provide a check on the data distribution and hence the validity of estimates. They are used as an aid in product reliability studies (Section 8.8) and in process control (Section 9.11).

However, this chapter explains that it is not necessary to have special graph paper in order to carry out the basic tests of suitability.

■ 8.2 General principles

The basic idea behind the graphical methods for identifying suitable distributions is that the observed cumulative relative frequency at successive data points should behave in much the same way as the underlying theoretical cumulative distribution function (cdf) for the distribution which generated the data. The properties of particular distributions can then be used to devise special ways of plotting the data so as to give a set of points that lie close to a straight line if the appropriate distribution has been chosen.

The general approach can be examined in more detail by considering the data in Table 8.1, which represents the lifetimes (in months) of 5 motor vehicle wing mirrors.

Table 8.1 Lifetimes of five wing mirrors

Item i	As observed x_i	Ranked x_i	Cumulative relative frequency
1	73	15	$1/5 = 0.2$
2	40	40	$2/5 = 0.4$
3	94	49	$3/5 = 0.6$
4	15	73	$4/5 = 0.8$
5	49	94	$5/5 = 1.0$

At first, the lifetimes are labelled according to the order in which they were observed, so that $x_1 = 73$ was observed first and $x_5 = 49$ was observed last. The next step is to rearrange the values in rank order (that is, smallest first and largest last, as would be done in order to find the median) and relabel them with their rank numbers. Having done this, $x_1 = 15$ (the smallest value) and $x_5 = 94$ (the largest value).

After the rearrangement, just one of the five ranked values is less than or equal to x_1, two out of five are less than or equal to x_2 and so on. This can be summarized by saying that the cumulative relative frequency at x_i (the ith ranked value) will be $i/5$. The summary is shown graphically in Figure 8.1, where it can be seen that the cumulative relative frequency rises in steps of height $1/5$ at each of the ranked values.

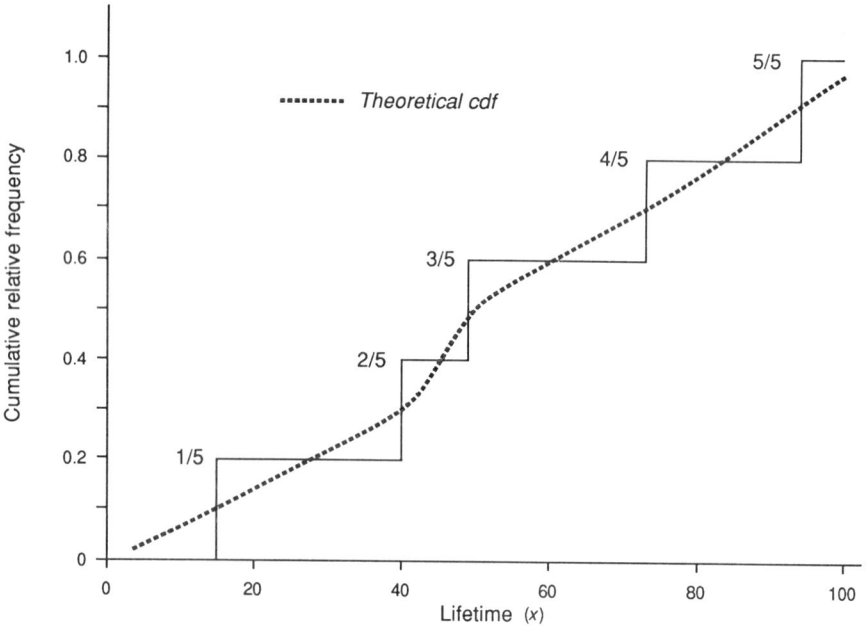

Figure 8.1 Cumulative relative frequency for wing mirror lifetimes

Since the lifetime of a wing mirror would be a continuous variable, its cumulative distribution function would be a continuous curve that rises smoothly from zero at the minimum possible lifetime to 1 at the maximum possible lifetime. This curve should pass close to the stepped graph that represents the cumulative relative frequency. It is usual to consider that the curve passes somewhere between the bottom and the top of the step at each x_i value, like the broken line in Figure 8.1. The height at the bottom of this step is $(i - 1)/5$ and the height at the top of it is $i/5$, so the estimated height of the broken line must lie somewhere between these values. The most simple assumption is that it lies *halfway between the bottom and the top of the step*, at height $(i - 1/2)/5$.

Plotting positions

In general, if there are n observed values, it can be assumed that the height of the cumulative distribution function curve (often called the *plotting position*) at the ith ranked value is

$$P_i = \frac{i - 0.5}{n}$$

This expression is called the *Hazen formula*, after an American civil engineer who introduced it in a study of water supplies. It provides the simplest means of getting a sensible plotting position for any kind of underlying distribution of x values. For example, with $n = 5$ and $i = 3$ it gives

$$P_3 = \frac{3 - 0.5}{5} = \frac{2.5}{5} = 0.5.$$

In other words, the third ranked value (which is the sample median in this case) corresponds to an estimated cumulative distribution function of 0.5, so it provides a natural estimate of the median of the underlying distribution.

More complicated mathematical reasoning, involving specific assumptions about the form of the underlying distribution, has led to other formulae for plotting positions. The most widely used are of the form

$$P_i = \frac{i - c}{n - 2c + 1}$$

where c is a constant with a value between 0 and 0.5. All formulae of this type give the sample median as the estimate of the population median.

The Hazen formula has $c = 0.5$ and is recommended for most sample sizes and most kinds of underlying distribution. Other commonly used values are $c = 0$, $c = 0.3$ and $c = 0.375$. For example, *Benard's formula* uses $c = 0.3$:

$$P_i = \frac{i - 0.3}{n + 0.4}.$$

Benard's formula and the formula having $c = 0.375$ specifically assume that the underlying distribution is normal, but they are often, mistakenly, used in testing for Weibull distributions.

For n greater than about 10, all the formulae give very similar values for most of the plotting positions. They differ most about the positions for the first and last points (corresponding to the smallest and largest observed values of x). In practice, whichever formula is being used, the extreme points should be treated as being less reliable than the rest.

Transformations

Values of P_i plotted directly against x_i will usually produce a set of points that seem to lie near some S-shaped curve. A particular curve is expected if the x values have been generated by some particular theoretical distribution. However, it is often difficult to decide whether the plotted points lie sufficiently close to the expected curve, and therefore it is not easy to confirm the x-value distribution.

The art of probability plotting is to find a way in which to *transform P_i* to another quantity, say y_i, so that plotting y_i against x_i gives a set of points that lie on (or near) a straight line when the correct distribution is chosen. To achieve this straight line with some distributions it may also be necessary to transform the x_i values.

When checking for a *normal distribution* the transformation for P_i is

$$y_i = 4.91 \ [(P_i)^{0.14} - (1 - P_i)^{0.14}]$$

and no transformation is needed for x_i.

When checking for *Weibull* distribution the transformation for P_i is

$$y_i = \ln\{\ln[1/(1 - P_i)]\}$$

and x_i is transformed to $x_i = \ln(x_i - \gamma)$, where γ is the smallest possible value of x.[1] As explained below, some trial and error may be required to determine an appropriate value of γ, the location parameter.

Although it is fairly easy to calculate y_i from these formulae, especially if a computer spreadsheet is used, it is often more convenient to use a specially printed probability paper on which there is a non-linear scale labelled with values of P_i (usually expressed as percentages). The paper with the transformation for normal distributions is often called 'normal probability paper'. The paper for Weibull distributions may include extra scales for parameter estimation. Both of these types of paper are described in more detail below. There are also special probability papers for some other types of distribution.

---

[1] x denotes the basic Weibull variable here, rather than t which was used in Chapter 6.

■ 8.3 Normal probability plotting

The transformation $y_i = 4.91 \; [(P_i)^{0.14} - (1 - P_i)^{0.14}]$ is an approximation to the value of the standard normal deviate $z_i = (x_i - \mu)/\sigma$ which would give $P(z < z_i) = P_i$. For example, the transformation of $P_i = 0.95$ gives $y_i = 1.647$, but the standard normal tables (Appendix B) show that the exact z value corresponding to a cumulative probability of 0.95 is $z_i = 1.645$. However, in the context of scales generally used for probability plotting, the difference between y_i and z_i is negligible.

If the plotted points do lie close to a straight line and the line of best fit is drawn through them by eye, the value on the x axis corresponding to $y = 0$ will be an estimate of μ, the mean (and median) of the underlying normal distribution. The difference between this x value and that corresponding to $y = 1$ will be an estimate of σ, the standard deviation of the underlying normal distribution.

i	1	2	3	4	5
x_i	15	40	49	73	94
P_i	0.10	0.30	0.50	0.70	0.90
y_i	-1.28	-0.52	0	0.52	1.28

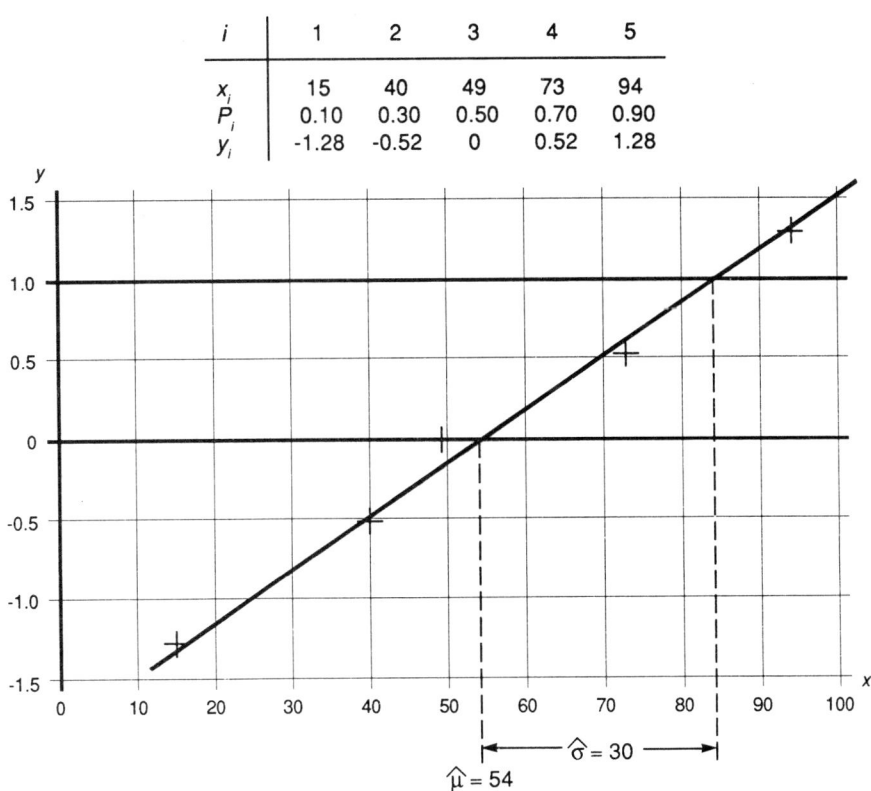

Figure 8.2 Normal probability plot of wing mirror data

Figure 8.2 shows the calculations and the normal probability plot for the wing mirror data from Table 8.1. Note that the graphical estimates of the mean and standard deviation ($\hat{\mu} = 54$ and $\hat{\sigma} = 30$) are the same as those calculated from the original data. However, graphical estimates can be affected by the exact position of the line and, especially if this is fitted by eye, they should be used only as rough guides to the true values of μ and σ.

The x axis in a normal probability plot need not be of individual x values; the method can be extended to checking whether sample means or sample medians are normally distributed. Figure 8.3 shows a plot of the sample means from Table 7.2, below a table of their rank order and plotting positions. In this plot, cumulative percentages ($100P_i$) are shown on the vertical scale, as on many proprietary brands of normal probability paper.

Note that, when several values tie for the same rank order, a single point is plotted at the highest rank only.

Most industrial papers show cumulative percentage frequency ($\Sigma f\%$) on the horizontal axis (Figure 9.8). The largest observed value corresponds to 100%, but plots cannot be made at 100% on probability papers. Therefore, it is usual to plot a single point at the average of the two highest values with the idea of making some use of all data. Also, it is common practice to plot a single point at the average of any values that tie for the same rank order. There are theoretical objections to these industrial conventions, but they find acceptance because they are considered to be 'user friendly' and the inaccuracies that result are usually negligible.

Apart from the first point (at 2%), all the points plotted in Figure 8.3 lie near a straight line. This indicates that the $\bar{x}$ values from Table 7.2 are normally distributed, as the central limit theorem would suggest.

When a percentage scale is shown on a normal probability paper axis, the graphical estimate of the underlying mean of the quantity plotted on the other axis is the value of this quantity that corresponds to 50%, and the graphical estimate of its standard deviation is the difference between the values corresponding to 84% and 50%.

Thus for $\bar{x}$ in Figure 8.3, the estimate of the mean (μ) is approximately 78.8 and that for the standard deviation ($\sigma/\sqrt{n}$) is approximately $82.8 - 78.8 = 4$. These agree quite well with 78.9 and $7.6/\sqrt{5} = 3.4$ obtained by the more accurate methods described in Section 7.4.

Although it might appear to be secondary in some industrial applications (Section 9.11), the main reason for normal probability plotting is to check for normality. If there is insufficient data to draw a histogram and see whether this approximates to the characteristic bell shape, then a normal probability plot provides the easiest check. The Kolmogorov–Smirnov test (Section 11.11) is a more exact method.

154

Rank = i, mean = $\bar{x}_i$, plot = $100P_i$ where $P_i = (i - 0.5)/n$ and $n = 25$

Rank	1	2	3	4	5	6	7	8	9	10	11	12	13
Mean	73.5	73.8	73.8	74.2	74.9	75.4	75.8	76.3	77.0	77.1	77.7	78.2	78.5
Plot	2	(6)	10	14	18	22	26	30	34	38	42	46	50

Rank	14	15	16	17	18	19	20	21	22	23	24	25
Mean	79.7	79.7	79.9	80.6	80.9	81.2	83.1	83.1	83.2	84.4	85.1	87.6
Plot	(54)	58	62	66	70	74	(78)	82	86	90	94	98

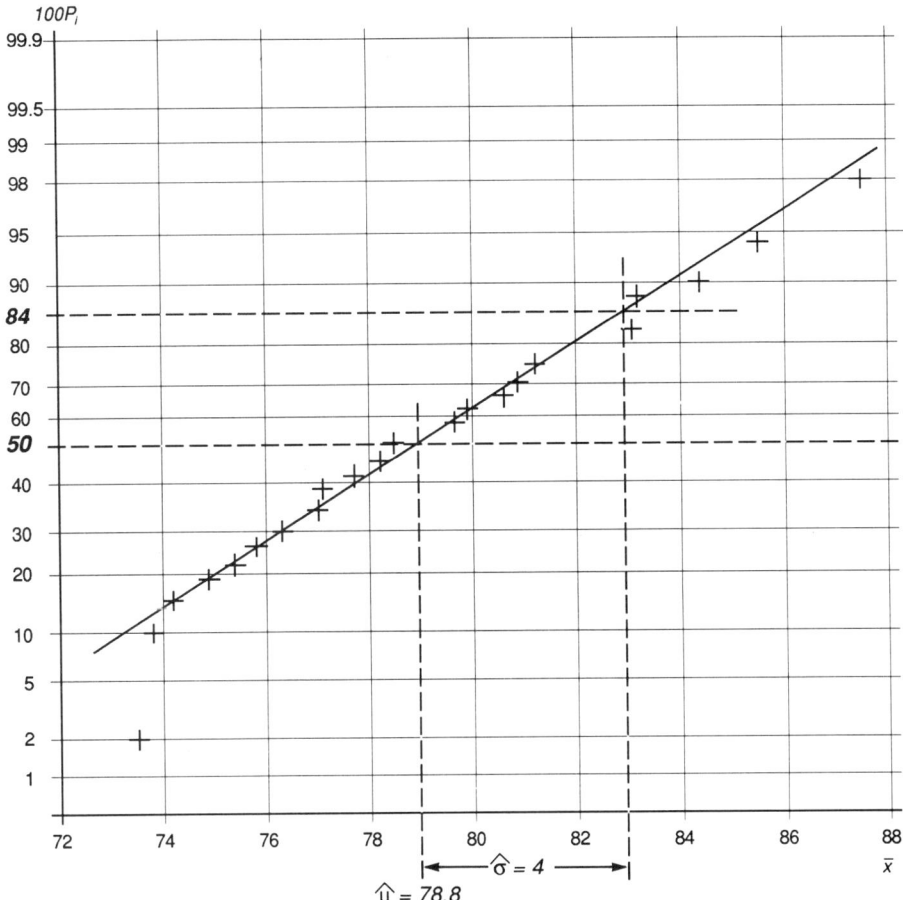

Figure 8.3 Normal probability plot of sample means

It is important to remember that if there is very little data, such as only 5 wing mirror lifetimes, several different kinds of distribution might provide a reasonable model. In fact this particular set of data (being fitted here to a normal distribution) could just as well have come from a Weibull distribution (as shown below).

155

■ 8.4 Probability plots for grouped data

If the data is available only in grouped form it can still be used for probability plotting, regardless of whether there is enough to construct a reasonable histogram. The essential difference from plotting all the individual observations is that the only x values considered are the *upper boundaries of each class*. The

Class = i, x = upper boundary x_i, plot = $100P_i$ where $P_i = (F_i - 0.5)/n$ and $n = 125$

Class	1	2	3	4	5	6	7	8	9
x	19.5	29.5	39.5	49.5	59.5	69.5	79.5	89.5	99.5
F_i	3	10	22	43	82	105	116	124	125
Plot	2.0	7.6	17.2	34.0	65.2	83.6	92.4	98.8	99.6

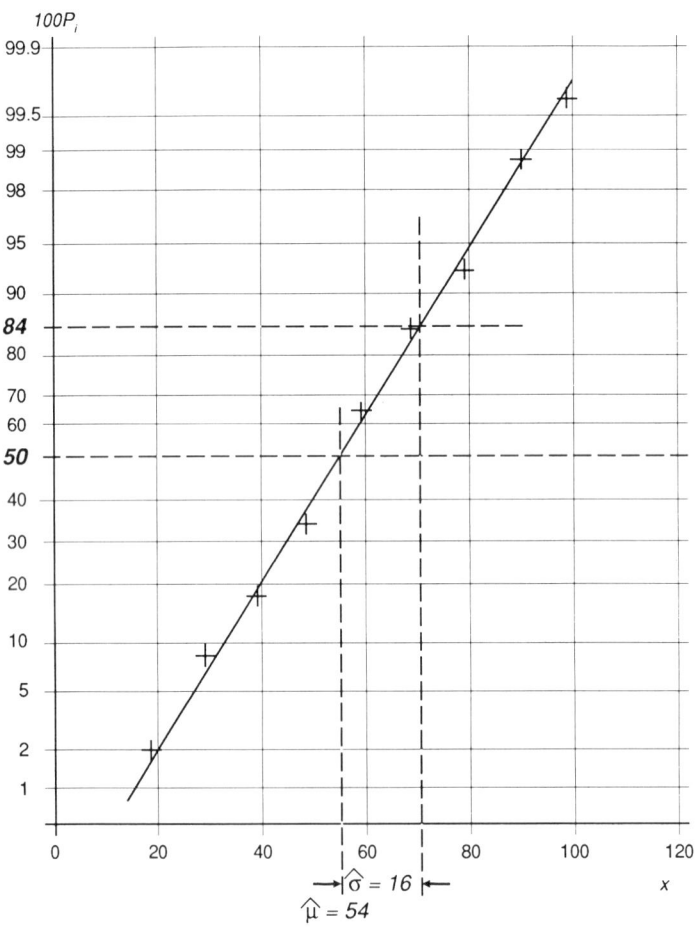

Figure 8.4 Normal probability plot of flange separation data

plotting positions for these are calculated using the cumulative frequency for the class concerned: that is, the total number of observations which are less than or equal to the value at the upper boundary of the class.

For a set of n observations which have been grouped so that the cumulative frequency for class i is F_i, the Hazen plotting position corresponding to the upper class boundary x_i is

$$P_i = (F_i - 0.5)/n.$$

It is a common mistake to plot at the midpoint rather than the upper boundary of each class. This distorts the linearity of the plot and gives incorrect graphical estimates of the mean and standard deviation.

Figure 8.4 shows the normal probability plot for the flange separation data that was discussed in Chapter 3, and is based on the grouping used in Table 3.6. A good linear plot is obtained, confirming other indications that the separations were normally distributed. Graphical estimates of the mean and standard deviation are approximately 54 and $70 - 54 = 16$ respectively. These agree well with 54.1 and 16.1 calculated from the raw data or with 54.1 and 16.4 calculated from the grouped data by the methods described in Chapter 3.

■ 8.5 Weibull probability plotting

When checking to see whether a set of observed x values could have come from a Weibull distribution, Hazen plotting positions are calculated in the same way as for normal probability plotting. If specially printed Weibull probability paper is not available, the double-logarithmic transformation below is used to obtain the values that are plotted on the y axis.

$$y_i = \ln\{\ln[1/(1 - P_i)]\}$$

Note that natural logarithms (ln), not logarithms to the base 10, must be used.

The x values also need to be transformed. This requires making a guess at an appropriate value for the location parameter, γ, which is the smallest possible value of x. Unless there is some good reason to do otherwise, it is best to make an initial guess that $\gamma = 0$. This γ is then subtracted from each x_i value and the natural logarithm taken:

$$x_i^* = \ln(x_i - \gamma)$$

Values of y_i are then plotted against these values of x_i^*. The method is illustrated in Figure 8.5, which is the Weibull plot for the wing mirror data from Table 8.1, using $\gamma = 0$. All the points lie reasonably close to a straight line, indicating that the data could have come from a Weibull distribution. Note that, as shown in Section 8.3, this data could have come from a normal

x_i	15	40	49	73	94
$x_i^* = \ln x_i$	2.71	3.69	3.89	4.29	4.54
P_i	0.10	0.30	0.50	0.70	0.90
y_i	-2.25	-1.03	-0.37	0.19	0.83

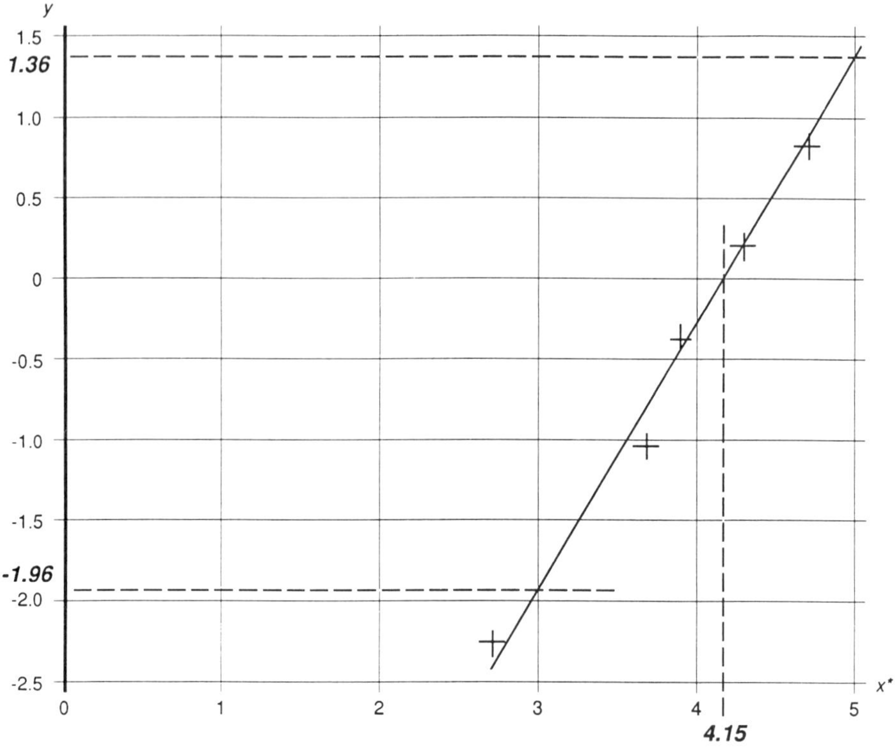

Figure 8.5 Weibull probability plot of wing mirror data

distribution; because of the small sample size, there are other reasonable models besides these two.

If the Weibull plot gives a reasonable straight line, graphical estimates can be obtained for the characteristic life (η) and the shape parameter (β).

The estimate of the characteristic life, $\hat{\eta}$, equals exp(x_0) which is the anti-log of x_0^*, where x_0^* is the value of x^* at which y is zero (in other words, the value at which the line crosses the horizontal axis).

In Figure 8.5, the line crosses the horizontal axis at $x^* = 4.15$ approximately, so $\hat{\eta} = \exp(4.15) = 63.4$.

The estimate of the shape parameter, $\hat{\beta}$, is the slope of the graph of y against x.

Table 8.2 A table of $\hat{\beta}$, g and P_μ values for Weibull estimation

$\hat{\beta}$	g	P_μ	$\hat{\beta}$	g	P_μ
0.5	2.000	75.7	2.0	0.886	54.4
0.6	1.505	72.1	2.2	0.886	53.5
0.7	1.266	69.3	2.4	0.887	52.7
0.8	1.133	66.9	2.6	0.888	52.0
0.9	1.052	64.9	2.8	0.891	51.5
1.0	1.000	63.2	3.0	0.893	50.9
1.2	0.941	60.5	3.5	0.900	49.9
1.4	0.911	58.4	4.0	0.906	49.1
1.6	0.897	56.8	4.5	0.913	48.4
1.8	0.889	55.5	5.0	0.918	47.9

In Figure 8.5, a convenient estimate of the slope can be found by dividing the difference between the heights of the line at $x^* = 3$ and $x^* = 5$ by the corresponding difference between the values of x^*. In other words, $\hat{\beta} = [1.36 - (-1.96)] \div [5 - 3] = 1.66$.

The estimate of the mean of the Weibull distribution, $\hat{\mu}$, can be found by multiplying $\hat{\eta}$ by a factor g that depends upon $\hat{\beta}$. A table of values of g is shown in Table 8.2. This table also lists values of P_μ, the percentage of the distribution lying below the mean; this will be referred to again in Section 8.8 below.

By interpolation from the table, g is approximately 0.895 when $\hat{\beta} = 1.66$, so the estimate of the mean is $\hat{\mu} = 63.4 \times 0.895 = 57$ to the nearest whole number. The discrepancy between this estimate and the sample mean, $\bar{x} = 54$, is slightly greater than that found for the corresponding estimate obtained from the normal probability plot.

The various transformations involved in Weibull plotting usually result in rather wide margins of error for any of the parameter estimates. Since the estimate of the mean depends upon estimates of the characteristic value and the shape parameter, it will include errors due to inaccuracies in both of these other estimates.

■ 8.6 Estimating the Weibull location parameter

In Figure 8.5, a reasonably straight line was obtained for the Weibull probability plot of the wing mirror data by assuming that the location parameter (γ) was zero. However, making the same assumption for other sets of data which could be explained by a Weibull model may result in a plot such as one of those shown in Figure 8.6.

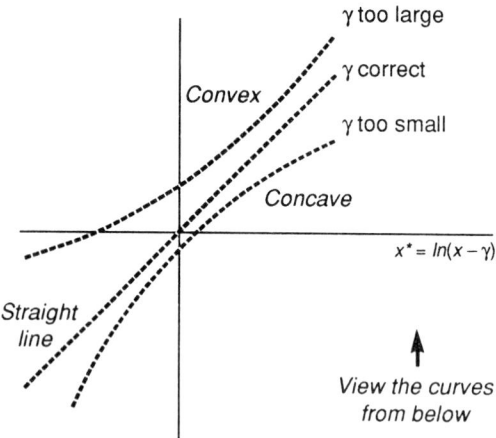

Figure 8.6 Effect of location parameter (γ) upon shape of Weibull probability plot

If the best freehand curve drawn through the points *bends downwards at the left*, so that the curve is *concave* when viewed from below, it indicates that *the assumed value of γ is too small*.

- When *x* represents the lifetime of a manufactured item, measured from the time that it was brought into use, it indicates an initial failure-free period of time or some delay in the onset of ageing.
- In this case the data should be replotted using a value of the location parameter (γ) that represents a reasonable guess at the length of the delay (expressed as a reasonable round number). Points corresponding to the higher values of *x* should stay much the same as in the original plot, but those corresponding to the lower values should move noticeably to the left, pulling the curve towards a straight line.

If the curve *bends upwards at the left*, so that it is *convex* when viewed from below, then *the assumed value of γ is too large*.

- This indicates that 'the clock was started too late'. In other words, failures or ageing had already started before the part was brought into use. It might reflect time spent in testing or the effects of storage.
- In this case an estimate of the equivalent pre-use ageing time must be added to the observed lifetimes for the revised probability plot.

Although a Weibull model can provide a reasonable explanation for many types of data, there is no guarantee that adjusting the value of the location parameter will produce the straight-line plot which indicates that Weibull is appropriate. However, some trial and error with different values of the location parameter is usually worthwhile if the initial Weibull plot does not give a straight line.

■ It is possible that a straight line is not produced because the data reflect two or more different failure mechanisms or ageing processes. This situation is discussed in the next section.

When a straight line has been obtained by using a non-zero value of γ, it is important to realize that the plot refers to the distribution of $x - \gamma$ and so estimates of the characteristic value and the mean must be adjusted for γ.

■ Adjusted $\hat{\eta} = \gamma + \hat{\eta}$ and adjusted $\hat{\mu} = \gamma + g\hat{\eta}$

■ 8.7 Fitting composite Weibull distributions

If a Weibull probability plot appears to consist of two linked straight lines with markedly different slopes, this indicates that the data may have been generated by two Weibull distributions with different shape parameters. A convenient way of modelling this is to fit a composite Weibull distribution of the type described in Section 6.9. The method is illustrated here by using the air-conditioning system data from Tables 6.2 and 6.3.

Since the data is grouped, the cumulative frequencies can be used to calculate Hazen plotting positions, P_i, for normal probability plots of grouped data (Section 8.4). These values of P_i may then be transformed to values of y_i in the same way as for Weibull plotting of ungrouped data.

Alternatively, if a table of hazard rates and cumulative hazards (such as Table 6.3) has already been constructed during the search for a suitable model, *values of y_i may be found by taking natural logarithms of the cumulative hazard estimates*. The reason for this is that $1 - P_i$ is an estimate of $R(t)$ at $t = x_i$, the upper boundary of class i, and the cumulative hazard is an estimate of $H(t) = \ln[1/R(t)]$ as stated in Section 6.7. Hence

$$\ln\{\ln[1/(1 - P_i)]\} \simeq \ln\{\text{cumulative hazard}\}$$

Values of y_i calculated by each method are shown in Table 8.3. The two sets of values are so close that effectively the same probability plot will be obtained whichever set is used.

Figure 8.7 shows the plot of y against x^*. The first three points appear to lie near one straight line whose slope is approximately 0.53. This is an estimate of β_1, the shape parameter for the first part of the composite distribution; since it is less than 1, it indicates a decreasing hazard rate, which is what the analysis in Section 6.7 has already shown.

The other points lie near a steeper straight line whose slope is approximately 1.7; this is an estimate of β_2, the shape parameter of the second part, and it indicates an increasing hazard rate.

Table 8.3 Calculation of y_i for air-conditioning system

x_i	x_i^* $= \ln x_i$	P_i	y_i from P_i	Cumulative hazard	y_i from cumulative hazard
6	1.79	0.095	−2.30	0.1053	−2.25
12	2.48	0.125	−2.01	0.1392	−1.97
18	2.89	0.165	−1.71	0.1862	−1.68
24	3.18	0.235	−1.32	0.2743	−1.29
30	3.40	0.305	−1.01	0.3708	−0.99
36	3.58	0.405	−0.66	0.5271	−0.64
42	3.74	0.495	−0.38	0.6922	−0.37
48	3.87	0.575	−0.16	0.8661	−0.14

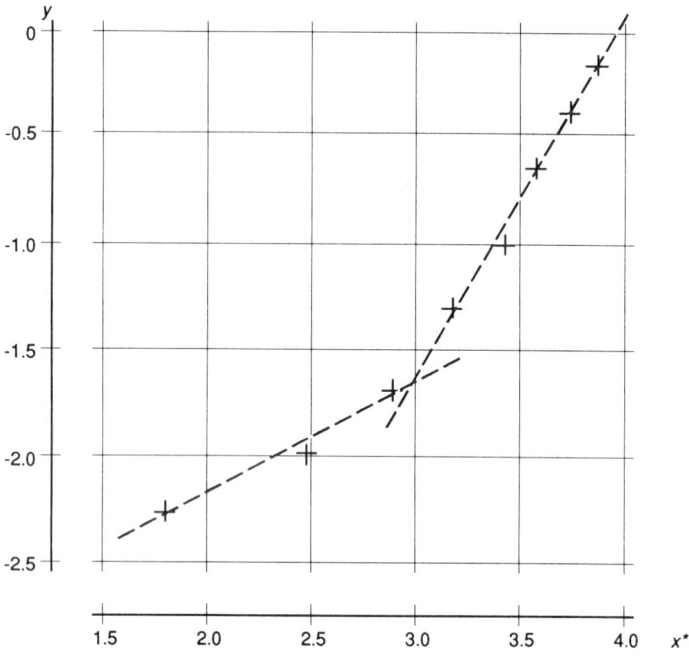

Figure 8.7 Composite Weibull probability plot for air-conditioning system

The two lines cross near $x^* = 3.0$; antilogging this gives $\exp(3.0) = 20$, which is a rough estimate of the transition time. A better estimate can be found when the characteristic values have been estimated.

To find the characteristic value for the second part of the distribution, read off the value of x^* where the steeper line cuts the horizontal axis. It cuts at $x^* = 3.95$ which is an estimate of $\ln \eta_2$; antilog this to get $\eta_2 = \exp(3.95) = 52$.

Rather than extend the lower line until it cuts the horizontal axis (which would require a very wide piece of graph paper), its intercept can be calculated by reading the x^* and y coordinates for some point already covered by the line and then using the slope to find the required additional distance along the horizontal axis.

For example, $y = -1.65$ at $x^* = 3.0$ and y increases by $\beta_1 = 0.53$ for each unit increase in x^*. Hence $y = 0$ at $x^* = 3.0 + (1.65/0.53) = 6.11 = \ln \eta_1$. Anti-logging gives $\eta_1 = exp(6.11) = 450$.

The transition time can now be re-estimated:

$$T = exp[(\beta_2 \ln \eta_2 - \beta_1 \ln \eta_1)/(\beta_2 - \beta_1)]$$
$$= exp[(1.7 \times 3.95 - 0.53 \times 6.11)/(1.7 - 0.53)] = 19.5.$$

Reverting to t instead of x as the variable representing lifetime in months, the model for hazard rate can then be written as

$$h(r) = \frac{0.53}{450} \left(\frac{t}{450}\right)^{-0.47} \quad \text{for } t < 19.5$$

$$\text{and } \frac{1.70}{52} \left(\frac{t}{52}\right)^{0.70} \quad \text{for } t > 19.5.$$

Figure 8.8 shows this $h(t)$ superimposed on the hazard rate histogram from Figure 6.7. The actual jump in the hazard rate at 19.5 months is unlikely to be as sudden as the model suggests, but it does seem likely that some new cause of failure begins to make itself felt around 20 months. Once this has been

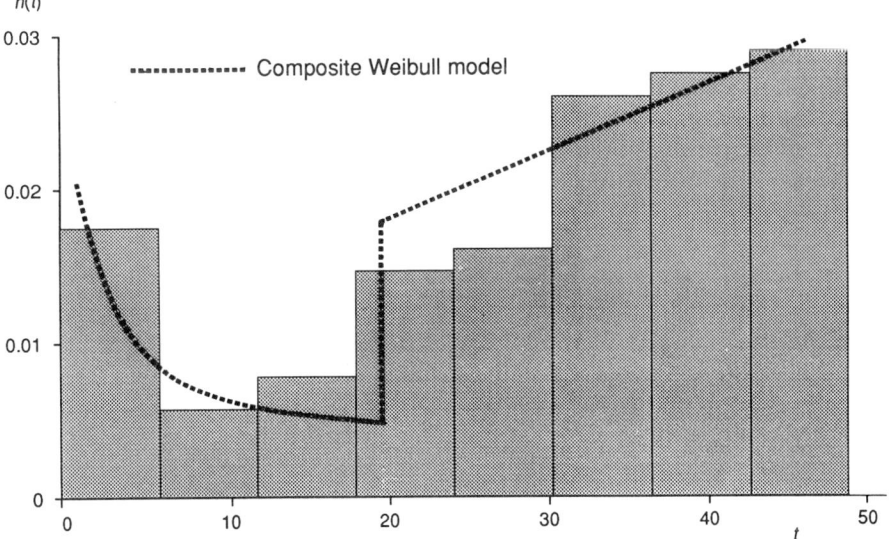

Figure 8.8 Hazard rates for air-conditioning system

highlighted by statistical analysis, thorough post-mortem examination of the failed air-conditioning systems should enable the separate causes of failure to be identified and the more realistic mixed Weibull model could be fitted. Even if this more accurate modelling is not required, the results of fitting the composite model give an indication of where to concentrate efforts on increasing the reliability of the air-conditioning systems.

■ 8.8 Weibull probability paper

There are several proprietary brands of probability paper which enable the user to carry out Weibull plots without needing to calculate logarithms or refer to tables such as Table 8.2. A paper that is widely used in the United Kingdom is the Chartwell 6572, produced by H.W. Peel & Co. of Greenford, Middlesex. This section explains how the paper is constructed and demonstrates how to use it by plotting the data for the failure of engine parts from Section 6.8.

The data referred to 15 engine parts, 9 of which had failed by 2015 hours when the remaining 6 were censored. Table 8.4 shows the calculations for Hazen plotting positions, using the methods described earlier in this chapter.

Figure 8.9 shows the failure times and the corresponding values of P_i plotted on a Weibull paper of the Chartwell type, assuming that the location parameter is zero. All the points lie fairly close to a straight line, which indicates that the Weibull is a satisfactory model.

The following points about the printed scales should be noted:

■ *The horizontal scale* (labelled *age at failure*). It is logarithmic (corresponding to $x^* = \ln t$ on a linear scale) and allows for values between 1 at the left and 100 at the right, with zeros omitted on the 10, 20, 30 and so on. Since the engine-part failure times vary between 468 hours and 2015 hours, the basic unit in this case is taken as 100 hours so that the plot covers the scale between 4.68 and 20.15.

Table 8.4 Plotting positions for engine-part failure data

Rank order i	Failure time t_i	Plotting position P_i
1	468	$(1 - 0.5)/15 = 0.033$
2	950	$(2 - 0.5)/15 = 0.100$
3	1177	$(3 - 0.5)/15 = 0.167$
4	1281	$(4 - 0.5)/15 = 0.233$
5	1313	$(5 - 0.5)/15 = 0.300$
6	1427	$(6 - 0.5)/15 = 0.367$
7	1768	$(7 - 0.5)/15 = 0.433$
8	1806	$(8 - 0.5)/15 = 0.500$
9	2015	$(9 - 0.5)/15 = 0.567$

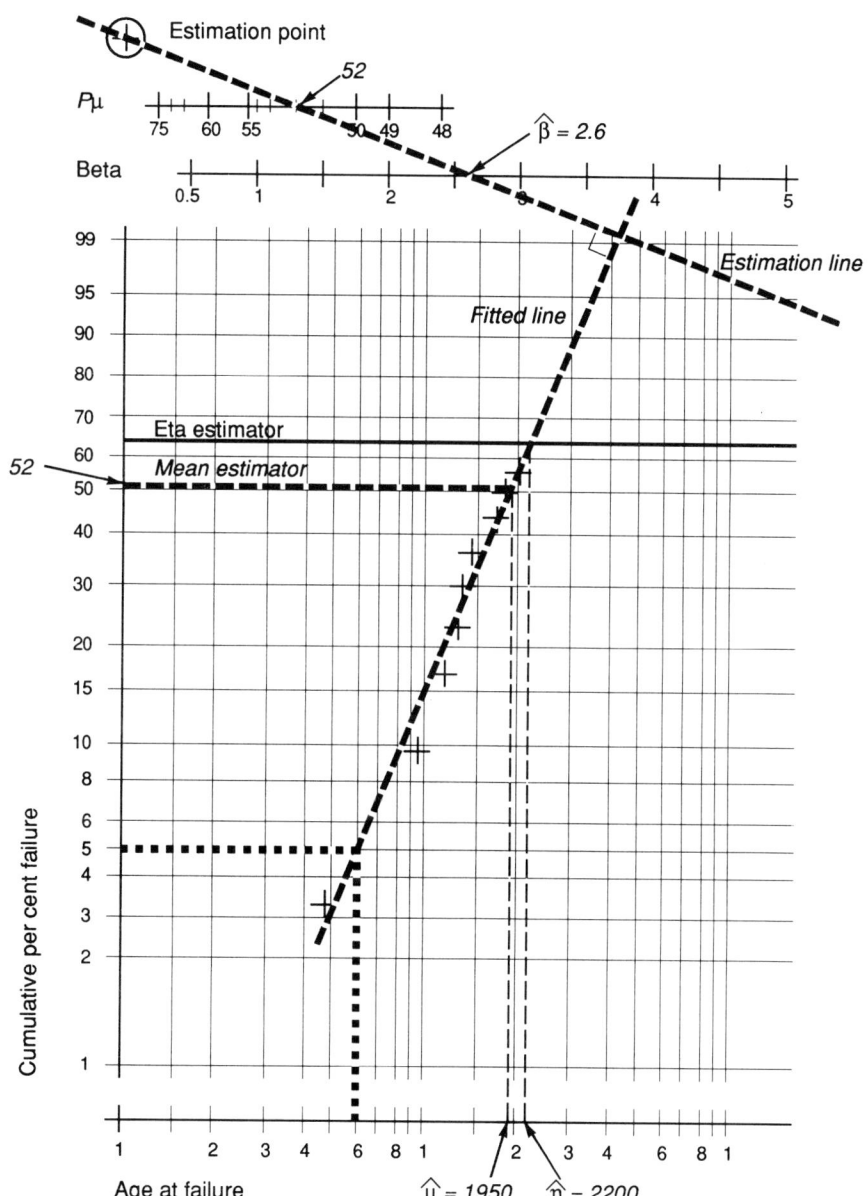

Figure 8.9 Weibull probability plot for engine-part failures

■ *The vertical scale (labelled cumulative per cent failure)*. It is non-linear and shows values ranging from 1 at the bottom to 99 at the top (most proprietary papers range from 0.1 to 99.9, to allow for larger samples). The calculated values of P_i are multiplied by 100 (so that they range from 3.3 to 56.7) and plotted on this non-linear scale, which corresponds to $y = \ln\{\ln[1/(1 - P)]\}$ on a linear scale.

Once it has been established that the Weibull is a satisfactory model, the characteristic life (η), the shape parameter (β) and the mean (μ) can be estimated graphically, using the line that has been fitted by eye in the usual way and various other supplementary devices printed on the Weibull probability paper and described below.

■ *Estimating the characteristic life.* The line fitted to the data points is extended to cut the line labelled *eta estimator*, which is drawn at 63.2 on the non-linear vertical scale (this corresponds to $y = 0$ on a linear scale). The age at failure where the two lines intersect is $\hat{\eta}$, the estimate of η.

In practice, parameters such as characteristic life are often distinguished from their estimates by different descriptions. For example, an estimate of characteristic life from development test data is sometimes called the expected in-service life.

■ *Estimating the shape parameter.* At the top left-hand corner of the Weibull probability paper is a small circle containing the *estimation point*. A set square should be used to draw an *estimation line*, which is perpendicular to the fitted line and passes through the estimation point. The estimation line will also pass through two supplementary scales which are printed between the estimation point and the top of the main graph. The lower of these scales shows values of $\hat{\beta}$. The estimate of the shape parameter is the value of $\hat{\beta}$ at which the estimation line cuts this scale.

The shape parameter that is estimated during development testing is often used as an indicator of the need for in-service maintenance. If $\hat{\beta} \leqslant 1$, any maintenance will have little or no effect; if $\hat{\beta} > 1$, maintenance is required and perhaps the design needs to be modified.

■ *Estimating the mean.* The upper supplementary scale shows values of P_μ, the percentage of failure times that are less than the mean; this percentage was tabulated in Table 8.2 above. Read off the value of P_μ where the estimation line cuts this scale, then draw a horizontal line (the *mean estimator line*) on the main graph at the cumulative per cent failure equal to P_μ and read off the age at failure where this mean estimator line cuts the fitted line; this is $\hat{\mu}$, the estimate of μ.

The cumulative per cent failure scale can be used also to estimate the time at which any percentage of parts will have failed. A practical application of this is in planning for servicing. For example, if it is economic to allow about 5% of parts to fail before the first routine service, the mean estimator line is drawn at 5% cumulative per cent failure and the age at failure is read off as before. This gives a rough idea of the interval before necessary servicing.

Because of difficulties in drawing the estimation line *perpendicular* to the fitted line, some proprietary brands of Weibull probability paper have the scales for β and P_μ at the side of the main graph and use an estimation line that is drawn *parallel* to the fitted line. Otherwise these papers are used in much the same way as the type described here.

In Figure 8.9, the fitted line cuts the eta estimator at 22 on the horizontal scale, so the estimated characteristic life, $\hat{\eta}$, is 2200 hours.

The estimation line cuts the $\hat{\beta}$ scale at 2.6, which is therefore $\hat{\beta}$, the estimate of the shape parameter.

The estimation line also shows that $P_\mu = 52$ (which can be verified by looking at Table 8.2 for $\hat{\beta} = 2.6$). Drawing the mean estimator line at 52 on the vertical scale makes it cut the fitted line just below 20 on the horizontal scale, so $\hat{\mu}$ is slightly less than 2000. This agrees with the result obtained by multiplying $\hat{\eta}$ by the g factor for $\hat{\beta} = 2.6$ obtained from Table 8.2 ($g = 0.888$ so $g\hat{\eta} = 0.888 \times 2200 = 1954$). This estimate of the mean takes account of the censoring and is appreciably greater than the average age at failure of the 9 parts which did fail, which is only 1356 hours.

Drawing the mean estimator line at 5 on the vertical scale makes it cut the fitted line at about 6 on the horizontal scale. In other words, about 5% of parts are likely to fail before about 600 hours in service.

These parameter estimates should be treated with considerable caution because they are all sensitive to small changes in drawing the fitted line and the estimation line, beside being subject to errors caused by trying to interpolate on the various non-linear scales. This sensitivity is made worse by the plotted points spanning such a narrow range on the horizontal scale. This is a price that has to be paid for not calculating logarithms.

The method described in Section 8.4 can avoid some of these problems by allowing the user to choose horizontal and vertical scales that are appropriate to the data in a particular problem. However, that method still gives parameter estimates with fairly wide margins of error. Rather than spend time on the complicated mathematics that is involved in quantifying these margins of error, it is better to use more exact mathematical methods for estimating in the first place. Even when these methods are considered, some kind of Weibull probability plot is a useful preliminary because it shows quite quickly whether the Weibull is an appropriate model to proceed with.

■ References to further reading on the topics of Chapter 8

Hahn and Shapiro (1968) and Barnett (1975) describe the general principles of probability plotting. Walpole and Myers (1993) give more details about normal probability plots.

See Bibliography for details of titles and publishers.

■ Self-assessment questions on Chapter 8

1. The following figures represent coded results from an investigation into the temperature of exhaust gases.

    72.8   69.8   64.6   71.8   73.7   75.4   70.2   79.6   66.9   76.4

    Rearrange the results in rank order and then carry out a normal probability plot, using any of the methods described in this chapter. Hence show that it is reasonable to conclude that the results could have come from a normal distribution with mean approximately 72 and standard deviation approximately 4.

2. As part of a survey of the performance of disposable batteries used in a certain kind of electrical appliance, information has been gathered about the lifetimes of these batteries. For 10 batteries selected at random, the numbers of weeks from installation to failure were as follows.

    132   186   90   49   70   108   149   73   113   78

    (a) Carry out a Weibull probability plot, assuming that the location parameter, γ, is zero. Confirm that this results in the kind of plot which indicates that there is a delay period before failure commences.

    (b) Subtract 30 weeks from each time to failure and then repeat the Weibull plot. Confirm that this does suggest that there is a delay period of about 30 weeks and that an appropriate distribution for the lifetimes would be a Weibull with β approximately 2, γ approximately 30 and η approximately 82.

Process Control

■ 9.1 A philosophy

Throughout this book there has been an emphasis on so-called *process control*. This might appear as a matter of semantics: the mathematical theories and statistical methods that have been explained are valid regardless of industrial philosophy. However, as was indicated in Chapter 1, the context in which they are used has changed and is continuing to change.

In keeping with the holistic concept of quality inherent in total quality management, the theories and methods are no longer the preserve of experts. Indeed, virtually none of the world's major companies employ 'policemen' in the form of departments devoted to inspection or to quality control. At the same time, for purposes of continuous improvement, there is more checking and measurement than ever before; this is carried out by employees on their own work.

Checking and measurement extends to all aspects of industrial processes: in other words, to people, materials, methods, facilities and the environment. The term 'process control' distinguishes this improvement activity from traditional *quality control*, which tended to focus on the result of processes in the context of a fixed specification.

In this chapter, process control is outlined and related to the statistical methods described elsewhere. Particular attention is given to the prerequisite for process control sometimes called a *stake-in-the-ground*. It is some measurement that provides a starting point for continuous improvement programmes based on reality rather than contractual specification.

■ 9.2 Terminology

Nominal

This refers to the single value of a product feature that is the supplier's target.

It might have been determined by calculation, by experiment or by experience, but it should be the same as the optimum value. It is the single value of a product feature that meets customers' needs with least whole-life cost, and for many industries it is a moving target.

Tolerance

This refers to the range of values about the optimum which give product performance that is satisfactory to the customer. Product performance tolerances are the criteria for specification of processes, rather than for day-to-day control. Control criteria are discussed in Chapter 10.

In most commercial industries, manufacturing industry in the Far East and increasingly in Western manufacturing industry, tolerances on details are not specified.

> Beware of 'standard tolerances' which are formalized in some manuals of drawing office practice. They were developed originally as a basis for contractual payments to piece-workers and external suppliers rather than as a basis for customer satisfaction.

Disturbances

Variation among the products of any process is inevitable. It arises from causes which create process *disturbances* that are generally classified as *common* or *special*.

Although the distinctions given below between common and special disturbances are clear, in the context of continuous improvement there is no distinction. Any cause of variation is a likely source of waste. In other words, the classifications help the assignment of responsibility for improvement, but they do not imply any priorities except that it is often easier to identify common disturbances when special disturbances have been eliminated.

Common disturbances

Causes that are supposedly inherent in a process are said to give rise to *common disturbances* which, randomly to a greater or lesser degree, *affect all products* of the process. Examples are an individual's alertness from moment to moment, uncontrolled ambient temperature, humidity, play in machines and inhomogeneous raw materials. Reducing the effect of common disturbances usually involves overall process management.

> Sometimes, common disturbances are said to arise from chance causes.

The total effect of all common disturbances in a process is to give results that are predictable within limits, for example, a data set that has a normal distribution (Figure 5.3). Processes that suffer only from common disturbances are said to be in *a state of statistical control*. The variation of in-control processes can usually be described by a simple statistical model that could be used to construct a chart for process monitoring (Section 9.6).

Special disturbances

Causes which *affect only some products* of a process and give unpredictable results that do not fit a simple statistical model are said to give rise to *special disturbances*. The causes of special disturbances are not inherent in the process, and they can sometimes be eliminated by local actions. Examples of the causes of special disturbances include deficiency in an individual's level of skill, material flaws, not observing specified procedures and power failures.

Sometimes, special disturbances are said to arise from assignable causes.

Processes that suffer from special disturbances are said to be *out of statistical control*. Usually, this state can be identified through use of a statistical model.

■ 9.3 Stages of process control

The stages of commissioning, monitoring and improvement in process control are illustrated in Figure 9.1. The first objective of control is to satisfy the customer. At the outset, when little or nothing is known about the process, its products are measured and compared with the customer's expectations. If it is likely that the expectations might not be met, the next step is to introduce some form of quality control such as screening, which amounts to inspection followed by acceptance or rejection.

The second objective of control is to identify and eliminate any special disturbances so that the process becomes consistent, predictable and able to be centred on specification. The elimination of special disturbances allows any information based on a mathematical model to be used with more confidence; also, it enables common disturbances to be more easily identified. At this point, it might still be necessary to have finished product 'quality control'.

Purists will debate the ordering of the first two objectives of control. The truth could be a matter of opinion. On one side 'customer rules' and on the other 'specification rules'. Certainly the former is true in an open market place. Under some contractual conditions, it might be the latter. In both cases the specification might be wrong: there can be no substitute for teamwork with the customer. Also affecting both cases is the simple fact that special disturbances might never be eliminated, especially where a process rather than one of its elements (such as a machine) is involved.

171

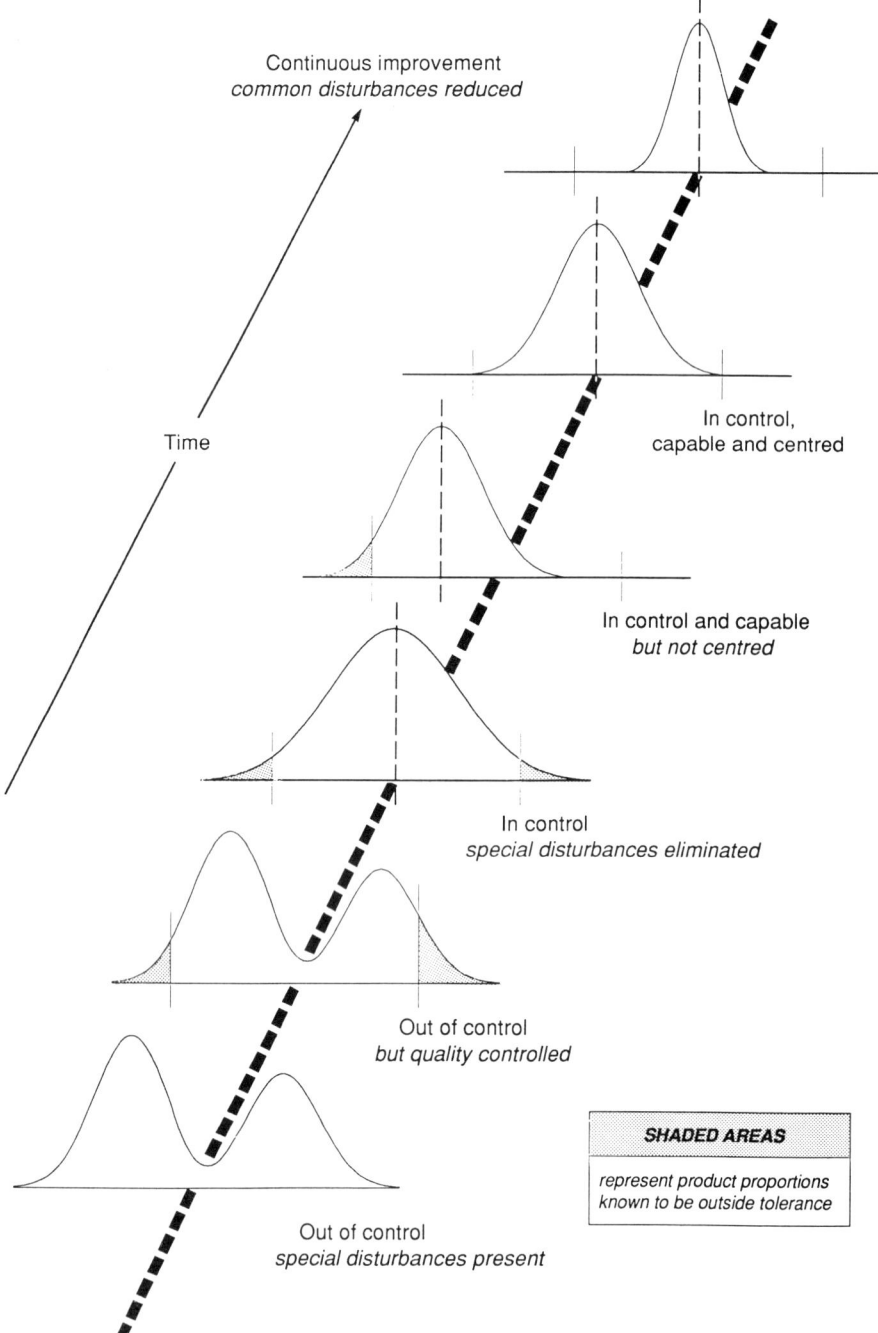

Continuous improvement
common disturbances reduced

Time

In control,
capable and centred

In control and capable
but not centred

In control
special disturbances eliminated

Out of control
but quality controlled

SHADED AREAS

*represent product proportions
known to be outside tolerance*

Out of control
special disturbances present

Figure 9.1 Stages of control

The third and final objective of control is to make the process as profitable as possible, which amounts to eliminating waste.

An obvious step is the removal of any quality control which does not add value to the product. Of course, such a step must be preceded by making sure that the process is *capable*: in other words, that the risk of not satisfying the customer is negligible. Process capability is considered in more detail later in this chapter.

> This step might seem contentious where quality is understood as total quality in the sense set out in Section 1.3. However, the step is aimed at activities that seek control independent of the work team and only through total or sample product inspection.

A further step is to set the process, to produce the majority of products at or close to the specified nominal, then to ensure that the specified nominal is the same as the customer's optimum. The final step is to approach the ideal of all products being at the customer's optimum.

At all stages, statistical models can be used to help achieve control. However, because these models usually assume an in-control state (which is rarely the case in practice), interpretation of objective statistical information requires process knowledge that is sometimes subjective. In other words, as

Mathematical description

$$f(x) = f(x_1) + f(x_2) + \ldots + f(x_n)$$

$$\text{where } f(x_i) = \frac{\beta_i}{\eta_i}\left(\frac{x - \gamma_i}{\eta_i}\right)^{\beta_i - 1} exp\left[-\left(\frac{x - \gamma_i}{\eta_i}\right)^{\beta_i}\right]$$

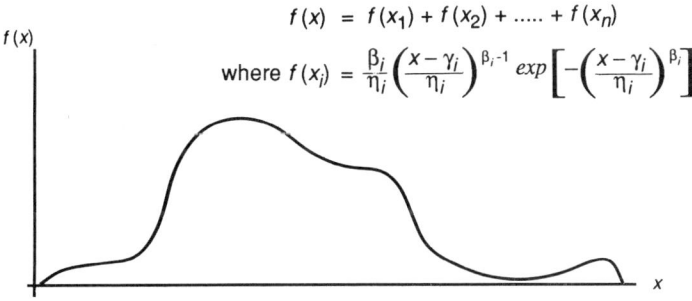

Industrial interpretation

*with acknowledgement for the idea
to Antoine de Saint-Exupéry*

Figure 9.2 Approaches to statistics

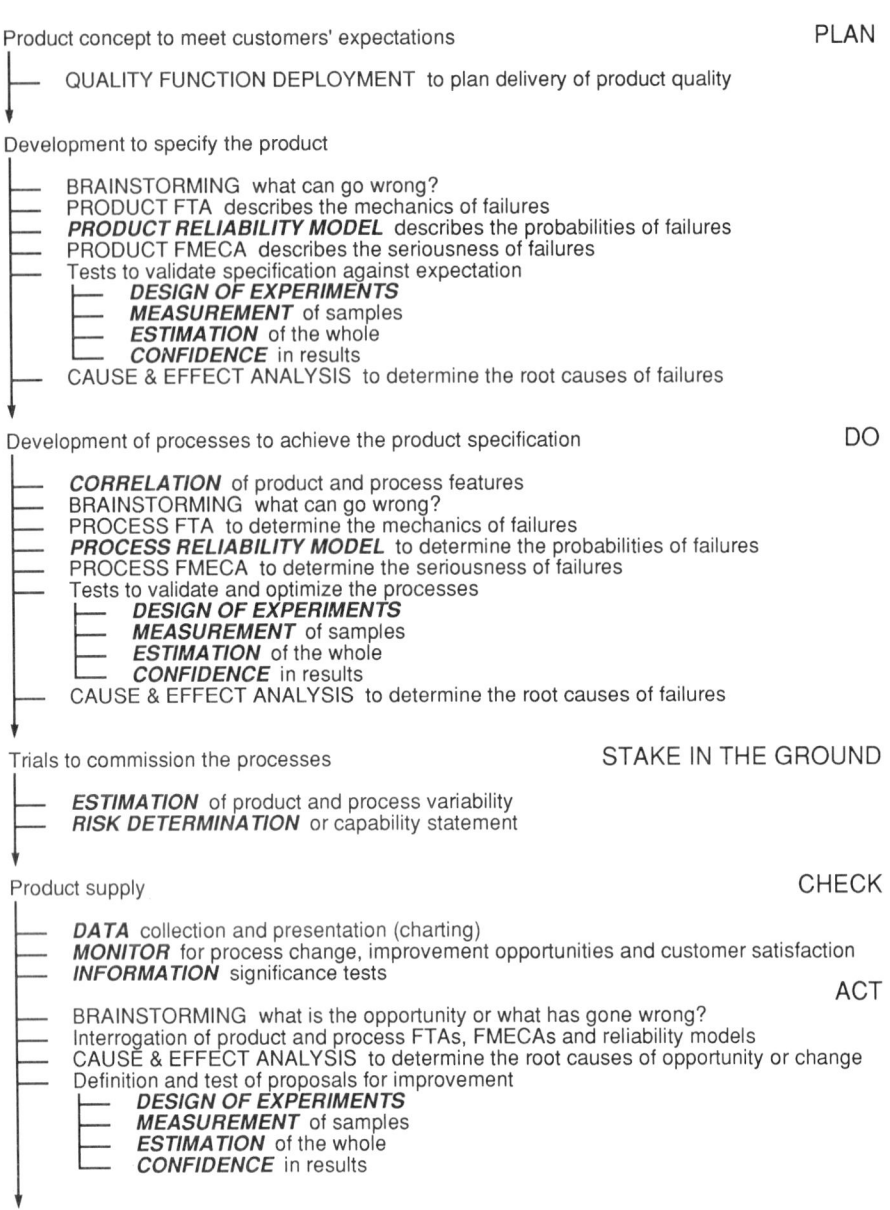

Product concept to meet customers' expectations　　　　　　　　　　PLAN

├── QUALITY FUNCTION DEPLOYMENT to plan delivery of product quality

Development to specify the product

├── BRAINSTORMING what can go wrong?
├── PRODUCT FTA describes the mechanics of failures
├── *PRODUCT RELIABILITY MODEL* describes the probabilities of failures
├── PRODUCT FMECA describes the seriousness of failures
├── Tests to validate specification against expectation
│　　├── *DESIGN OF EXPERIMENTS*
│　　├── *MEASUREMENT* of samples
│　　├── *ESTIMATION* of the whole
│　　└── *CONFIDENCE* in results
├── CAUSE & EFFECT ANALYSIS to determine the root causes of failures

Development of processes to achieve the product specification　　　　DO

├── *CORRELATION* of product and process features
├── BRAINSTORMING what can go wrong?
├── PROCESS FTA to determine the mechanics of failures
├── *PROCESS RELIABILITY MODEL* to determine the probabilities of failures
├── PROCESS FMECA to determine the seriousness of failures
├── Tests to validate and optimize the processes
│　　├── *DESIGN OF EXPERIMENTS*
│　　├── *MEASUREMENT* of samples
│　　├── *ESTIMATION* of the whole
│　　└── *CONFIDENCE* in results
├── CAUSE & EFFECT ANALYSIS to determine the root causes of failures

Trials to commission the processes　　　　　　　STAKE IN THE GROUND

├── *ESTIMATION* of product and process variability
├── *RISK DETERMINATION* or capability statement

Product supply　　　　　　　　　　　　　　　　　　　　　　　CHECK

├── *DATA* collection and presentation (charting)
├── *MONITOR* for process change, improvement opportunities and customer satisfaction
├── *INFORMATION* significance tests
　　　　　　　　　　　　　　　　　　　　　　　　　　　　　　ACT
├── BRAINSTORMING what is the opportunity or what has gone wrong?
├── Interrogation of product and process FTAs, FMECAs and reliability models
├── CAUSE & EFFECT ANALYSIS to determine the root causes of opportunity or change
├── Definition and test of proposals for improvement
│　　├── *DESIGN OF EXPERIMENTS*
│　　├── *MEASUREMENT* of samples
│　　├── *ESTIMATION* of the whole
│　　└── *CONFIDENCE* in results

Modification of the product and/or the process　　　　　　START AGAIN!

Figure 9.3 Techniques for quality achievement summarized by
sequence

illustrated in Figure 9.2, the mathematical description of a data distribution is of little practical use without some knowledge of the factors giving rise to variations in the data.

It must be emphasized that statistical methods only provide information. Control and improvement are achieved by people, who interpret and act upon the information in the context of their process. Figure 9.3 is a summary of some principle techniques at successive stages in time. Techniques that use statistical methods are written in ***BOLD ITALIC CAPITALS*** and are explained in detail elsewhere in this book. Some important qualitative methods are written in CAPITALS and are described briefly in Chapter 13.

■ 9.4 Process control planning

Processes involve people, materials, facilities, methods and the environment. They must be designed to meet product specifications, projected volume requirements and minimal whole-life cost of the product (in other words, the *total quality cost* of development, actual production, marketing and ownership).

Process control specifications, sometimes called stage criteria, are increasingly being developed using quality function deployment (Section 13.6). This technique focuses process controls on to those aspects of the process that are crucial to the achievement of customers' expectations. Often, the need for these controls is not always obvious in product specifications.

> For example, when a motor vehicle gear is changed, operation of the gear stick selects a drive wheel in the gear box. The specification of a drive wheel requires its diameter only to be within sizes that allow it to be selected smoothly. The specification also requires the wheel's gear teeth to be chamfered so that it does not become disengaged after selection. In the case of a particular drive wheel, the diameter is formed by forging and the chamfer is formed later by rolling. The range of diameters allowed for proper functioning in use is much greater than can be accepted by the rolling machine. In other words, the forging tolerance is determined by the rolling machine and is more crucial to quality than the finished product tolerance.

Once quality function deployment has identified crucial process features, other techniques are used during product and process development to find features that are easily measured as the product is being made. Statistical methods are used with data from trials to establish correlations and the significance of results.

■ 9.5 Process commissioning

Commissioning is the term usually given to the proving tests that are carried out when a new or changed process is introduced. The tests cover matters of health, safety, production rate and product conformance with requirements.

During the initial stages of commissioning, the process is adjusted and set so that the dimensional and material features of a product are produced on average at their specified nominal values.

Proving tests and process setting involve measurement of samples. The measurements are used in capability studies (Section 9.8) to assess the risk of not satisfying customers and to provide the stake-in-the-ground against which the process can be monitored.

At the commissioning stage, the process might or might not be in a state of statistical control. It is important to recognize that, even if the process is in a state of statistical control, out-of-specification products might still occur randomly: in other words, their occurrence cannot be predicted.

Where the risk of not satisfying customers is too great, because product variability is too large or because the process is not in a state of statistical control, the process needs to be changed.

In most cases the immediate change is to introduce the straightforward quality control activity of 100% checking of features at risk. This practice occurs frequently in industries where products or processes are at the forefront of technology. For example, in the silicon chip industry, 100% checking is necessary to ensure that only those products meeting the required standards are dispatched to customers; the remainder are scrapped.

It is unusual for 100% checking to result only in acceptance or rejection. More usually, its main purpose is grading (another process change) and items can then be sold by grade (for example, large, medium and small eggs) or used selectively (yet another process change). Selective use is common in industries that are manufacturers of complex assemblies, such as aircraft and motor vehicles. With the best processes in the world to make the parts, it is often not possible automatically to achieve satisfactory matching: for example, between holes and rivets in aircraft structures or between moving parts in engines.

Selective use amounts to assembling small with small, large with large, and so on. This practice makes best use of what is available and minimizes material and equipment costs. It could be argued that selective use opposes standardization and therefore ease of product maintenance. The counter-argument is that the customer seems to be satisfied; the evidence for this is in the performance of Japanese manufacturing industry, where selective use of parts is a norm.

Checking followed by acceptance, rejection, grading or selective use is not a substitute for the process monitoring described below. Also because it increases costs, it is an opportunity for process improvement through elimination of wasteful activity.

Throughout commissioning, suppliers, manufacturers and customers need to work as a team in order to achieve a satisfactory end product. In practice, this means that production develops detail with the downstream customer rather than relying upon upstream design. Expressed in a more popular phrase, the focus is on the wood rather than the trees. A result of this is that capability study data is used to set targets for the average and variability of component detail feature values. In other words, process commissioning puts a stake-in-the-ground of criteria for day-to-day monitoring.

As well as monitoring, other process control mechanisms are introduced during commissioning. They include fail-safe devices (or idiot-proofing, called Poka-Yoke in the Far East) and meticulous self-motivated observance of best practice. Also and especially bearing in mind that the customers' requirement often changes over time, there is the ultimate mechanism that reviews the effectiveness of process control itself.

■ 9.6 Process monitoring

Once a process has been commissioned and it is operational, information is needed to monitor its stability and to identify opportunities for improvement. Monitoring amounts to using statistical charting methods (Chapter 10).

The charting methods used are usually determined in particular cases by the dominant controlling features of the process. These are broadly classified as set-up dominant, materiel dominant and people dominant as described below.

Note that 'materiel' means all the hardware of a process; it includes all facilities (such as tools, machines and buildings) as well as materials (whether durable or consumable), but it excludes people.

Set-up dominant

Continuous processes are usually set-up dominant; they include most chemical processes such as foundries, oil refineries and motor vehicle paint plants. They operate virtually non-stop and are often able to self-compensate for tool wear or ambient temperature change. After setting, the inherent variation in most continuous processes is likely to be steady and (special disturbances excepted) slow to change. It is frequently the case that the product cannot be tested: for example, paint which is a corrosion protection coating would be destroyed by a corrosion test. In most cases, control is exercised by automated systems that are triggered through computerized charting of process variables (Section 10.7).

For set-up dominant processes, manual measurement and charting of variables can be useful during commissioning and later as an investigative tool. Otherwise, charting might be limited to periodic monitoring of attributes (Section 10.9) such as levels of customer satisfaction.

Materiel dominant

This is the group of processes where manual charting of process variables is most relevant; it covers many manufacturing situations where control of materiel is not automated. Regular charting can give information about setting and distribution changes, and when it is done by the production team, there is some motivation for them to interpret and act upon the information.

Note that the information is useful whether or not the process is in a state of statistical control and whether or not it is within the customer's tolerance.

People dominant

These are the processes that are largely controlled by the perceptions, skills, disciplines and vigilance of people. Examples are manual assembly operations and most personal service activities. Note that the facilities (such as buildings, machines and tools), utilities (such as natural gas, water, compressed air and electricity) and materials (such as lubricating oils, cleaning cloths and protective clothing) which are used in the course of these activities might be set-up or materiel dominant.

Occasionally, charting of variables might be appropriate but samples should be taken irregularly, so that results are not influenced by special efforts. More usually, some form of attribute charting provides an effective monitoring method.

■ 9.7 Process improvement

Quality achievement means satisfying customers and making a profit (Section 1.3). The target of process improvement is enhancement of quality achievement, but note that when the two elements, customers and profit, are being considered, one cannot be improved at the expense of the other.

Opportunities for process improvement can be found in information from various sources. For example:

■ Capability studies during commissioning which identify product variability and out-of-statistical-control situations.
■ Charting which highlights process changes and again any out-of-statistical-control situation.

■ The costs of failure represented by scrap, rework, overtime inside the company and warranty costs outside the company.
■ Customers comments, given freely or obtained through special surveys.

The techniques used to make the most of the information amount to the same that have been used already in planning, commissioning and monitoring the process. However, their application is sometimes not so easy; in the early stages (like building on a green-field site) the necessary steps can be clearly

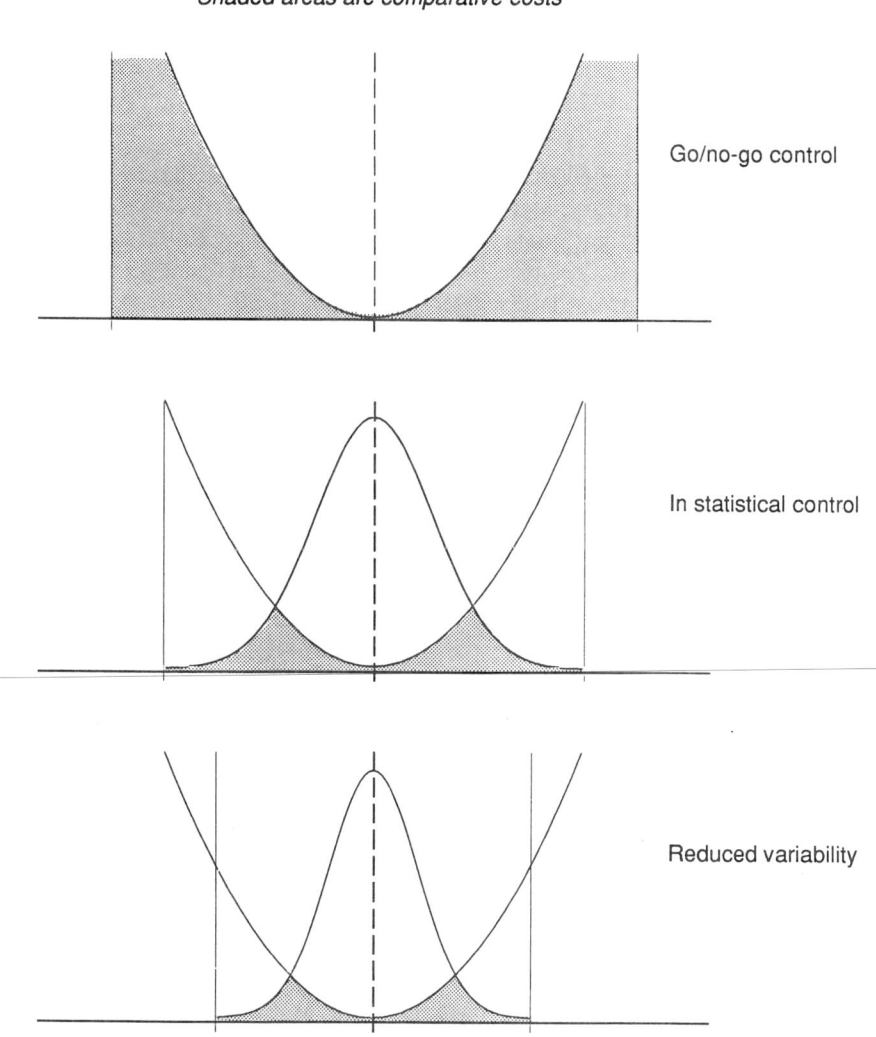

Shaded areas are comparative costs

Go/no-go control

In statistical control

Reduced variability

Figure 9.4 Total quality costs (example for a particular component)

179

prescribed and followed, but later on there are likely to be unseen obstacles. Perhaps the ideal is to get things right first time, but in any case, there is no substitute for process knowledge and experience.

Variability reduction makes the single most important process control contribution towards profitability. This is illustrated in Figure 9.4, where the shaded areas indicate the relative total quality costs of a simple gear wheel under different conditions of control. As explained below, this example merely illustrates the point; it follows no general mathematical rules.

> The top picture represents the situation when items anywhere within tolerance were made and sold. The middle picture represents the situation when controls were applied such that most items made and sold were at or near to nominal. The bottom picture indicates the effect of variability reduction.

The significance of the bell-shaped curve is explained in Chapter 5. The U-shaped curve is arbitrary, but it reflects the idea that divergence from optimum increases the total quality or whole-life costs. An empirical theory which explains the cost/size relationship more precisely (and results in different shapes for the shaded areas) is discussed in Section 12.10.

It will be appreciated that the consequence of improvement action will be changes to one or both of the product and the process, and also to criteria for control. Hence the slogan 'start again' at the foot of Figure 9.3: it emphasizes that improvement is a never-ending activity.

■ 9.8 Process capability

The capability of a process is a measure of its ability to satisfy customer requirements. Process capability studies are carried out to compare customer requirements with process performance and to identify the stakes-in-the-ground for process monitoring.

> Process capability assessments are not limited to obtaining information about product conformance. At the same time, information about significant process features can be obtained. For example, comparatively few of a particular drilled component might be produced, but the activity of drilling by a single machine might extend to many millions of holes. In this case, information about the machine (such as tool wear, lubrication intervals, setting precision and so on) is of more use than information about the single component.

Capability criteria

There is no evidence to suggest that any process can have the capability of producing zero defects; there is always some risk. The interaction of events (or process disturbances) cannot always be predicted and process design cannot always prevent the effects. An extreme example of this, taken out of its original context in the 'theory of chaos', is of a butterfly flapping its wings in China which can lead to a tornado in North America.

In practice, the criteria for a capable process vary and usually depend upon economics and an acceptable risk of not satisfying a customer.

For example, in mass production manufacturing industry, a benchmark of maximum risk is that the average item has a value of nominal and that no more than 6 in ten million items might be outside the limits of customer satisfaction. In practical terms for a statistically in-control process,

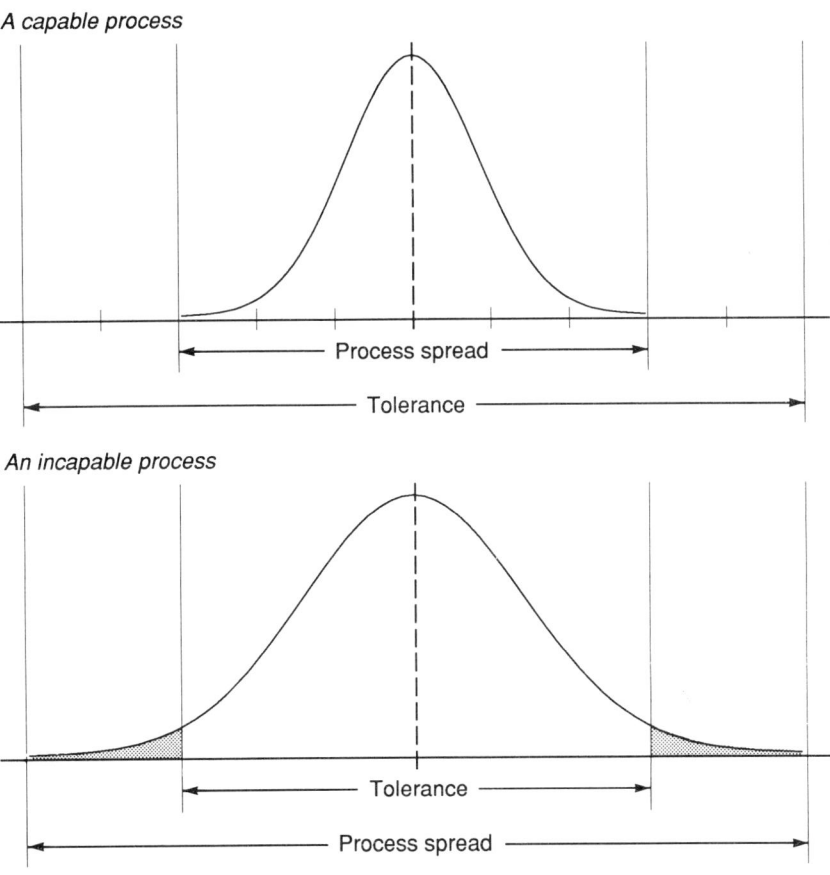

Figure 9.5 Capable and incapable processes

this means that there might be 2 unsatisfactory units among an item supplied at the rate of about 4000 per week over a 20-year period (see Figure 5.6).

The limits of customer satisfaction are defined by the tolerance band, which is mainly, but not always, centred on the nominal. Whatever its position, the idea of a *capable process* and an *incapable process* is expressed by Figure 9.5.

■ 9.9 Capability indexes (C_p, C_m, SPR, RPI)

Although capability criteria are sometimes expressed directly in terms of risk, it is much more usual to assume a normal distribution (Section 5.4) and to use an index which is defined as

$$C_p = \frac{\text{tolerance band}}{\text{process spread}}$$

where process spread is an estimate of six standard deviations. Process spread is sometimes called the inherent or natural process range.

For example, when specified tolerance limits are equally spaced about nominal and presuming that the process mean is equal to nominal, the acceptable risk of 6 in ten million being outside tolerance could be expressed as a requirement that an estimation of $\mu \pm 5\sigma$ (or ten standard deviations) must be less than or equal to the tolerance band. In other words

$$C_p = \frac{10\sigma}{6\sigma} \geqslant 1.66.$$

The value $C_p = 1.66$ is the most widely accepted expression of the minimum value for a process to be considered capable. However, as pointed out earlier, the criteria for some particular processes might be different, depending upon economics and acceptable risk.

By convention, the index C_p relates to *process* capability. Sometimes, attempts are made to estimate the capability of elements of the process: for example, a machine in isolation from its operator, its environment, its cutting tools and the material being worked. In such a case, the index might appear as C_m, indicating that it relates only to *machine* capability. The designation C_m is used also to distinguish a capability assessment that has been made over a short period of time, which does not allow for long-cycle common disturbances such as ambient temperature.

The terms, symbols and ways of expressing capability used in this book are widely accepted. However, the reader may need to adapt to local situations where other conventions are sometimes used to suit particular practical and theoretical work. They include the following:

- *Standardized precision ratio (SPR)*, which gives a value that is six times the C_p value (SPR = tolerance/one standard deviation) and gives rise to a concept of process precision which is expressed as $SPR \leqslant 4 =$ low precision, $4 > SPR < 8 =$ medium precision, $SPR \geqslant 8 =$ high precision.

   Note that $C_p = 1.66$ is equivalent to $SPR = 10$.

- *Relative precision index (RPI)*, which is the ratio of tolerance to the average sample range (R) that is determined in some charting. It can be converted to a C_p value through multiplication by a constant (that is, $C_p = \text{RPI} \times 6/d_2$), where d_2 is dependent upon sample size and its values are given in Appendix F.

Use of the capability index C_p (and the setting index C_{pk} below) requires tolerance to be specified. Without a tolerance, the supplier can determine process spread or $\mu \pm 5\sigma$ (when μ = nominal) and have a basis to work with the customer towards product acceptance.

It is important to recognize that capability indexes should be used as targets rather than specifications. They express risk or the probability of future failure, rather than absolute achievement or not of customer expectations. Failure to meet a target C_p value is a signal to management. Perhaps it signals that the customers' requirements have been misstated, perhaps that the process is incomplete.

■ 9.10 Setting indexes (C_{pk}, Z_U, Z_L)

The capability index C_p is calculated without reference to nominal; its interpretation assumes that the process mean is equal to nominal. The setting index C_{pk} is designed to confirm this assumption, but *only when nominal is equally distant from the tolerance limits*. Ideally,

$$C_{pk} = C_p.$$

The requirement for no more than 6 in ten million outside tolerance should appear as

$$C_{pk} = C_p \geqslant 1.66.$$

It must be emphasized that C_{pk} is only an index; it is a mathematical way of saying that process mean is or is not the same as nominal. Of far greater importance is the estimated difference in measurable units between process mean and customer's optimum, because this value is of more use to the process setter. This fact is reflected in the design of some industrial probability papers for capability assessment (see Figures 9.8 and 9.9).

Determination of C_{pk} involves calculation of two indexes, Z_U and Z_L, which relate to the standard deviates between the process mean and the upper and lower tolerance limits.

$$Z_U = \frac{\text{upper tolerance limit} - \text{process mean}}{\text{estimate of three standard deviations}}$$

$$Z_L = \frac{\text{process mean} - \text{lower tolerance limit}}{\text{estimate of three standard deviations}}$$

C_{pk} *is the lower value of* Z_U *and* Z_L.

When $Z_U = Z_L$, the process is set on nominal and $C_{pk} = C_p$. When they are not the same, they indicate a process deviation from nominal as illustrated in Figure 9.6.

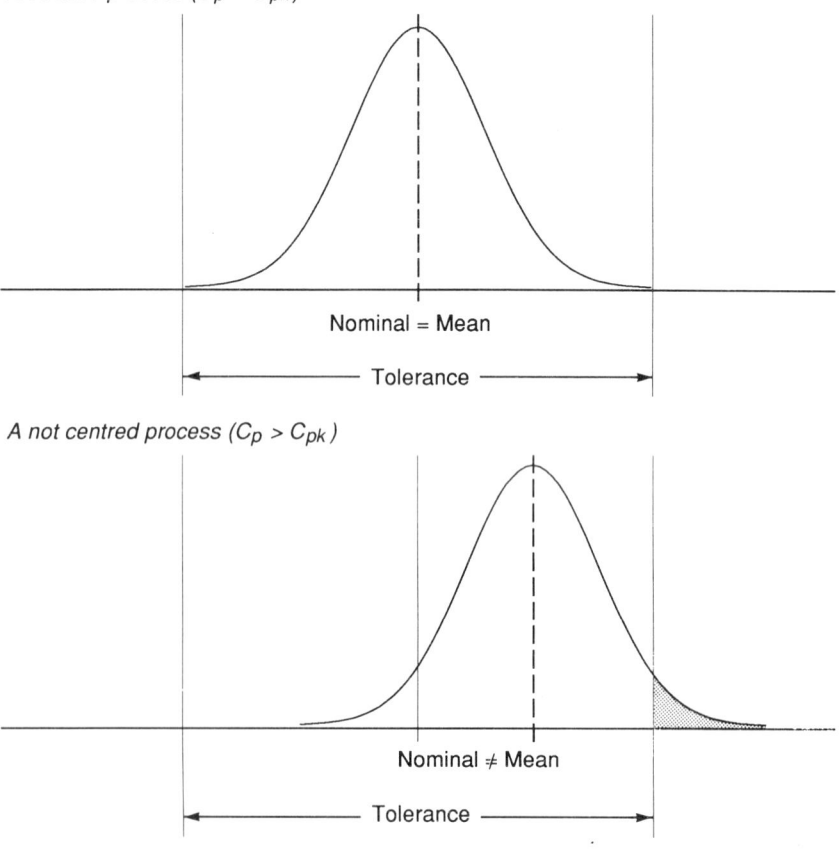

Figure 9.6 Centred and not centred processes

Table 9.1 Z value interpretation

Z_U or Z_L	0	0.1	0.2	0.3	0.4	0.5	0.6	0.7	0.8	0.9	1.0
% outside limits	50	38	27	18	12	7	4	2	1	0.4	0

When a Z value is less than 1, it is likely that there will be a proportion of components above (when $Z_U < 1$) or below (when $Z_L < 1$) the tolerance band.

Table 9.1 indicates the likely proportions for a few values of Z. The values are much simplified and rounded derivations from the standard normal probability table (in particular for Z_U or $Z_L = 1.0$, when the true value is 0.00135% outside limits – there can never be an absolute zero probability of failure).

■ 9.11 Capability assessments

The accuracy of capability assessments hinges on estimation of the underlying process mean. The estimate is affected in particular by sample size, by the randomness of sample selection, by the process' state of control and by the data distribution.

Sample size

The influence of sample size is discussed in Chapter 7. In practice, the quantity taken relates to the index calculation method that is used (commonly used methods are summarized at the end of this section).

Initial estimates made from measurements of about 125 random units are generally accepted as sufficiently accurate. Accuracy is less with smaller sample sizes. Estimates can be misleading with a sample size of below about 30.

Random sample selection

Although they might reflect the best capability of an element of the process (C_m), it is unlikely that 125 units will provide accurate information about overall process capability (C_p), especially if they have been produced consecutively. In other words, they are unlikely to be a random sample from the process. In practice, attempts are made to obtain more honest values for the index: for example, by taking 25 random samples of 5 consecutive units from

185

no fewer than about 300 units. The calculated index is designated C_p when the data is thought to represent the effects of all common disturbances.

State of control

Ideally, the process should be in a state of statistical control before capability is assessed. This state is rarely achieved by the time that commercial decisions must be made.

In practice, data is checked for any statistical outliers which indicate that the process is suffering from special disturbances. Again ideally, the causes of outliers are eliminated and new data is obtained. Otherwise, the outliers are discarded for purposes of capability calculations. Note that discarding outliers does not mean ignoring them; their causes should still be determined and appropriate action taken.

Data distribution

Capability requirements are usually expressed assuming that data has a normal distribution. Therefore, before use, the data should be checked to confirm that its distribution is approximately normal. Some methods used to deal with non-normal distributions are indicated in Section 9.12.

The most simple checks are given in Section 5.2. An alternative is probability plotting (Section 8.3), which gives simple indications of unusual and non-normal distributions. The pictures in Figure 9.7 represent best-fit lines through plots of data on probability paper such as that illustrated in Figure 9.8.

> Figure 9.7(a) is a straight diagonal line. It represents the best-fit line through a plot of data from a normal distribution.
>
> Figure 9.7(b) exhibits a kinked line or two off-set lines. It is typical of data being taken from two distributions: for example, from measurements of products that have been mixed after production by machines having different settings. This particular example has been drawn from the data used to construct Figure 2.5 and is evidence to support hypothesis 2 (two leak sources) because it suggests that there are two distributions.
>
> Figure 9.7(c) has a line that starts straight and diagonal but bends to the vertical. This line suggests that some data is missing at higher values and is indicative of a truncated distribution. For example, the data might be of the weights of eggs in a small-grade batch; it does not reflect the capability of the hens that produced them (large and medium sizes are not in this batch).
>
> Figure 9.7(d) shows a line that starts as a vertical, bends to a diagonal then bends back to vertical, and indicates a doubly truncated distribution.

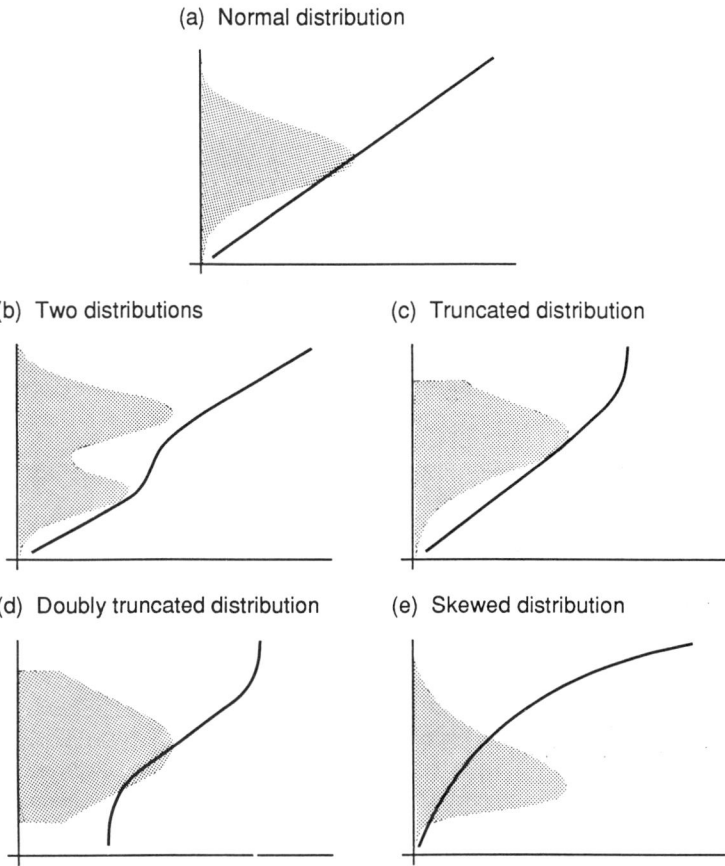

(a) Normal distribution

(b) Two distributions

(c) Truncated distribution

(d) Doubly truncated distribution

(e) Skewed distribution

Figure 9.7 Best-fit lines on normal probability paper

It shows the effect of inspection being used to control the products from an incapable process; items that are too small or too large are not present in the batch, which was measured on a shipping deck.

Figure 9.7(e) is a curved line that is typical of a skewed distribution. The example has been drawn from measurements of the indentations in an as-cast surface; it reflects sand grain agglomeration in the casting's mould.

Index calculation methods

There are three commonly used methods for calculating capability statistics. Respectively, they use probability plotting, charted data and first principles.

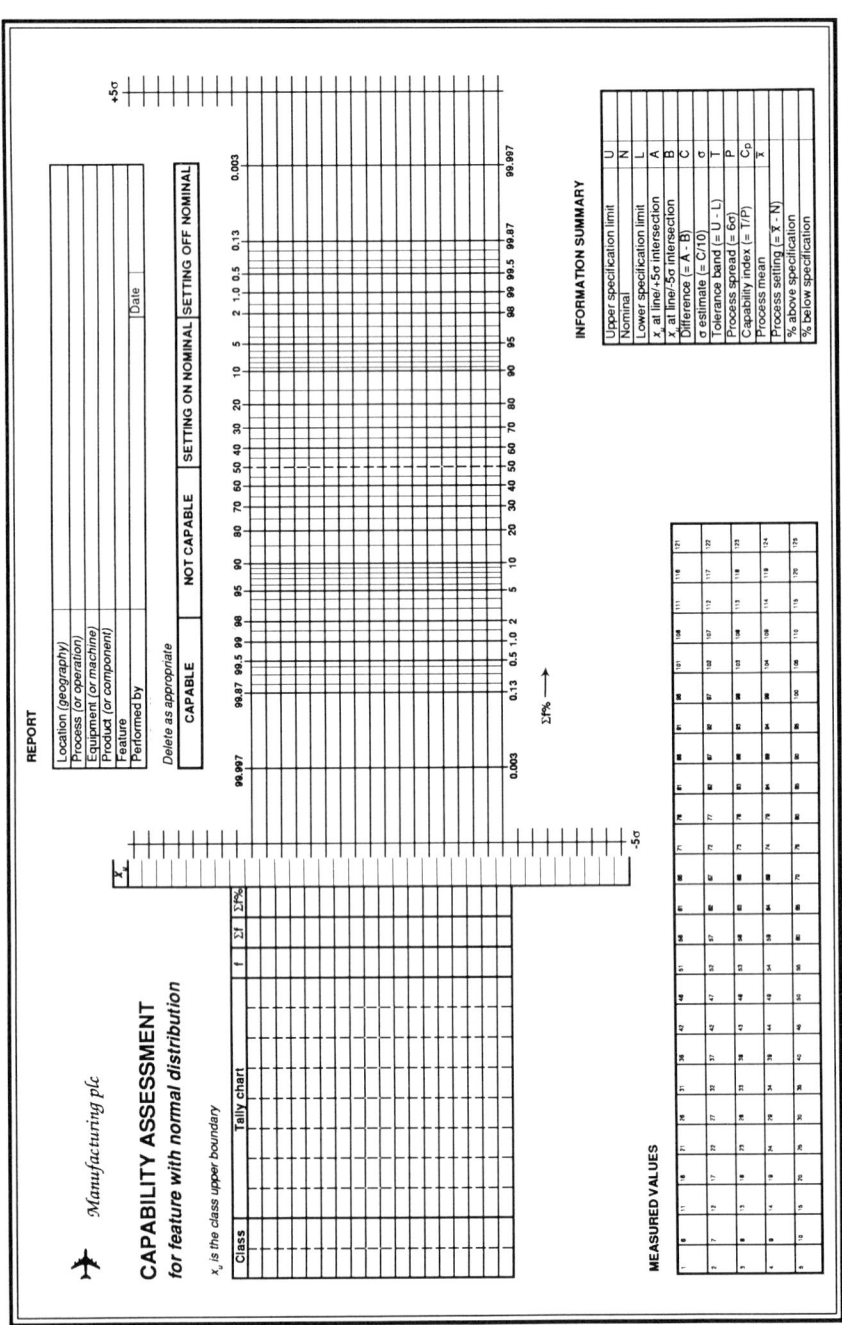

Figure 9.8 Industrial normal probability paper (reduced from A3 size)

Probability plotting (explained fully in Section 8.3) is used most frequently for 'snap-shot' assessments: in other words, when the data is obtained over a short time, effectively from consecutive events.

The disadvantages of probability plots are that they do not readily identify special disturbances (outliers) and they give no idea of common disturbances occurring over time. Also (as explained in Chapter 8), statistics produced by this method must be treated with some caution. However, the method gives usually adequate first-sight estimations even when the data base is small: that is, essentially above 10, preferably above 30 and ideally about 125 measurements. Of course, it is always advisable to confirm the results when more data becomes available.

A typical industrial paper for normal probability plotting is illustrated in Figure 9.8, and Figure 9.9 is a step-by-step example of its use.

Industrial probability papers are designed to be used by people who often do not have (and sometimes would not wish to have!) a knowledge of the mathematical principles involved. The result is a layout which has been found to work well in practice, provided there is somebody in the background to whom reference can be made when there are unusual results.

Step 1:   Collect and record the data, usually on the paper – but for purposes of the example in Figure 9.9, the data set out in Table 3.6 have been used.

Step 2:   Complete a cumulative frequency table of the data as shown in Figure 9.9(a). The tally chart is similar to that of the same data in Table 3.6, but it will be noticed that the class order is inverted; this follows a logic that higher values should be at the top of the table. Class limits are put into the 'class' column and upper boundaries are shown without the words 'just less than'.

Step 3:   From the values in the frequency table, plot x_u against $\Sigma f\%$ on the probability chart and draw a best-fit line through the plots as shown in Figure 9.9(b). Note that a plot cannot be made at $\Sigma f\% = 100$ and, as explained in Section 8.3, a plot is made at the average of the two highest classes' x_u and $\Sigma f\%$ values.

Step 4:   Complete the information summary as shown in Figure 9.9(c). Specification details are as given (U, N and L). Values of x_u at the intersections of the best-fit line with the vertical lines designated $\pm 5\sigma$ and x ($\Sigma f\% = 50$) are read off the chart (A, B and x), as are values of $\Sigma f\%$ at intersections of the best-fit line at $x_u =$ upper and lower specification limits (% above and below specification). The remainder of the summary involves simple calculations. The results in this example agree with those obtained by other methods (Section 8.4).

(a) Frequency table

(c) Information summary

x_u is the class upper boundary

Class		Tally chart	f	Σf	Σf%	x_u
90	99	/	1	125	100	100
80	89	#### ///	8	124	99.2	90
70	79	#### #### /	11	116	92.8	80
60	69	#### #### #### #### ///	23	105	84.0	70
50	59	#### #### #### #### #### #### #### ////	39	82	65.6	60
40	49	#### #### #### #### //	21	43	34.4	50
30	39	#### #### //	12	22	17.6	40
20	29	#### //	7	10	8.0	30
10	19	///	3	3	2.4	20

Upper specification limit	U	100
Nominal	N	50
Lower specification limit	L	0
x_u at line/+5σ intersection	A	134
x_u at line/-5σ intersection	B	-26
Difference (= A - B)	C	160
σ estimate (= C/10)	σ	16
Tolerance band (= U - L)	T	100
Process spread (= 6σ)	P	96
Capability index (= T/P)	Cp	1.04
Process mean	$\bar{x}$	54
Process setting (= $\bar{x}$ - N)		4
% above specification		~0.2
% below specification		~0.05

(b) Probability plot

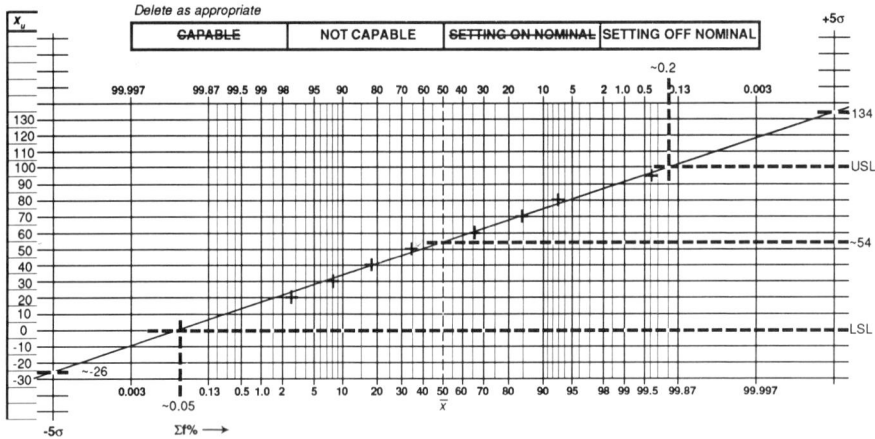

Figure 9.9 Capability assessment using normal probability paper

> *Step 5:* Report the result. The calculated C_p value is 1.04 and, using the capability criteria explained in Sections 9.8 and 9.9, the process is sentenced as 'not capable'. Also, the process setting is 4 (above nominal) and therefore the process is sentenced as 'off nominal'.

The other two methods for calculating C_p are as follows:

■ Using a reasonable number of small samples to estimate standard deviation (an example is given in Section 10.8). This is the preferred method when there is time and opportunity to take samples at intervals from a continuous series of events. Ideally, samples should be taken to reflect the effects on the process of all likely common disturbances. Outliers are fairly easy to identify, but unusual and non-normal distributions are not readily identified.

■ Using first principles to estimate standard deviation (examples are given in Section 3.20). This method requires more calculation effort. Calculators can be used to minimize the effort but should be supplemented with pictorial representation, otherwise some important factor might be missed.

■ 9.12 Non-normal capability assessments

If preliminary work indicates that a distribution is non-normal or grossly out of control, there are four approaches which might be adopted.

First and most important, investigate the data more thoroughly. Many apparently out-of-control and non-normal distributions only reflect measurement practice (Chapter 2) and are often a result of the following:

■ Not considering the polarity of measurements (for example, the so-called one-sided distributions described below).
■ Not reporting results above or below particular values.
■ Reporting results beyond the precision of the measurement method.
■ Having differing standards of measurement (perhaps a problem of people).
■ Reporting combined results off differently set machines.

These phenomena should be investigated and resolved, in the context of the process, before the data is interpreted. Investigation is sometimes helped by the use of probability papers. The effect of investigations is often to improve process consistency and to determine that the underlying distribution is in fact normal.

Second, where reasonable, treat all or part of the distribution as normal. Normal criteria are given in Section 5.1 and approximations are discussed in Section 6.5; also, the central limit theorem (Section 7.3) can be used.

In addition, there are the special cases, often referred to as *one-sided distributions*, that occur when nominal is zero, and which are common features of many machined components.

Perfectly round holes and external surfaces are a rarity; most machines introduce some degree of ovality. Tolerance and measurements are often expressed as positive numbers which are the absolute difference between major and minor diameters. Similarly for adjacent parallel or perpendicular surfaces, tolerances and measurements of taper or run-out are expressed as positive numbers. These measurements give a one-sided distribution, ideally with the mode at zero as in Figure 9.10(a).

If tolerances and measurements are expressed with reference to a particular plane or direction, there can be positive and negative numbers and a

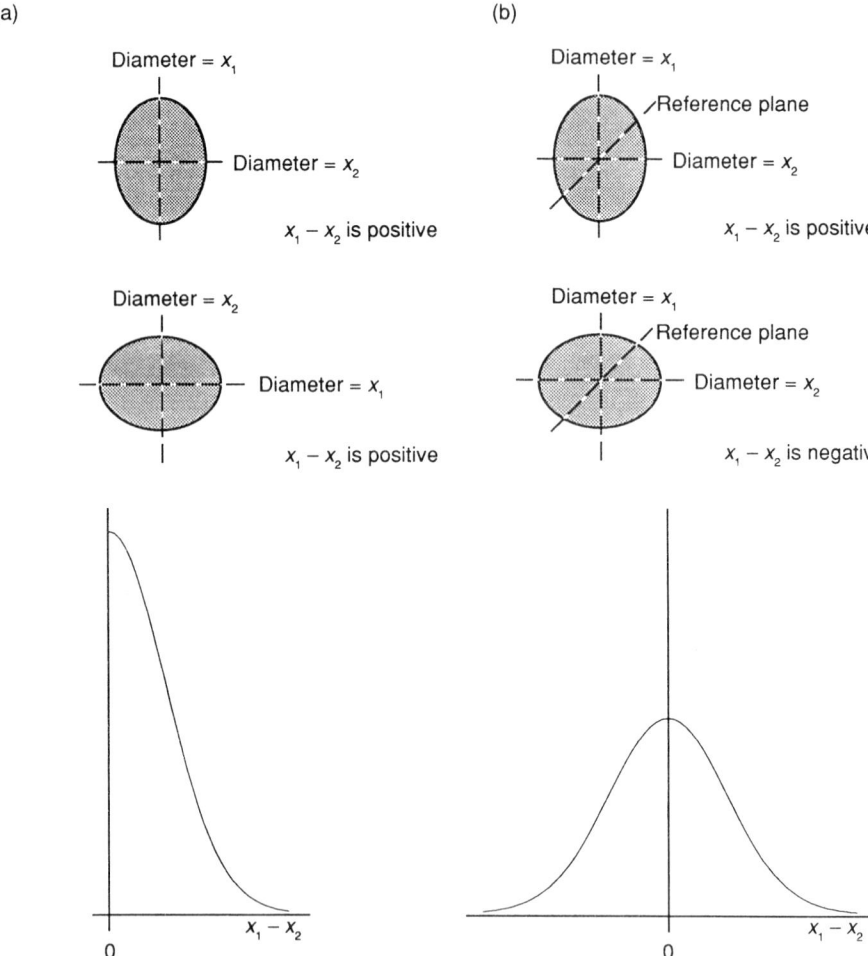

Figure 9.10 One-sided distribution

two-sided distribution, ideally with mode, median and mean at zero as in Figure 9.10(b).

The distribution shown in Figure 9.10(b) is said to be truncated at zero. It is a particular example where part of a distribution might be usefully treated as normal. If one tail of the distribution approximates to normal, sufficiently accurate calculations can usually be made using deviations from the mode instead of the mean and only the values in the approximately normal tail of the distribution.

- ■ When the mode of the distribution is at or very close to zero, the statistic representing process spread is effectively half that of a normal distribution: in other words, three standard deviations.

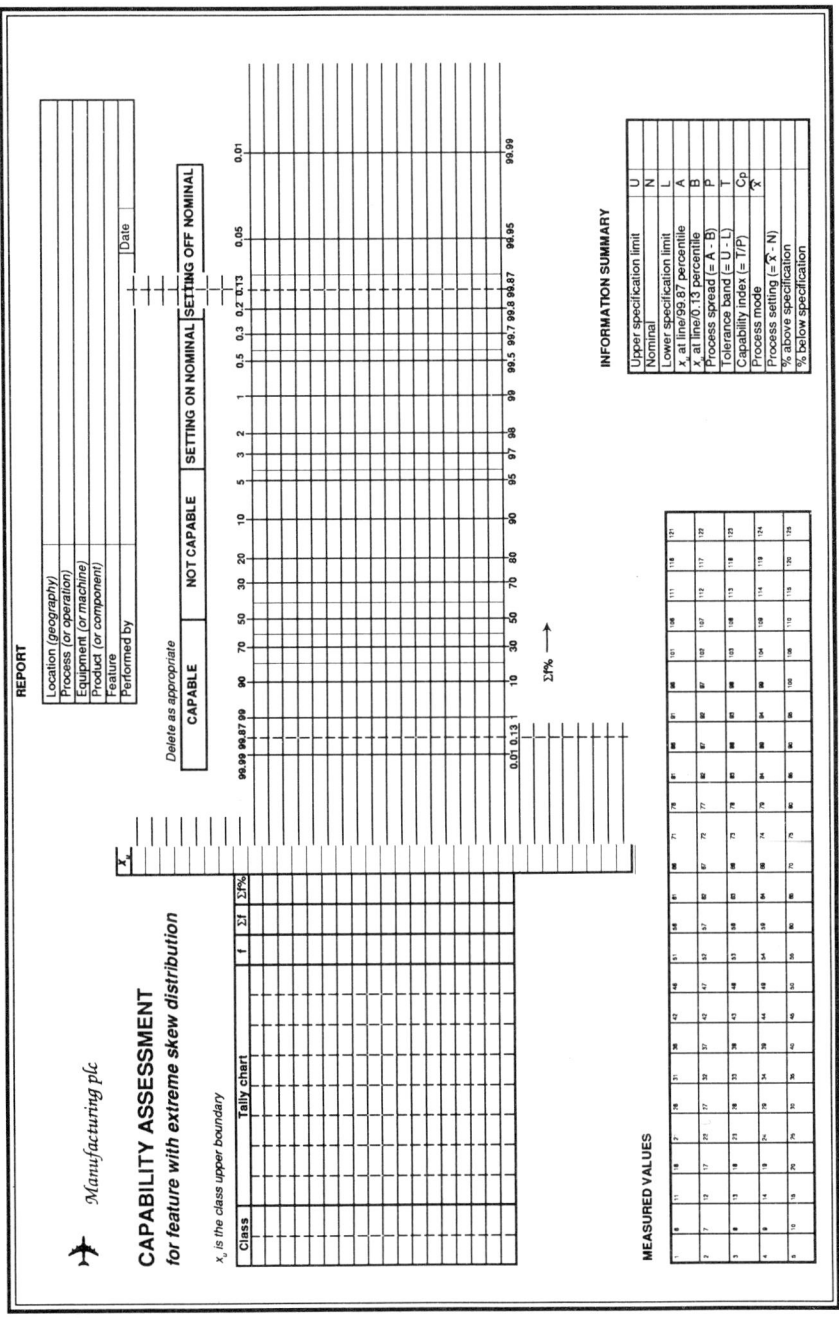

Figure 9.11 Industrial probability paper for extreme skew distribution (reduced from A3 size)

- When the distribution is truncated at zero but the mode is not at zero, the process spread is effectively three standard deviations plus the width of the non-normal tail (zero to the mode).

Third, if appropriate, confirm another distribution such as Weibull. The mathematical principles set out in Chapter 6 could be applied, but probability plotting usually provides the most simple method of confirming a distribution and of determining process spread.

In this situation, the methods set out in Chapter 8 are recommended in preference to most pre-printed industrial papers. Many of the latter are customized or prepared for specialized applications, and they are often complex and difficult to use. However, the paper illustrated in Figure 9.11 is used occasionally to obtain an estimation of process spread for extreme skew distributions.

When process spread is determined for a non-normal distribution, it is taken usually and for convenience as the value of the interval between the 0.13 and 99.87 percentile lines. This is the same interval as that of six standard deviations in normal distributions. These values are shown as broken lines in Figure 9.11.

Finally, if there is a very large amount of data (that is, thousands of results), simply studying a histogram will usually give sufficient information about process spread and its relationship to the tolerance band.

■ 9.13 Attribute capability statistics

All of the previous sections in this chapter apply to continuous data (Section 2.6). When the only data available is discrete, capability may be expressed as the proportion of satisfactory events among the total events.

Note that in the context of process control, attributes amount to defectives and defects for which the target is zero without any tolerance.

For the other form of discrete data (numbers of occurrences such as accidents, live births or aircraft movements), a capability statement may be made if the data can be reduced to an attribute form. A statement is not appropriate when the quantity of non-occurrences is not known.

■ References to further reading on the topics of Chapter 9

Oakland (1986) gives a brief overview of process capability. Montgomery (1985) and Mitra (1993) give more detailed treatments, and Bissell (1990) considers the accuracy of various capability indices.

See Bibliography for details of titles and publishers.

10 | Control charts

■ 10.1 Control chart types

There are two basic types of control chart, one to deal with variables (Section 10.4), the other for attributes (Section 10.9). There are many versions of each type. Most of those described in this book are known as Shewhart charts after W.H. Shewhart, who developed their use during the mid-1920s.

The so-called British charts are described briefly in Section 10.19, the differences between them and Shewhart charts are discussed in Sections 10.19 and 10.20.

■ 10.2 Control chart design

The working part of any control chart, on paper or on an electronic display screen, is essentially a graph that shows variation from time to time in measured or counted data. Most chart designs also provide for supplementary information. The chart in Figure 10.1 provides for the following:

- Basic instructions for the operator, including subject, sample size and frequency.
- Reminders for measurement equipment calibration.
- The raw data and identifying references.
- Calculation instructions and a table of appropriate constants.
- Remarks on relevant actions and subjective observations at the time of measurement.

Charts are constructed using data that is collected while the process is operating. Sometimes the data is converted to another more convenient form, such as an average or proportional value. It is then plotted, conventionally against an appropriate scale shown on the vertical axis in a sequence that is

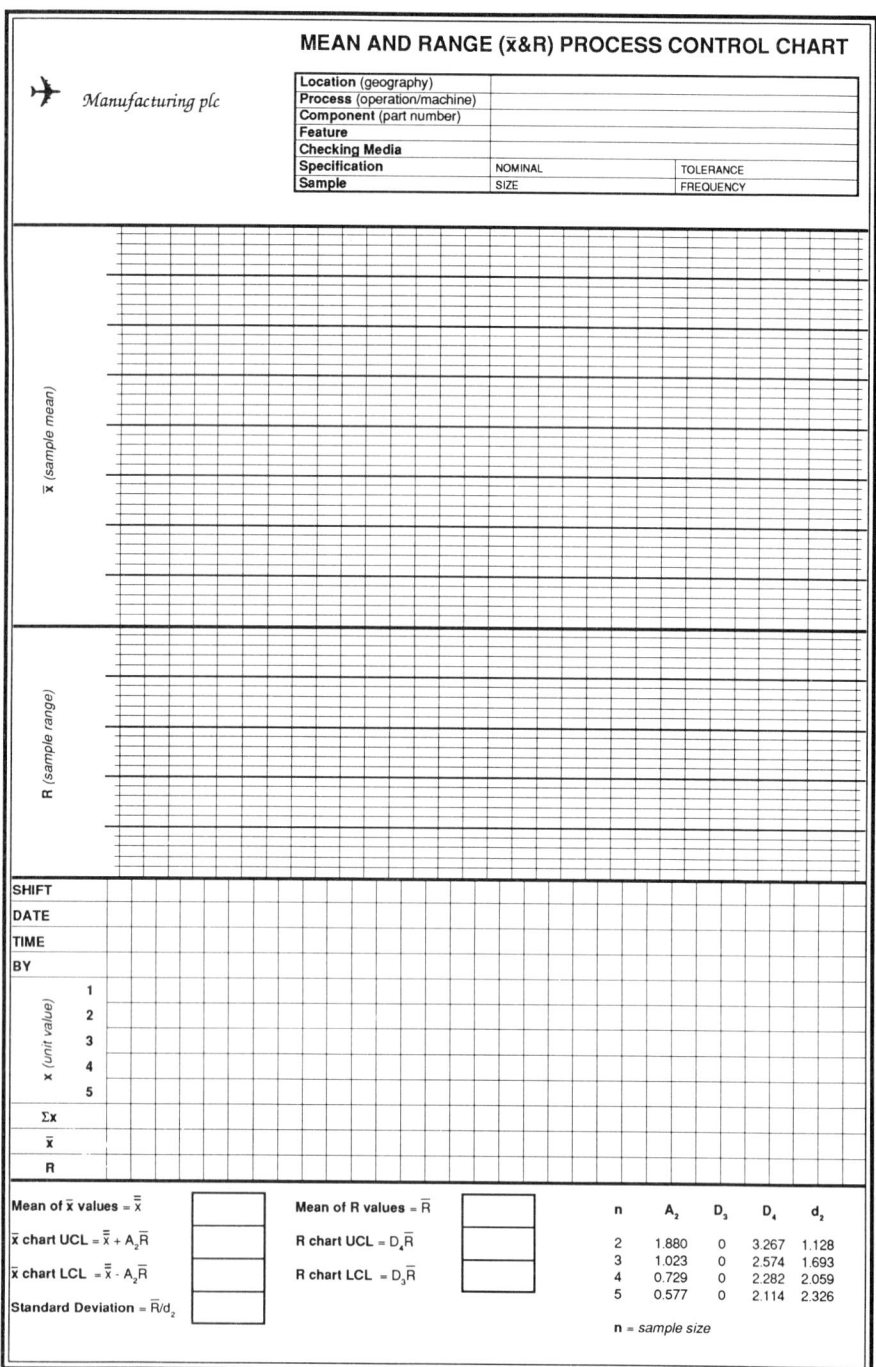

Figure 10.1 An industrial control chart (reduced from A3 size)

shown on the horizontal axis. The sequence usually, but not always, relates to the time of data collection.

- The chart should display sufficient plots to give a coherent picture of the process. This often means that the horizontal scale must accommodate a minimum of 25 plots.
- The vertical scale should be selected so that extreme values can be plotted. Also, to help interpretation (Section 10.18) there should be about 12 but not fewer than 8 steps at convenient equal intervals between the control lines introduced in the next section.

■ 10.3 Control lines

Control lines are sometimes called control limits. The word 'lines' is preferred because it distinguishes the parameters for process control from the tolerance limits for quality control. The lines are often well inside the tolerance limits.

The purpose of control and process mean lines on the chart is to help the identification of plots which represent occurrences that are sufficiently unusual to warrant special attention.

When there is sufficient data available, a process mean line and control lines are calculated and drawn on the chart. Conventionally, means are drawn as broken lines and control lines are solid (see Figure 10.2).

- Practical experience suggests that there should be data from about 25 samples of the feature under review before the lines are calculated. Sample sizes are discussed as the principle chart types are explained. The recommended quantities offer a generally acceptable level of accuracy to chart interpretation; the theoretical basis and implications of different quantities are discussed in Chapter 7.

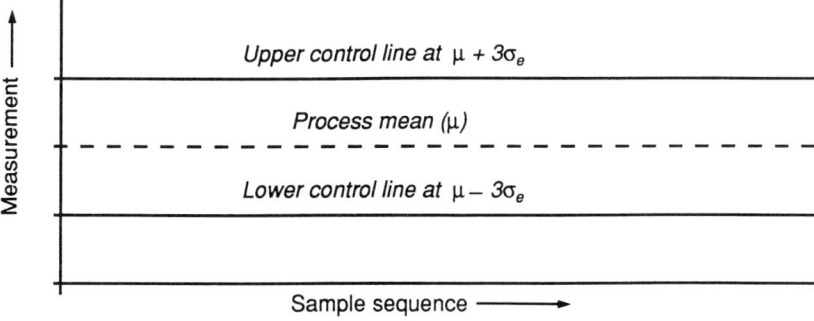

Figure 10.2 Chart layout

197

■ Control line calculations are made using the data in simple formulae which sometimes involve constants that vary according to chart type and sample size. The formulae are given for the principle chart types in following sections.

The calculated lines are designed to enclose most expected sample values. For Shewhart charts, they are set at $\mu \pm 3\sigma_e$, where μ is the process mean and σ_e is the standard deviation of the distribution of plotted values. When the plotted value is the mean ($\bar{x}$) of a sample size n, $\sigma_e = \sigma/\sqrt{n}$ and is the standard error of the mean. Of course, σ is the standard deviation of the distribution of x values.

Other control line settings are discussed in Sections 10.19, 10.20 and 10.21.

■ 10.4 Charts for variables

Charts for variables are used in pairs: one to monitor process setting (or the middle of the distribution), the other to monitor variability (or the spread of the distribution).

The underlying principles of formulae used for calculation of control lines are explained for the mean and range chart in Section 10.5. The formulae for other charts are similar although the theory is more complicated in most cases.

Sample size

It is perfectly feasible to use a sample size of one and to measure variation as the sample-to-sample difference (see Section 10.17). However, there are frequent occasions when the distribution of individuals' measurements is not normal and this leads to complications in chart interpretation. The central limit theorem states that, with increasing sample size, the distribution of sample means tends towards normal; hence a reason for having sample sizes greater than one. In practice, a minimum sample size of 5 has been found to give reasonable results and is the most commonly used, although a larger sample is recommended for $\bar{x}$ & s charts (Section 10.7).

Sample selection

As discussed in Chapter 9, the initial calculation of control lines might be made from commissioning data when the samples are obtained over a comparatively short production run or time span. More usually, samples are taken periodically at regular intervals when processing is dominant, and randomly when people are dominant.

The sampling interval should reflect known process stability. In other words, it should take account of all likely process disturbances, such as start-up/shut-down, material batches and shift changes. It could be every 500 produced or every hour or four times a day, and so on. Whatever the interval, wherever practicable, the individuals in a sample should have been produced consecutively. Calculations from such samples give better estimates of inherent variability and trends than samples where the individuals are chosen randomly.

■ 10.5 Mean and range chart (or x̄ & *R* chart)

This type of chart is the most commonly used chart for variables because it avoids complicated arithmetic calculations.

Data collection

As indicated already, about 25 samples of 5 units are taken and measured. The quantities are not absolute, they often depend on what is available, but they have been found to work well in practice.

During commissioning, 125 consecutive units might be taken, measured and then treated as if they were 25 samples of 5. In this case, it is probable that the process includes the best operator, a single material batch and constant environmental conditions; the results are unlikely to represent the true behaviour of the process. Especially for high-volume industrial processes it is usual to make some compensation: for example, by taking the initial 125 units randomly from a much longer production run.

The measurements of each individual in the 25 samples are recorded; the mean ($\bar{x}$) and range (R) of each sample are calculated, recorded and plotted on the chart. Lines are drawn to join successive plots. Then the process mean and control lines are calculated and drawn on the chart.

Mean and range chart formulae

The general formulae for control lines is given in Figure 10.2. For example:

upper control line = $UCL = \mu + 3\sigma_e$.

for the means chart, this general formula becomes

upper control line for means = $UCL_{\bar{x}} = \mu_{\bar{x}} + 3\sigma_{\bar{x}}$

and from Chapter 7, $\mu_{\bar{x}} = \mu$ and $\sigma_{\bar{x}} = \sigma_e = \sigma/\sqrt{n}$. Therefore

$UCL_{\bar{x}} = \mu + 3\sigma/\sqrt{n}$

As indicated above, μ or σ will not be known at the outset. They have to be estimated from the sample data. The following methods are a summary of the theory given fully in Section 7.7.

The *process mean*, μ is estimated from the mean of the 25 sample means, which is denoted by $\bar{\bar{x}}$ and is sometimes called the *grand mean* or *grand average*.

The parameter σ is estimated using the formula $\sigma = \bar{R}/d_2$ where $\bar{R}$ is the mean of the 25 sample ranges and d_2 is a constant. In fact, there is no direct mathematical relationship between σ and $\bar{R}$, but the constant gives the best possible conversion if the samples come from a normal distribution.

Hence for a sample size of 5, $UCL_{\bar{x}} = \mu + 3\sigma/\sqrt{n} = \bar{\bar{x}} + 3\bar{R}/(d_2\sqrt{5})$ where $3/(d_2\sqrt{5})$ is another constant which is written as A_2. Therefore

$$\text{upper control line for means} = UCL_{\bar{x}} = \bar{\bar{x}} + A_2\bar{R}$$

Similarly

$$\text{lower control line for means} = LCL_{\bar{x}} = \bar{\bar{x}} - A_2\bar{R}$$
$$\text{upper control line for ranges} = UCL_R = D_4\bar{R}$$
$$\text{lower control line for ranges} = LCL_R = D_3\bar{R}$$

The constants A_2, D_4 and D_3 depend upon sample size and are given for a range of sample sizes in Appendix F. Constants for commonly used sample sizes are often given on industrial charts such as Figure 10.1.

The process mean, process range and control lines should be recalculated from about the first 25 samples following any significant process change.

Example of mean and range chart calculations

In order to set up a control chart 25 samples were taken, each sample consisting of measurements from 5 consecutive units. The 125 individual measurements were those already shown in Table 7.2. They are repeated in Table 10.1, which shows how the grand mean and average range are calculated.

From Table 10.1, the grand mean, $\bar{\bar{x}} = 1963.4/25 = 78.5$, and the average range, $\bar{R} = \Sigma\, R/25 = 17.60$.

From Appendix F for $n = 5$, $A_2 = 0.577$, $D_3 = 0$ and $D_4 = 2.114$; hence

$$UCL_{\bar{x}} = \bar{\bar{x}} + A_2\bar{R} = 78.5 + 0.577 \times 17.6 = 88.7$$
$$LCL_{\bar{x}} = \bar{\bar{x}} - A_2\bar{R} = 78.5 - 0.577 \times 17.6 = 68.3$$

$$UCL_R = D_4\bar{R} \quad = 2.114 \times 17.6 \quad = 37.2$$
$$LCL_R = D_3\bar{R} \quad = 0 \times 17.6 \quad\quad = 0.$$

Table 10.1 Mean and range control chart

Sample	Measurements					Mean	Range	Median
j	x_1	x_2	x_3	x_4	x_5	$\bar{x}_j$	R_j	$\tilde{x}_j$
1	81.5	66.6	84.2	70.1	72.2	74.9	17.6	72.2
2	72.3	83.6	76.3	84.5	76.0	78.5	12.2	76.3
3	78.3	99.8	88.9	83.0	87.9	87.6	21.5	87.9
4	75.9	66.7	94.2	93.2	85.7	83.1	27.5	85.7
5	69.0	71.8	81.6	86.8	94.0	80.6	25.0	81.6
6	66.4	81.4	63.3	87.3	70.6	73.8	24.0	70.6
7	75.6	85.9	84.4	89.8	89.7	85.1	14.2	85.9
8	73.8	83.6	77.5	75.4	78.1	77.7	9.8	77.5
9	82.3	77.3	84.9	80.2	73.7	79.7	11.2	80.2
10	76.9	74.6	88.2	83.7	67.4	78.2	20.8	76.9
11	83.1	81.8	72.4	85.8	75.6	79.7	13.4	81.8
12	71.0	79.5	81.9	69.6	83.7	77.1	14.1	79.5
13	71.3	86.4	70.5	77.3	76.2	76.3	15.9	76.2
14	80.0	72.4	53.9	75.6	85.8	73.5	31.9	75.6
15	65.1	68.5	75.0	80.0	80.3	73.8	15.2	75.0
16	78.6	72.1	85.2	80.8	68.4	77.0	16.8	78.6
17	80.8	86.8	79.5	78.9	80.0	81.2	7.9	80.0
18	76.8	85.6	65.0	70.2	81.5	75.8	20.6	76.8
19	88.8	64.6	74.5	72.6	70.5	74.2	24.2	72.6
20	74.9	91.4	91.2	72.3	85.6	83.1	19.1	85.6
21	77.5	78.2	81.9	70.1	76.8	76.9	11.8	77.5
22	74.8	88.7	76.7	80.7	78.6	79.9	13.9	78.6
23	69.1	76.0	78.3	72.3	81.5	75.4	12.4	76.0
24	72.3	76.1	87.8	85.7	94.0	83.2	21.7	85.7
25	85.5	70.4	81.4	79.9	68.3	77.1	17.2	79.9
	Totals					1963.4	439.9	1974.2
	Averages					78.5	17.6	79.0

The control chart containing the above information is shown in Figure 10.3. The mean and range of each sample have been plotted on their separate graphs, and lines have been drawn to join successive plots. Also, the estimated process mean, process range and control lines have been drawn at their calculated values.

Control charts are prepared as above to provide means of process monitoring, and ideally the process should be in control. An in-control situation is suggested by the pattern of plotted points on Figure 10.3.

However, charts often have greater value when they highlight an unsuspected out-of-control situation that requires investigation. The chart might even suggest an avenue of investigation: for example, there might be a pattern over time that can be correlated with other events. Out-of-control situations are suggested by some of the patterns described in Section 10.18.

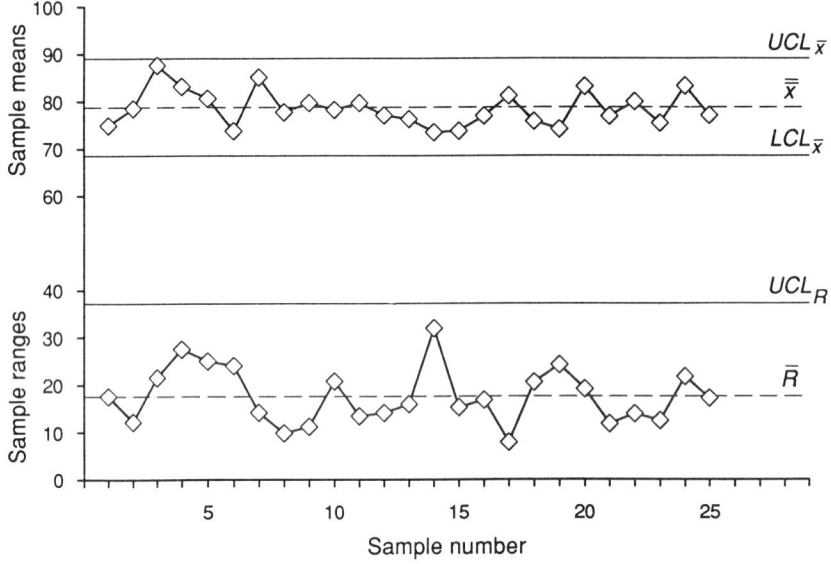

Figure 10.3 Mean and range control chart: data plot

Figure 10.4 is a chart used for monitoring a liquid dispensing process. The raw data is the weight of filled containers. The control lines were calculated from the first 25 samples. Among subsequent samples, one point appears outside the upper control line for ranges. This plot triggered investigation that found it was necessary to clean a release catch.

New sample values are added to the chart as they are obtained, whether or not the process is in control. The chart will monitor continuing in-control or the effectiveness of improvement action. In practice, the latter use is the most frequent reason for manual charting. Usually, it is more economic to automate the monitoring of consistently in-control processes.

■ 10.6 Median and range chart (or $\tilde{x}$ & R chart)

This type of chart is used widely because it avoids day-to-day arithmetic calculations and it minimizes the recording procedure; therefore, it finds more ready acceptance than $\bar{x}$ & R and $\bar{x}$ & s charts. If the medians are normally distributed, the medians graph is an effective method for detecting changes in the process mean. However, it is not an estimator of the process mean. In other words, when the actual value of the mean is important, more useful information will be obtained from $\bar{x}$ & R and $\bar{x}$ & s charts.

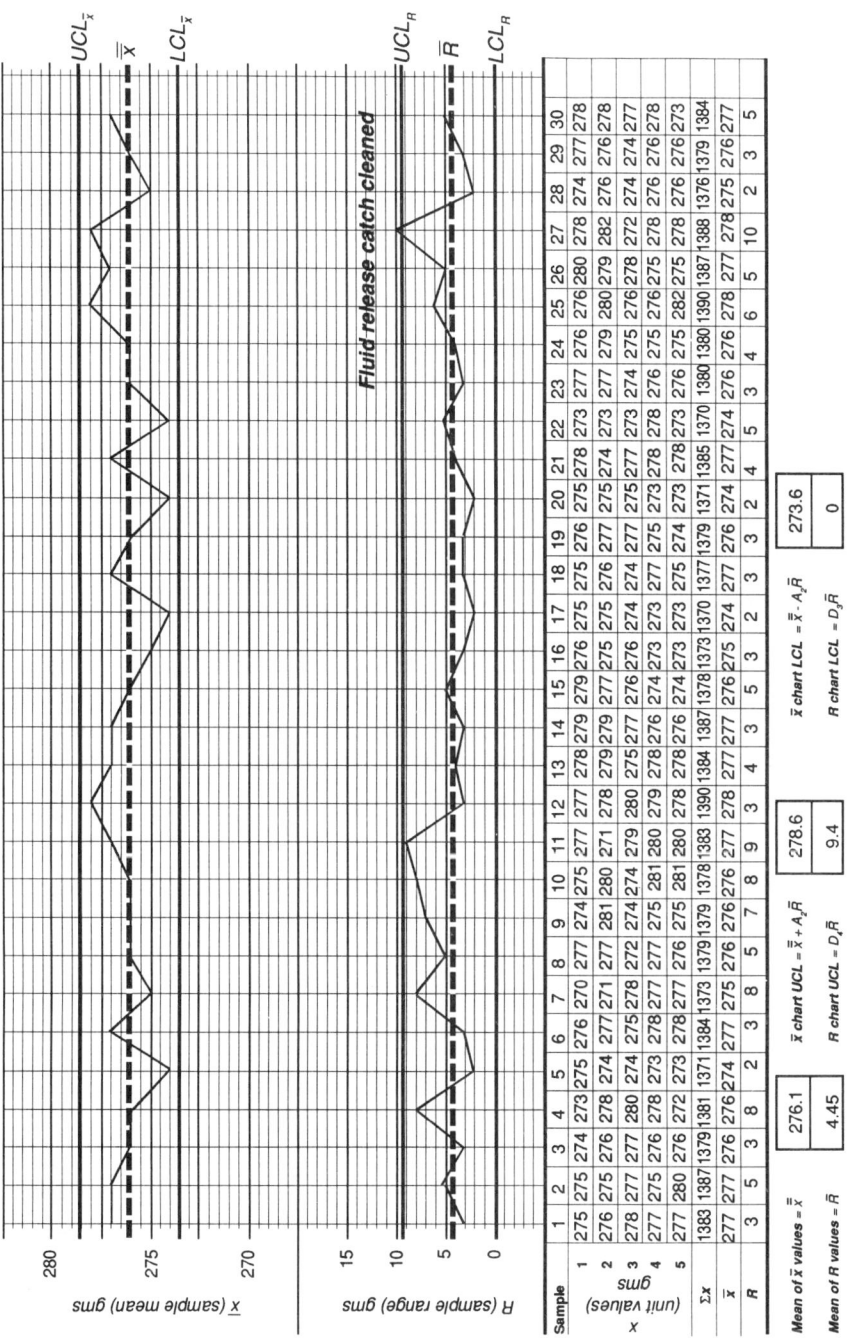

Figure 10.4 Mean and range chart for liquid dispensing process

203

The constants, used in medians control line calculations, have been calculated assuming a normal distribution for medians; this might not be true because there is no equivalent of the central limit theorem for medians. A normal probability plot (Section 8.3) of sample medians can be used as a check. The resulting control lines for medians are further apart than those for means, but this merely reflects the differences between the sampling distributions of means and medians.

Apart from two exceptions, the calculations for this chart are the same as those for the mean and range chart in Section 10.5.

■ The measurements of each individual in a sample are recorded as plots on the chart rather than as the written numbers on the mean and range chart. Then for each sample, instead of calculating the mean, the median value ($\tilde{x}$) is merely highlighted on the chart and lines are drawn to join successive medians.

■ Control lines are calculated from $\bar{\tilde{x}}$ and $\bar{R}$, using the following formulae.

$$\text{upper control line for medians} = UCL_{\tilde{x}} = \bar{\tilde{x}} + \tilde{A}_2 \bar{R}$$
$$\text{lower control line for medians} = LCL_{\tilde{x}} = \bar{\tilde{x}} - \tilde{A}_2 \bar{R}$$
$$\text{upper control line for ranges} \quad = UCL_R = D_4 \bar{R}$$
$$\text{lower control line for ranges} \quad = LCL_R = D_3 \bar{R}$$

Note the different constant $\tilde{A}_2$ in the medians control lines formulae and that the formulae for ranges control lines is the same as that on the mean and range chart. The constants $\tilde{A}_2$, D_4 and D_3 depend upon sample size and are given for a range of sample sizes in Appendix F.

Figure 10.5 is a median and range chart used to monitor a metal turning process. The raw data is the diameter of finished turned shafts. The control lines were calculated from the first 25 samples. Among subsequent samples, two points appear outside the control lines for medians. The appearance of the first led to the cutting tool being deburred, the second (a result of deburring) led to the tool being reset.

Example of median and range chart calculations

For the median and range chart, data can be laid out in the same way as for the mean and range chart in the previous section, but instead of a column of sample means $\bar{x}$, there would be a column of sample medians $\tilde{x}$ like that at the right of Table 10.1.

The median is the middle value when data is arranged in order of magnitude. The easiest way to pick out the median in a small sample of 5 is mentally (or physically) to cross out the smallest value, then the next smallest; the smallest of the three remaining is the median.

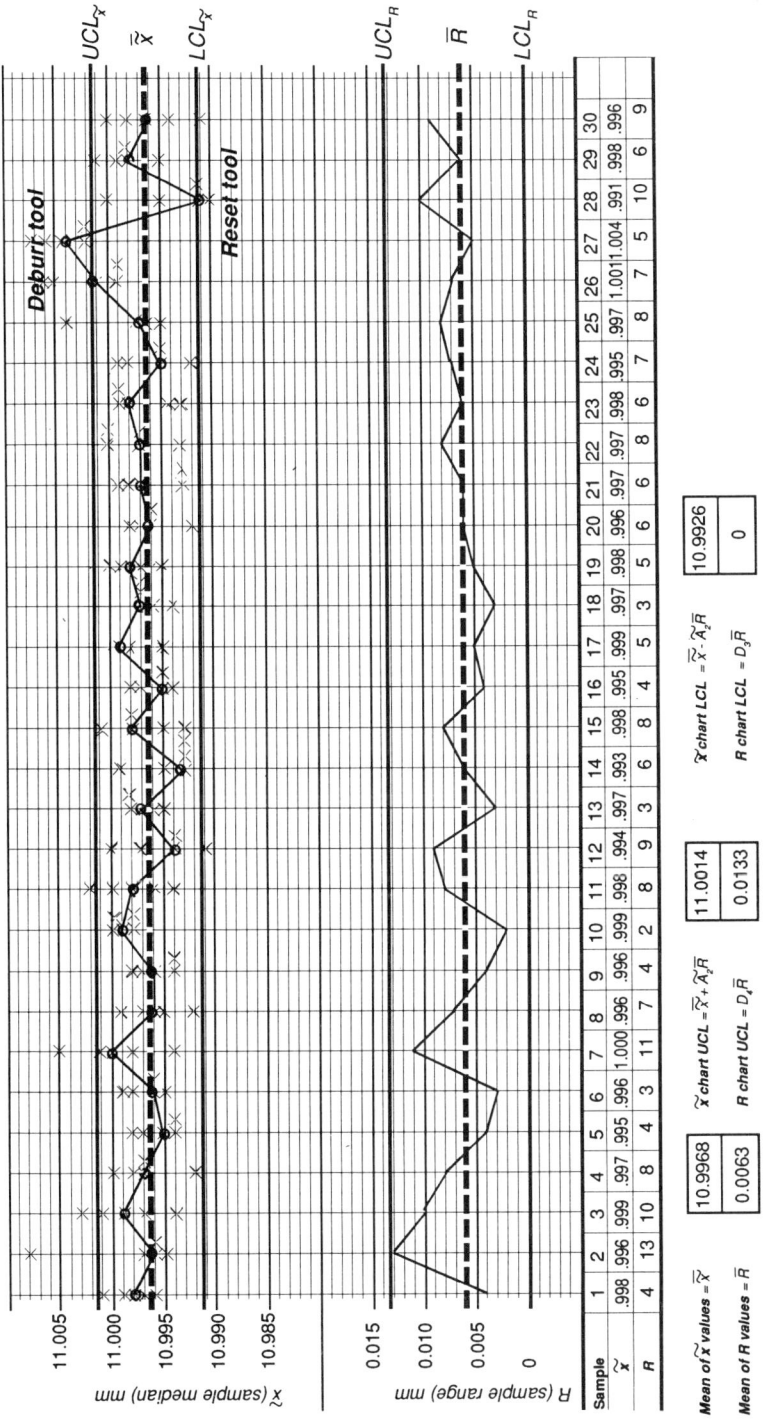

Figure 10.5 Median and range chart for a metal turning process

For example, take the first sample from the data in Table 10.1 (81.5, 66.6, 84.2, 70.1, 72.2), delete the smallest and the next smallest (81.5, ~~66.6~~, 84.2, ~~70.1~~, 72.2), and then the next smallest (72.2) is the median.

The mean of all the medians in Table 10.1 is found by summing the 25 values and dividing by 25. This gives $\bar{\bar{x}} = 79.0$ and, from Appendix F, $\tilde{A}$ is 0.691 so

$$UCL_{\tilde{x}} = \bar{\bar{x}} + \tilde{A}_2\bar{R} = 79.0 + 0.691 \times 17.60 = 91.2$$
$$LCL_{\tilde{x}} = \bar{\bar{x}} - \tilde{A}_2\bar{R} = 79.0 - 0.691 \times 17.60 = 66.8.$$

The control lines for ranges will be the same as in the previous section.

■ 10.7 Mean and standard deviation chart (or $\bar{x}$ & s chart

The description of this chart is the same as that for the mean and range chart except that the standard deviation, instead of the range, is calculated for each sample and control lines are calculated from $\bar{\bar{x}}$ and $\bar{s}$, using the following formulae.

upper control line for means	$= UCL_{\bar{x}} = \bar{\bar{x}} + A_3\bar{s}$
lower control line for means	$= LCL_{\bar{x}} = \bar{\bar{x}} - A_3\bar{s}$
upper control line for standard deviations	$= UCL_s = B_4\bar{s}$
lower control line for standard deviations	$= LCL_s = B_3\bar{s}$

Note the different constants, B_3 and B_4 in the standard deviations control lines formulae and A_3 in the formulae for means control lines, which are similar to those for the mean and range chart. The constants A_3, B_4 and B_3 depend upon sample size and are given for a range of sample sizes in Appendix F.

The s graph has narrower control lines than the R graph, which simply reflects the sampling distribution of s as opposed to R. The two graphs are equally sensitive in detecting changes in their representation of spread, but the s graph is less sensitive in detecting single outliers in a sample.

However, for purposes of process control and particularly when sample sizes are above 8, the s graph is a better indicator of trends. Over long periods of time, maximum and minimum values of many features tend to be normally distributed: in other words, the R values are fairly constant. Process improvement or deterioration usually changes the distribution of values towards maximum or minimum, and this is reflected in s value changes.

Manual calculations of s are somewhat more difficult than R, but most pocket calculators have functions that make the job easy. The $\bar{x}$ & s chart is best suited to applications where measurements, calculations and plots can be automated.

■ 10.8 Standard deviation calculation

It was explained in Section 7.7 how an estimate of standard deviation can be made from R values using the formula

standard deviation estimate $= \hat{\sigma} = \bar{R}/d_2$

and a similar estimate can be made from the mean of sample standard deviations, s:

standard deviation estimate $= \hat{\sigma} = \bar{s}/c_4$.

The constants d_2 and c_4 depend upon sample size. They are given for a range of sample sizes in Appendix F and often for commonly used sample sizes on control charts. For example, on Figure 10.4, where $\bar{R} = 4.45$ and $d_2 = 2.326$ for $n = 5$, the estimate of σ is $4.45/2.326 = 1.91$.

■ 10.9 Charts for attributes

In many situations, the nature of a process or the complexity of expectations inhibits collection of measured data. Where measurement is not possible, there is usually qualitative information about product or process features. This amounts to 'yes' or 'no' against some predetermined standard. For example, a valve functions or it does not, a seal leaks or it does not. Such processes can be monitored using attributes charts.

The target nominal for attributes is zero. On the way to achieving this ideal, it is preferable to have consistent performance rather than widely variable performance. Information about yes/no attributes can be used in a single chart, to obtain a picture of opportunities for improvement.

It must be stressed that in any situation where go/no-go gauges are used in *periodic process checks*, they should be replaced wherever practicable by actual measurement and variables control charts. In situations where screening is part of the process, the quality control activity should be supplemented by process control charts.

The four principal types of attribute chart are summarized in Figure 10.6.

Sample size

There is one simple guideline for attribute chart sample size: it should be large enough to allow the feature being monitored (usually a defect) to appear in the majority of samples taken. Successive sample sizes need not be constant, although there is an advantage of less calculation with constant sample sizes.

Chart	Plot	Process mean		Control lines **
p	p	$\bar{p}$ =	$\dfrac{f_1 + f_2 + ... + f_m}{n_1 + n_2 + ... + n_m}$	$\bar{p} \pm 3\sqrt{\dfrac{\bar{p}(1-\bar{p})}{n}}$
np	f	$\bar{f}$ =	$\dfrac{f_1 + f_2 + ... + f_m}{m}$	$\bar{f} \pm 3\sqrt{\bar{f}\left(1 - \dfrac{\bar{f}}{n}\right)}$
c	c	$\bar{c}$ =	$\dfrac{c_1 + c_2 + ... + c_m}{m}$	$\bar{c} \pm 3\sqrt{\bar{c}}$
u	u	$\bar{u}$ =	$\dfrac{c_1 + c_2 + ... + c_m}{n_1 + n_2 + ... + n_m}$	$\bar{u} \pm 3\sqrt{\dfrac{\bar{u}}{n}}$

Key to symbols

c	Defects (faulty features) in sample	
f	Defectives (faulty units) in sample	
n	Sample size	
m	Number of samples	
$\bar{n}$	Average sample size	$= \dfrac{n_1 + n_2 + ... + n_m}{m}$
u	Defects per sample unit	$= \dfrac{c}{n}$
p	Proportion of defectives in sample	$= \dfrac{f}{n}$
**	*The lower control line is drawn at zero when the calculation gives a negative number.*	

Figure 10.6 Attributes charts statistics and formulae

This guideline presupposes knowledge of defect rate and requires application of Poisson or binomial probability theory at an acceptable confidence level. When the objective is to monitor a frequently occurring situation, comparatively small samples (say 10) might be used. However, if the objective is to confirm the effectiveness of action to eliminate a defect that previously occurred at a rate of about 1%, the required sample size will be well over 1000. In general, sample sizes for attribute charts are much larger than for variables charts.

Sample selection

Similar to variables, samples should be taken randomly to reflect all likely process disturbance, but where practicable, the individuals in a sample should have been produced consecutively. Calculations from such samples give better estimates of inherent variability and trends than samples where the individuals are chosen randomly.

Ideally in all attributes charts, at least 25 samples should be obtained in order to calculate the control lines. Sometimes in practice a smaller number has to be taken.

■ 10.10 Proportion of defectives chart (*p* chart)

| Defectives are sometimes called non-conforming or faulty units.

The *p* chart is used to monitor defective units, regardless of the number of defects in any unit, when the data is expressed as the fraction or proportion of defectives in the sample.

Sample sizes should vary by not more than ± 12.5% of the average sample size. Standardized charts can be used where there is a larger sample size variation (Section 10.15).

Example of *p* chart calculations

Control lines are calculated in Table 10.2 for data which gives the number of scratched windscreens (defectives) in samples taken from daily deliveries.
Hence

$$\bar{p} = \frac{15 + 12 + 11 + \cdots}{126 + 134 + 115 + \cdots} = \frac{384}{3207} = 0.120$$

and

$$\bar{n} = \frac{126 + 134 + 115 + \cdots}{25} = \frac{3207}{25} = 128.3.$$

$$\text{control lines} = 0.120 \pm 3 \sqrt{\frac{0.120(1 - 0.120)}{128.3}}$$

$$= 0.120 \pm 0.086 = 0.034 \text{ and } 0.206.$$

Clearly no self-respecting company would claim that the process of windscreen delivery is under control! However, mathematically, the process is in a state of statistical control because all the plotted values are within the control lines, as can be seen on the control chart in Figure 10.7.

The message from the chart to management is that the process is totally predictable, and also that any improvement will require radical change in one or more of the facilities, methods, people, materials and environment that make up the process.

Table 10.2 Data for control chart in Figure 10.7

Sample number	Sample size	Defective units	Proportion defective	Sample number	Sample size	Defective units	Proportion defective
1	126	15	0.119	14	128	21	0.164
2	134	12	0.090	15	130	14	0.108
3	115	11	0.096	16	127	11	0.087
4	121	9	0.074	17	119	18	0.151
5	142	18	0.127	18	135	10	0.074
6	133	21	0.158	19	140	16	0.114
7	115	15	0.130	20	129	20	0.155
8	131	7	0.053	21	125	13	0.104
9	126	20	0.159	22	133	14	0.105
10	119	11	0.092	23	125	18	0.144
11	132	7	0.053	24	129	26	0.202
12	131	17	0.130	25	137	21	0.153
13	125	15	0.120				

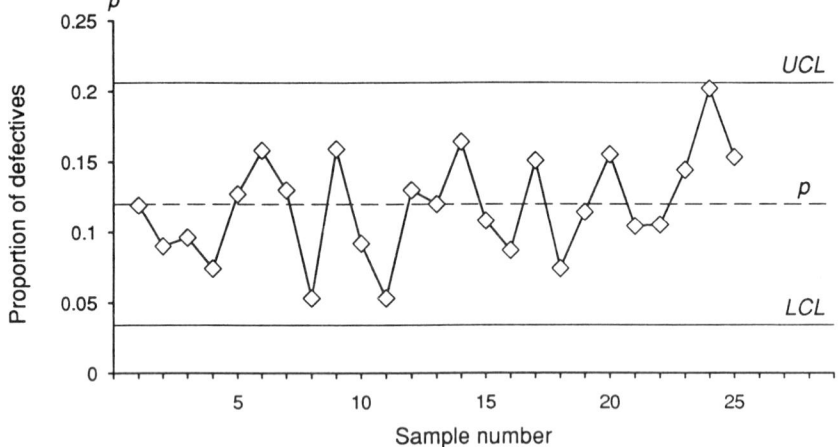

Figure 10.7 p chart illustration

Example of a p chart used outside manufacturing

Historically, control charts were used only as an aid to improving manufacturing production processes. More recently, with the recognition that processes are not special to production, control charts are being used in other situations. Whatever the activity, it can be expected that results will vary in a similar way to a manufacturing production process: for example, with points outside the control lines occurring due to special disturbances. Figure 10.8 is a proportion of defectives chart where late arrivals are recorded. It highlights unusual occurrences; the reasons were not difficult to find, as indicated on the chart.

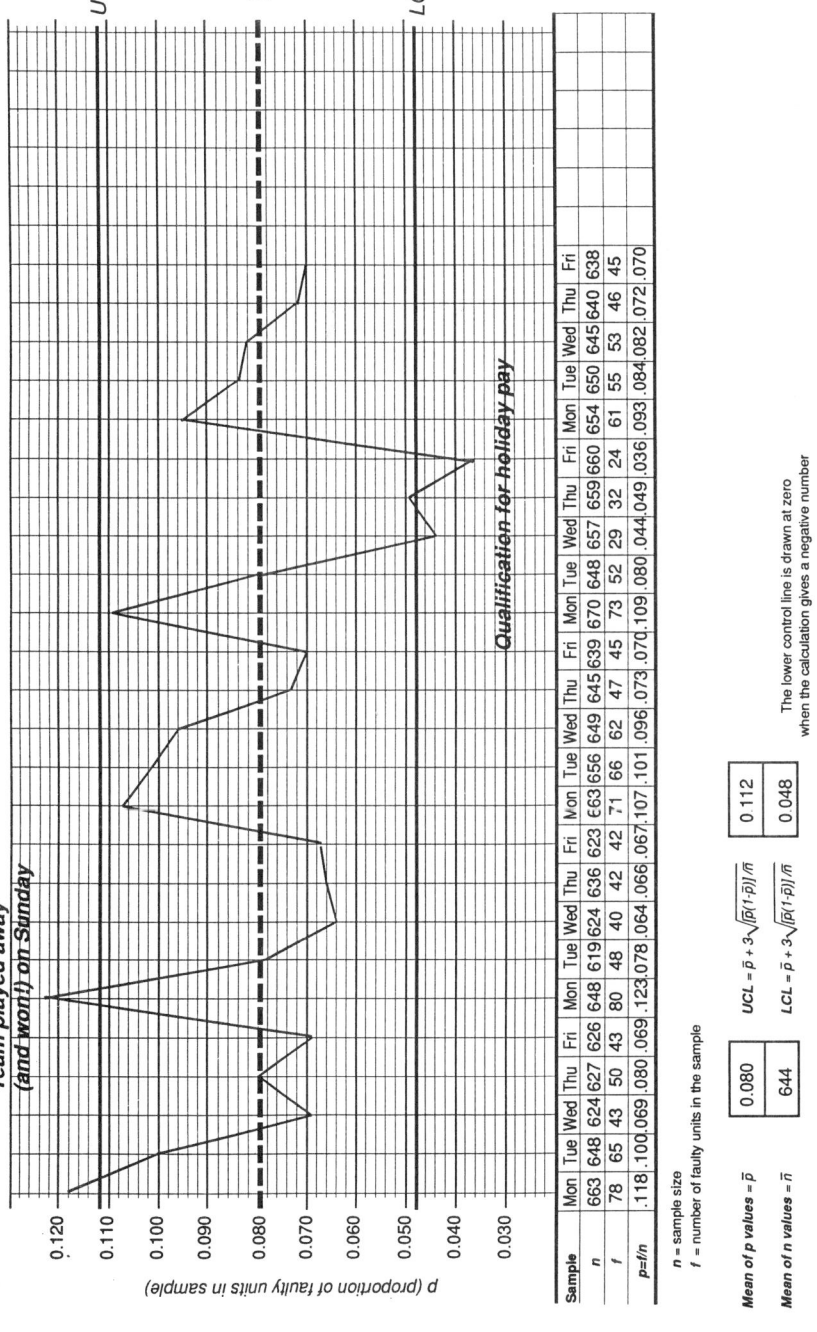

Figure 10.8 Proportion of defectives chart for late arrivals

The control lines might provide a better reference if they were calculated after discarding results influenced by special disturbances (or over a period when special disturbances do not occur). Even so, the charts are useful since they still highlight opportunities for process improvement.

The proportion of defectives charts can be used, of course, when sample sizes are all the same, but the chart described in the next section is more appropriate.

■ 10.11 Number of defectives chart (*np* chart)

The *np* chart is similar to the *p* chart. The difference is that the sample size, *n*, must be constant because the actual number of defectives is plotted. The two charts look exactly the same if plotted for the same data when there is constant sample size, except that the respective vertical scales will be labelled differently.

Example of *np* chart calculations

The number of defectives in 25 samples each of size 120 were

7,11,3,13,17,9,7,9,10,7,2,8,6,9,16,8,16,13,5,14,7,12,9,5,7.

The estimate of the process mean is

$$\bar{f} = \frac{7 + 11 + 3 + \cdots}{25} = \frac{230}{25} = 9.20.$$

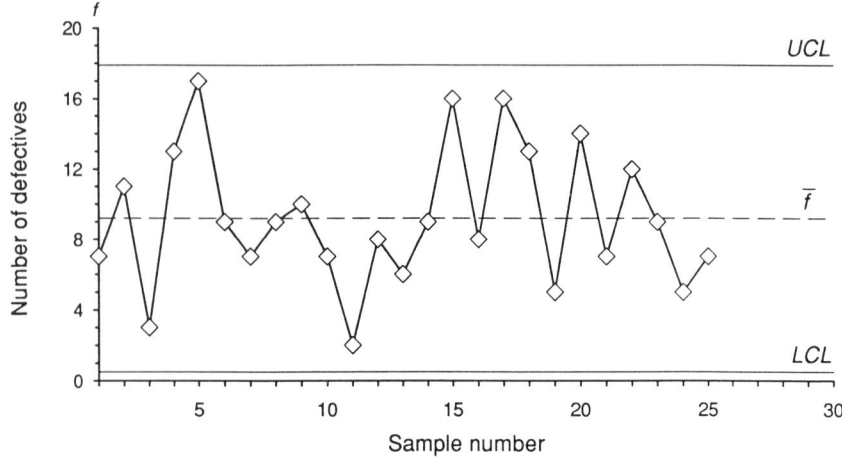

Figure 10.9 *np* chart illustration

212

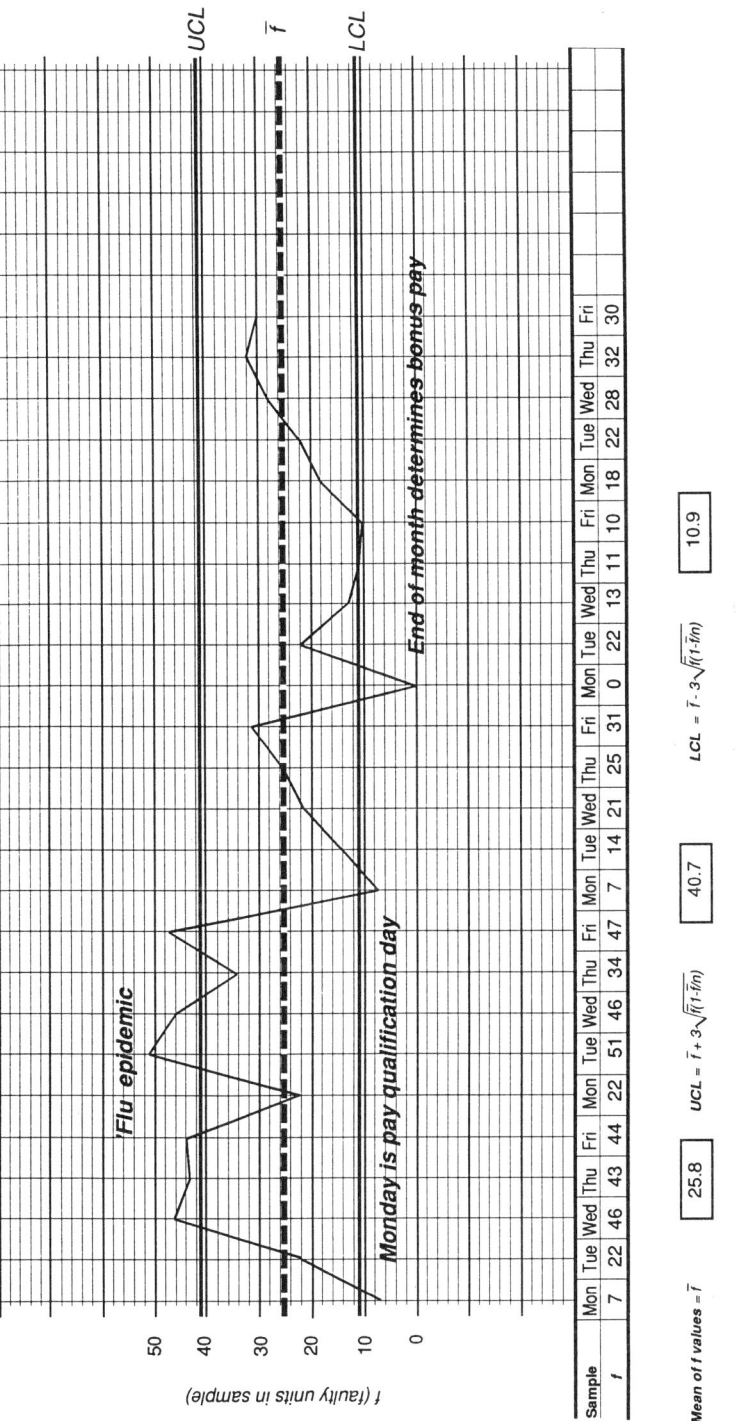

Sample	Mon	Tue	Wed	Thu	Fri	Mon	Tue	Wed	Thu	Fri	Mon	Tue	Wed	Thu	Fri	Mon	Tue	Wed	Thu	Fri
f	7	22	46	43	44	22	51	46	34	47	7	14	21	25	31	0	22	13	11	10	18	22	28	32	30

Mean of *f* values = *f̄* $\boxed{25.8}$

$UCL = \bar{f} + 3\sqrt{\bar{f}(1-\bar{f}/n)}$ $\boxed{40.7}$

$LCL = \bar{f} - 3\sqrt{\bar{f}(1-\bar{f}/n)}$ $\boxed{10.9}$

The lower control line is drawn at zero when the calculation gives a negative number

Figure 10.10 Number of defectives chart for absences

213

The control lines are at

$$9.2 \pm 3 \sqrt{9.2(1 - 9.2/120)} = 0.5 \text{ and } 17.9.$$

The chart is drawn in Figure 10.9, where the plots show that the process is in a state of statistical control. Figure 10.10 is a similar chart, but it is not in a state of statistical control for the reasons shown.

■ 10.12 Number of defects chart (*c* chart)

| The *c* chart is sometimes called a simple run chart.

The *c* chart is similar to the *np* chart except that it describes defects rather than defectives. Also, it is the most commonly used chart for illustrating other discrete data, such as occurrences of accidents, live births or aircraft movements. Samples might be a single unit or a constant-sized group of units.

Example of *c* chart calculations

The number of blemishes caused by factory fall-out on 30 car roofs were recorded as

2,5,3,7,1,6,1,3,3,4,2,7,3,4,2,5,4,2,6,4,3,9,4,2,5,2,4,7,2,3.

The estimate of the process mean is

$$\bar{c} = 115/30 = 3.83.$$

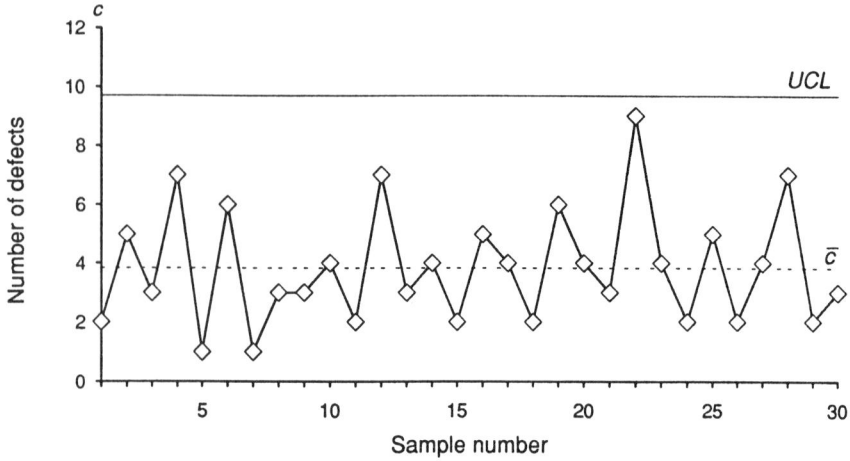

Figure 10.11 *c* chart illustration

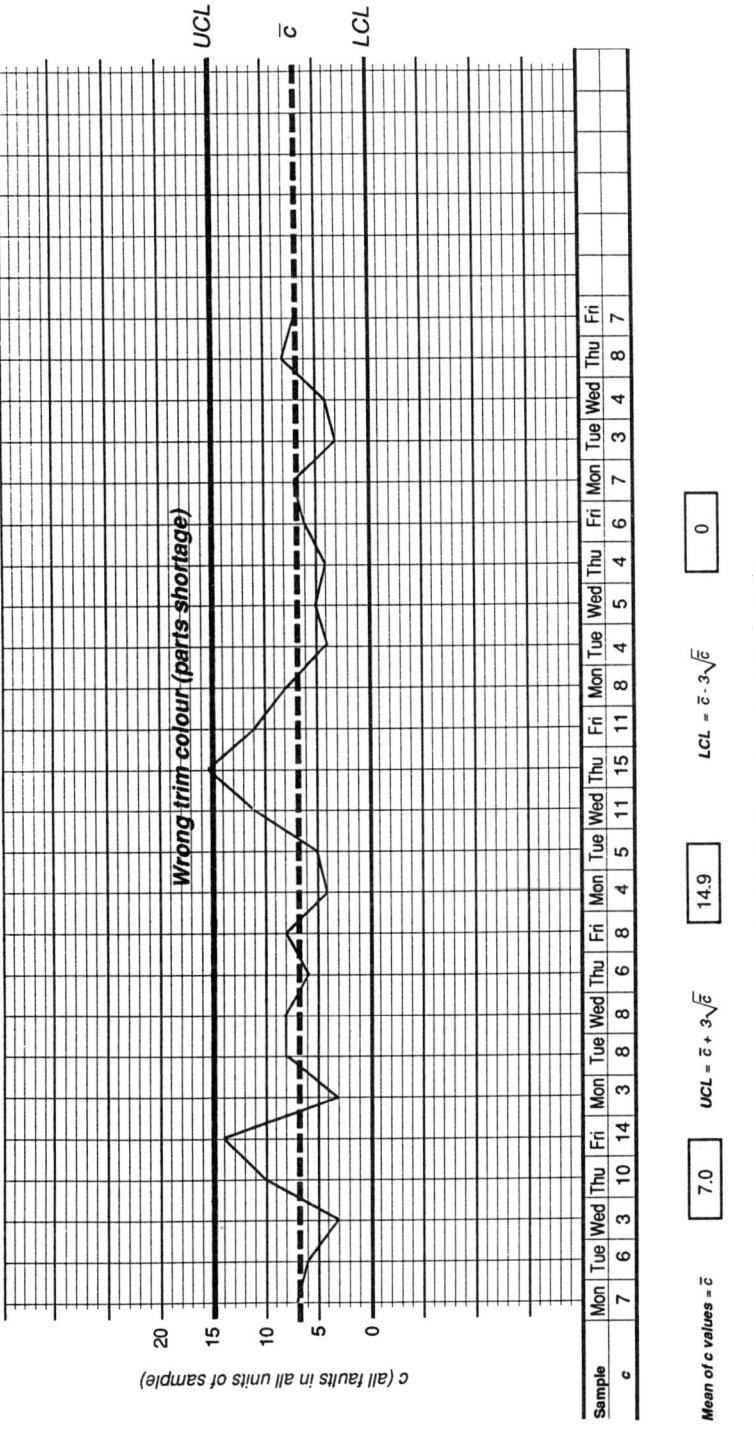

Figure 10.12 Number of defects chart for vehicle faults detected at audit

Sample	Mon	Tue	Wed	Thu	Fri	Mon	Tue	Wed	Thu	Fri	Mon	Tue	Wed	Thu	Fri	Mon	Tue	Wed	Thu	Fri
c	7	6	3	10	14	3	8	8	6	8	4	5	11	15	11	8	4	5	4	6	7	3	4	8	7

Mean of c values = $\bar{c}$

$UCL = \bar{c} + 3\sqrt{\bar{c}}$ $\boxed{14.9}$

$LCL = \bar{c} - 3\sqrt{\bar{c}}$ $\boxed{0}$

$\boxed{7.0}$

The lower control line is drawn at zero when the calculation gives a negative number

215

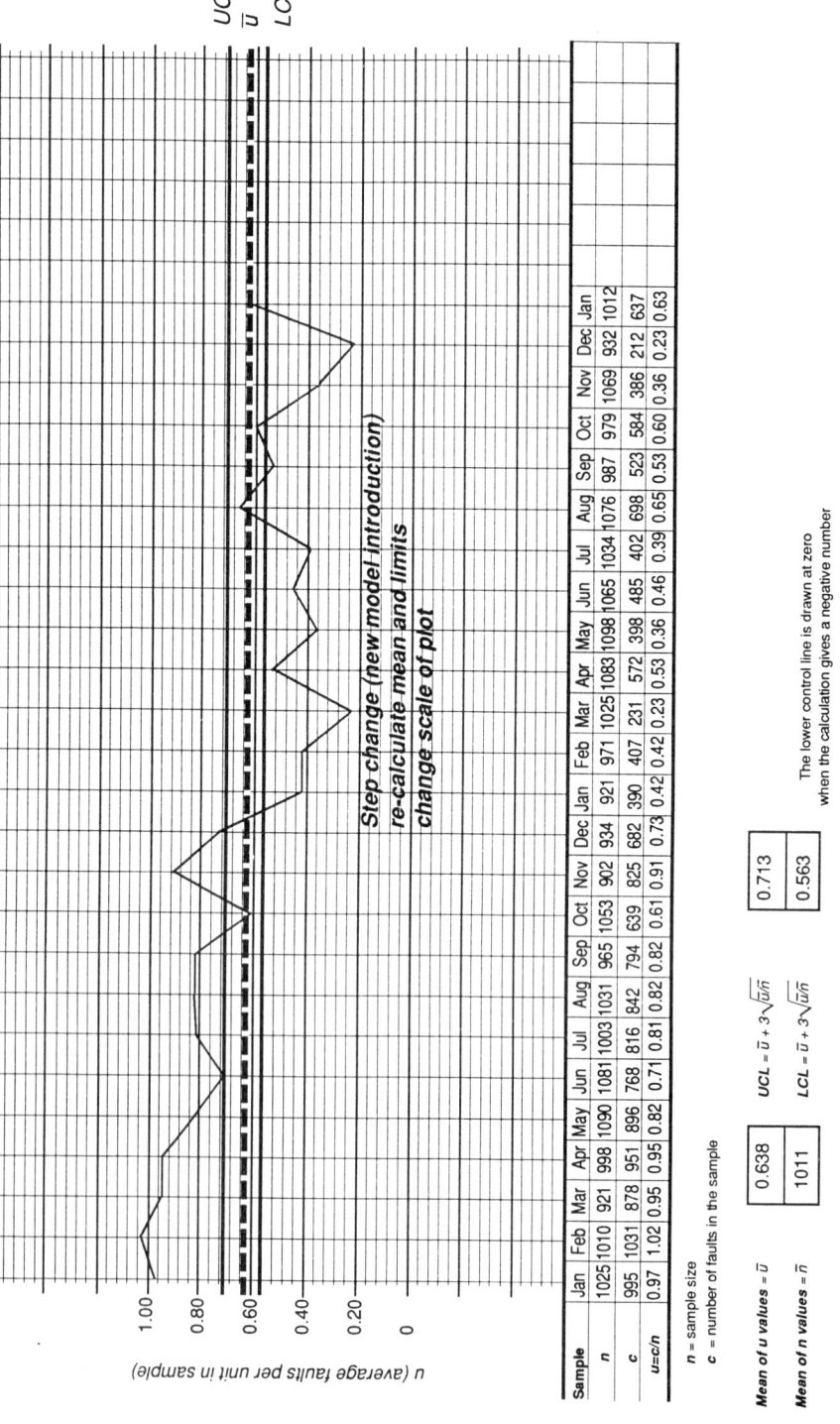

Figure 10.13 Proportion of defects chart for vehicle faults detected by dealers

The control lines are at

$$3.83 \pm 3 \sqrt{3.83} = 9.7 \text{ and } 0.$$

The chart is plotted in Figure 10.11, where the plots are in a state of statistical control. Figure 10.12 is a similar chart where the sample size is 5 and the total number of faults in each sample is recorded.

■ 10.13 Proportion of defects chart (u chart)

Figure 10.13 is an example of the u chart, which is similar to the c chart. The differences are that the u chart can be used when sample size varies and the average number of defects per unit in the sample is plotted. Similar to the p chart, sample sizes must not vary by more than $\pm 12.5\%$ of the average sample size.

Although the lines calculated and shown on the chart in Figure 10.13 clearly emphasize the step change, they are not appropriate for monitoring either the old or the new model. In practice, once enough samples have been collected (about 25 after the change), new lines will be calculated from this data alone.

■ 10.14 Moving mean charts

Sometimes a drift in the mean is an inherent characteristic of a process: for example, when wear of equipment is not automatically compensated and one or more parts (e.g. tool bits) are periodically replaced or reset. Where this situation exists, more information can be obtained from a chart that recognizes the trend. The $\bar{x}$ & R chart for moving means is such a chart. Sample ranges are dealt with in the same way as in the basic $\bar{x}$ & R chart, but the means are treated differently.

Moving mean charts are comparatively complex and it is unusual to see a hand-drawn example. However, the principles explained below are in use in many automated machines where calculations and signals are computerized.

Initially, data is collected over at least one process cycle (for example, between tool resets or replacements) and plotted in the same way as for a basic $\bar{x}$ & R chart. Provided the ranges are relatively stable (ideally in a state of statistical control), a best-fit line is drawn through the means plots (see Figure 10.14).

In the context of common and special disturbances, the best-fit line is effectively the process mean. Control lines are constructed parallel to it and at vertical distances from it of $\pm A_2\bar{R}$ (values of the constant A_2 are given in Appendix F). These sloping mean and control lines provide references for

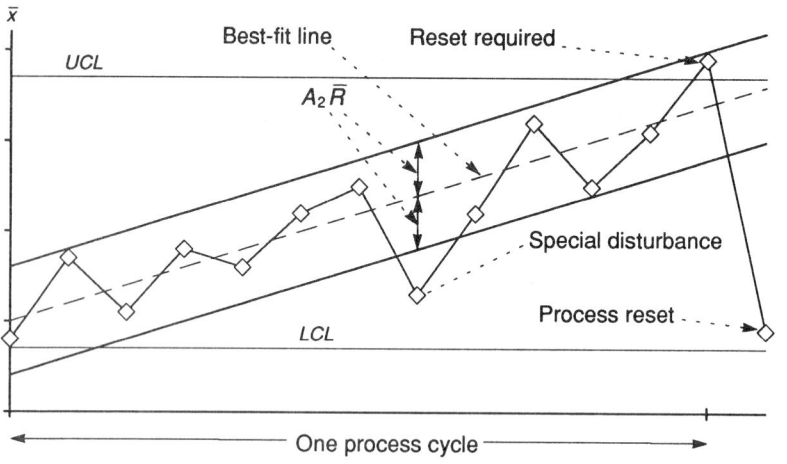

Figure 10.14 $\bar{x}$ chart for moving mean

interpretation of plots in the same way as the more usual horizontal lines. They must be repositioned on the chart each time the process is reset or when there is a tool change.

After several process cycles, horizontal control lines are constructed and used to monitor the need for resetting. This technique often results in extending the useful life of replaceable parts beyond theoretical fixed change points. The horizontal control lines are set at

$$UCL_{\bar{x}} = \bar{\bar{x}} + 0.5(\bar{x}_{max} - \bar{x}_{min}) + A_2\bar{R}$$

$$LCL_{\bar{x}} = \bar{\bar{x}} - 0.5(\bar{x}_{max} - \bar{x}_{min}) - A_2\bar{R}.$$

The quantity $(\bar{x}_{max} - \bar{x}_{min})$ is known as the *average movement of the mean* (AMM).

■ 10.15 Standardized charts

In general terms, standardized charts are used to monitor a basic process when it gives rise to two or more different populations of data and when the reasons for data differences are not a consequence of the process. In other words, the technique is based on the assumption that the particular features being measured are influenced by factors which are independent of the process. For examples:

■ On a country-wide basis, it is probable that the inherent ability of people is normally distributed. However, academic examination results show different distributions from area to area. Given that the examination is constant, it is reasonable to assume that the results (as a measure of ability) have been influenced by external factors such as the local examiner and the local preparation for the examination.

■ The cosmetic condition of motor vehicles from any single assembly line over particular periods of time is normally distributed. However, the distribution of customer reaction varies from country to country. From time to time, actions are taken to improve the condition; this causes change in the reaction distribution and the manufacturer needs a mechanism to standardize feedback information, so that the overall effectiveness of actions can be assessed.

■ Some processes involve small batch production, often of different components. Again, the manufacturer needs to standardize information to monitor the process.

■ Certain attribute data sample sizes vary beyond that which is acceptable to standard charting techniques (by greater than 12.5% from the average).

Monitoring in all the above examples can be carried out through estimation of the standardized deviate or z value of the sample average. The z value is plotted instead of sample average on a chart that has a mean at zero and control lines at ± 3 (see Figure 10.15).

An example of the use of a standardized chart appears in the subjective area of motor vehicle noise, vibration and harshness of ride. The subject is complex with many influencing factors, not least of which is human sensitivity to the effects. Charting is used to assess the compound effect of the frequent actions that are taken to improve all aspects of vehicles.

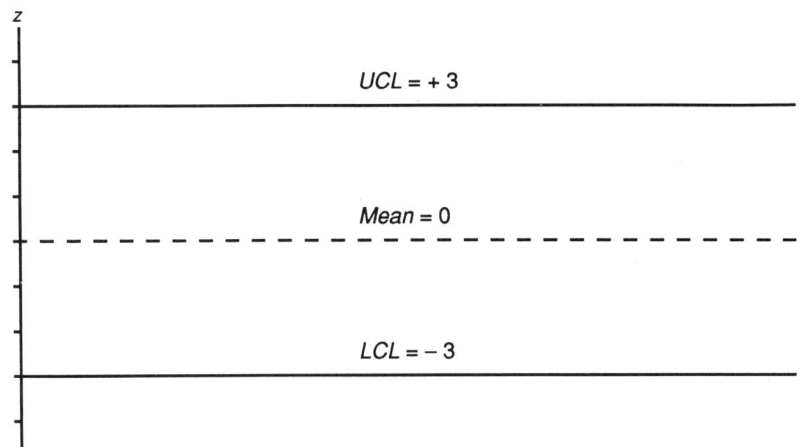

Figure 10.15 Standardized chart

About 5 vehicles are taken as a weekly sample for each model. They are driven for about 17 miles around a standard test route and each vehicle is rated by the driver for noise on a subjective 1 to 10 scale. Rating 1 is appalling, rating 10 is the other extreme and a rating of 6 is the level that the average car buyer would find acceptable, especially if his or her previous vehicle was more than one year old. There is more than one test driver.

The results over several weeks from two drivers, expressed as the sample average rating, are plotted on the chart in Figure 10.16(a) with control lines calculated from the data. The picture is of an out-of-control process (see chart interpretations Figure 10.29 in Section 10.17). The reason was found to be the naturally different perceptions of the drivers: in other words, the picture is of two processes having different distributions. When the results are separated by driver as in Figure 10.16(b) each process is in control, showing that the individual drivers are consistent in their assessments. Because the human element is not relevant to the purpose of this assessment, the results are presented on a standardized chart that eliminates driver sensitivity.

From the data presentation in Figure 10.16(b), each driver's ratings can be scaled to a value in the range -3 to $+3$ ($= \bar{x} \pm 3\sigma_e$ = the separation of the control lines). These values are plotted on the standardized chart in Figure 10.16(c), which offers a clearer picture of the overall effect of product modifications on noise, vibration and ride harshness trends.

For each new result from either driver, the z value is calculated by subtracting that driver's mean rating ($\bar{x}$) from the new rating and then dividing by σ_e for that driver.

If the purpose of charting in this example were to assess variations in human sensitivity rather than the effect of product actions, the picture offered by Figure 10.16(a) and the statistics derived from Figure 10.16(b) would be of more help than the standardized chart in Figure 10.16(c).

In practice, for ease of interpretation, the plots on Figure 10.16(a) would be made in sequence of driver 1, driver 2, driver 1, driver 2 and so on, rather than in strict time sequence.

■ 10.16 Charting data not over time

Most charting tracks a single process over a period of time, but this is not its only use. Charting can be used to compare the performance of similar elements in a process: for example, of different machines carrying out the same task or of different batches of material. In the following example, it is people engaged in the same activity who are compared.

(a)

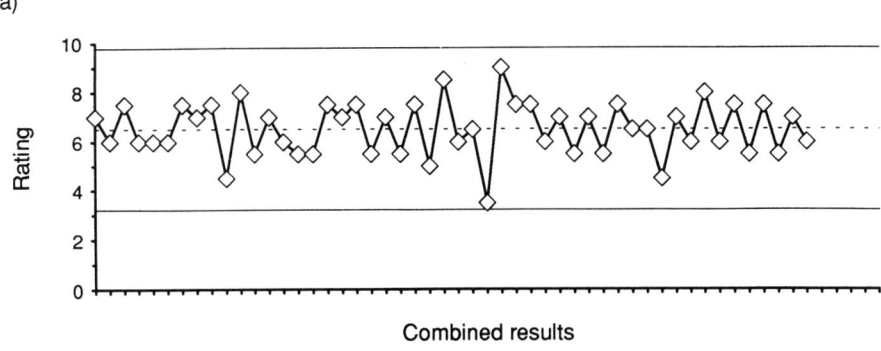

Combined results

(b)

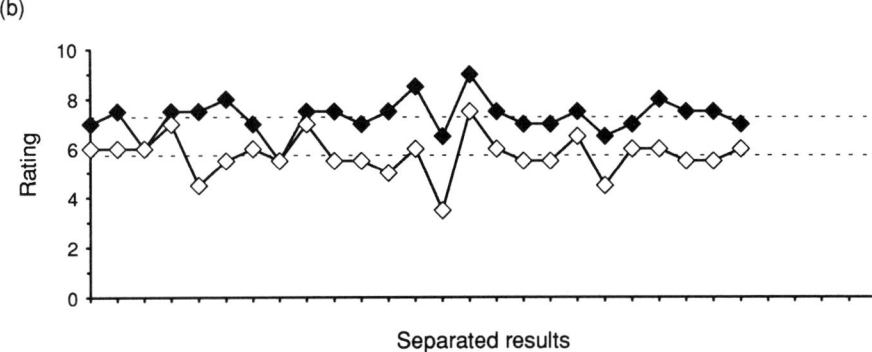

Separated results

(c)

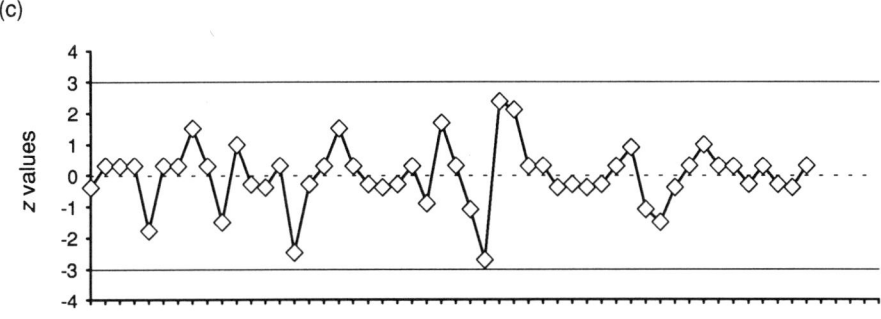

Figure 10.16 Standardizing

Figure 10.17 shows the number of defective ornaments made by 12 crafts-men where 100 ornaments were sampled from each of them. It is effectively a number of defectives (np) chart with an upper control line at 7.66 and a mean ($\bar{f}$) of 2.75.

221

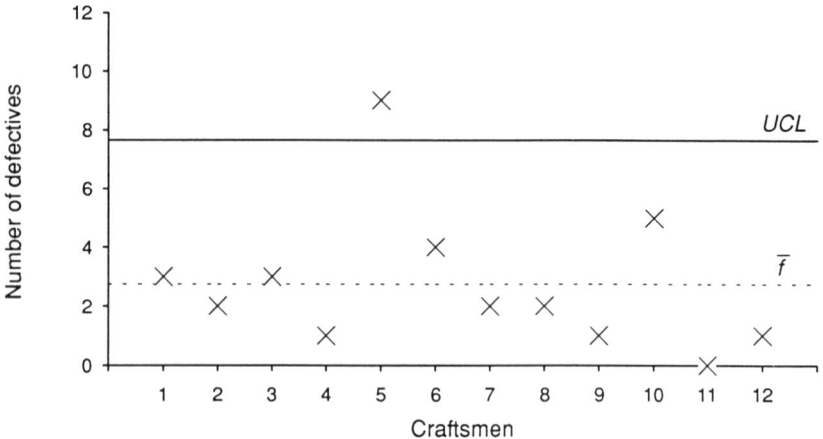

Figure 10.17 An *np* chart not over time

The chart shows that one craftsman is producing defectives outside the control line. If the production of ornaments is considered to be a process, then natural sources of variation between craftsmen can be regarded as common disturbances, and values outside control lines might be due to special disturbances. Put differently, the craftsman producing defectives outside the control line is in some way different to the others.

In this example, the actual difference was that the particular craftsman was the only trainee. It might have been that the work was being affected by illness or injury.

The use of a chart in this way can objectively highlight situations where there is a lesson to be learnt or action needs to be taken. Often it is a more acceptable basis for pursuing continuous improvement than subjective verbal observations.

■ 10.17 Charts with sample size 1

There are times when it is inappropriate to consider samples with size greater than 1. In this case the ranges are calculated by taking the difference between one value and the next, as illustrated in the following example. Effectively, each sample size for range is 2.

The plotted values are the individual measurement.
The estimate of the process mean is the mean of all measurements, $\bar{x}$.
The control lines for individual measurements are based on $\mu \pm 3\sigma$.

upper control line for measurements $= UCL_x = \bar{x} + 2.660\bar{R}$
lower control line for measurements $= LCL_x = \bar{x} - 2.660\bar{R}$
upper control line for ranges $\quad = UCL_R = 3.267\bar{R}$
lower control line for ranges $\quad = LCL_R = 0.$

For example, the weekly income for sales of exhausts at a garage were as shown in Table 10.3, where ranges are the difference between the income for that week and the income for the preceding week. A control chart, based on this data, has been drawn in Figure 10.18.

$$\bar{x} = \frac{256 + 302 + 243 + \cdots}{15} = \frac{4515}{15} = 301.0$$

$$\bar{R} = \frac{46 + 59 + 43 + \cdots}{14} = \frac{562}{14} = 40.14$$

Table 10.3 Weekly income from exhaust sales

Week	Income(£)	Range	Week	Income(£)	Range
1	256	–	9	327	95
2	302	46	10	292	35
3	243	59	11	281	11
4	286	43	12	305	24
5	281	5	13	333	28
6	277	4	14	294	39
7	294	17	15	322	28
8	422	128			

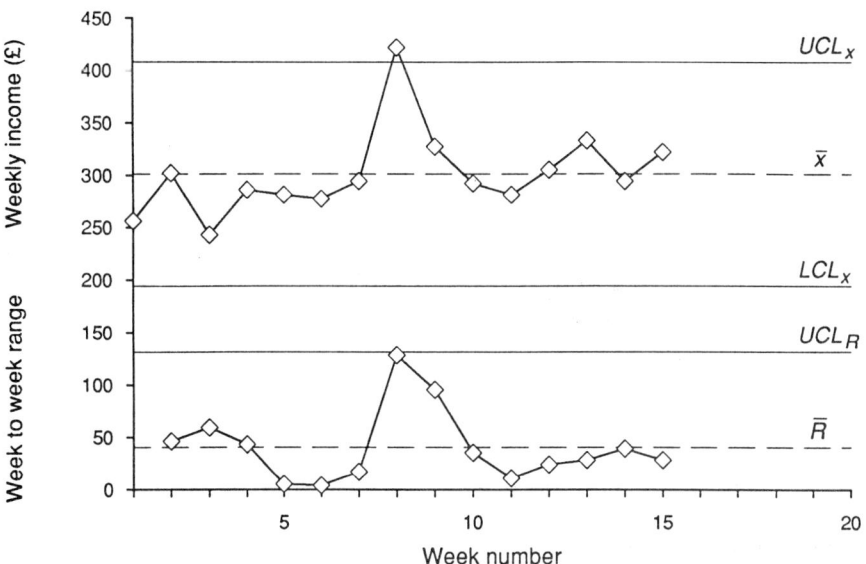

Figure 10.18 Mean and range chart for sample size 1

$$UCL_x = 301.0 + 2.660 \times 40.14 = 407.8$$
$$LCL_x = 301.0 - 2.660 \times 40.14 = 194.2$$
$$UCL_R = 3.267 \times 40.14 \qquad = 131.1$$
$$LCL_R \qquad\qquad\qquad\qquad = 0$$

The chart drawn on Figure 10.18 has one value outside the control lines (after a spell of bad weather). As always, it is preferable to use at least 25 readings, but this is not always possible. Note that in this particular chart the range values are not independent, in that given a large range, probably due to an unusually high or low value, the next one is also likely to be large.

■ 10.18 Chart interpretation

Chart interpretation requires a blend of process knowledge, experience and appreciation of probability. This section offers guidelines that amount to pattern recognition. The starting point is the expected distribution of plots set out in Figure 10.19 (see Figure 5.5) and illustrated in Figure 10.20.

Examples of unusual patterns and their interpretation are illustrated in Figures 10.21 to 10.30. These patterns and some noted variations are widely used in the world's motor vehicle and household durables industries.

The guidelines assume random results within a normal distribution when the process is stable, but the same pattern interpretations can be equally and usefully applied to data from non-normal distributions. This is particularly

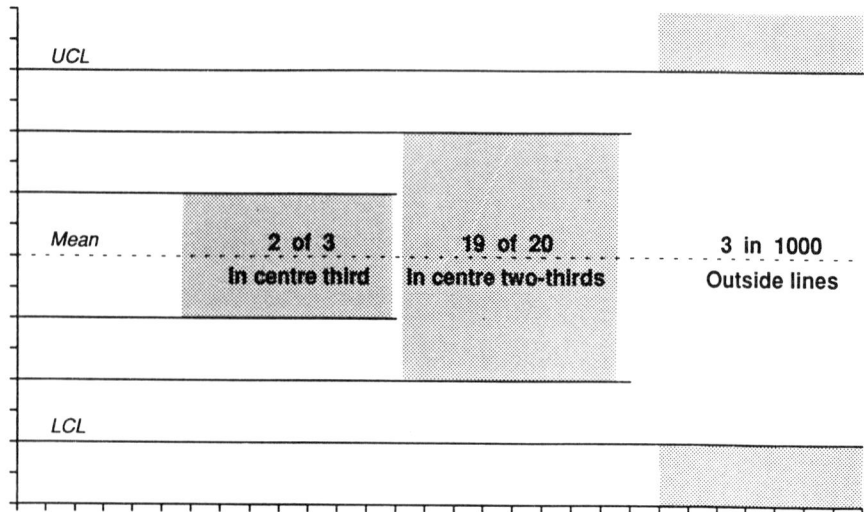

Figure 10.19 Normal plot proportions

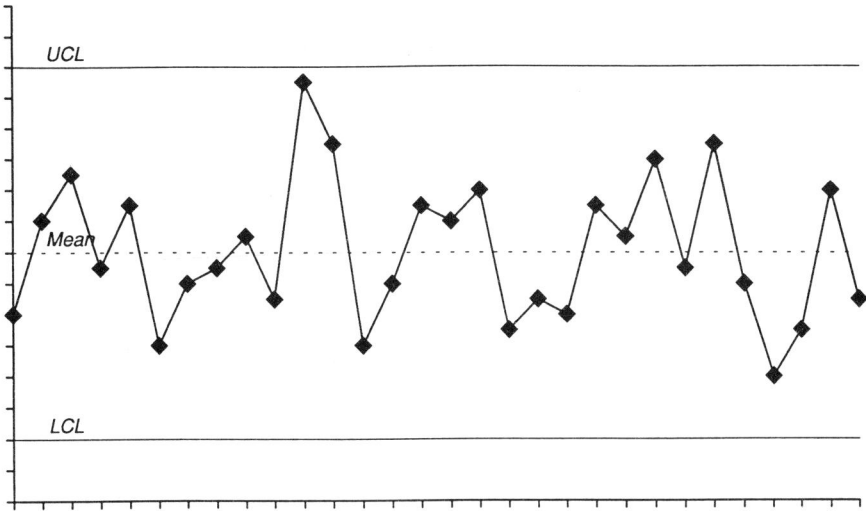

Figure 10.20 Normal plot

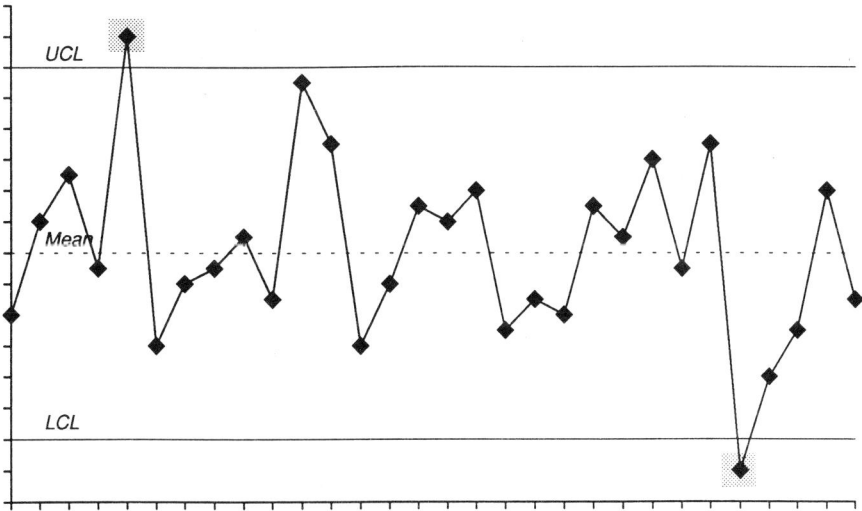

Figure 10.21 One or more plots above *UCL* or below *LCL*

true when sample means are being charted because the plotted values will have an approximately normal distribution, as shown by the central limit theorem. However, the quoted probabilities apply only to normally distributed data: in other words, the quoted probabilities do not apply to range, standard deviation and attribute charts.

225

The occurrence of an unusual pattern is a signal that the process requires investigation, including of course a check that the signal is not false. The illustrated patterns are by no means all that can occur. In practice, anything that looks unusual should be investigated if it persists.

> The probability of each of the results in Figure 10.21 occurring by chance is $\leqslant 0.00135$ (value from the standard normal probability table for $z = 3$) or less than once in every 740 plots.
>
> Most often, an occurrence indicates a single special disturbance, such as flawed material, a broken tool, an operator mistake or a power surge.
>
> The probability of each of the results in Figure 10.22 occurring by chance for any 3 consecutive plots is $\leqslant 0.00134$ ($= 0.0214 \times 0.0214 \times 0.9759 \times 3$, using the multiplication rule for three independent events, where 0.0214 is the probability $2 \leqslant z \leqslant 3$ and 0.9759 is the probability $-3 \leqslant z \leqslant 2$) or less than once in every 746 plots.
>
> Most often, it indicates a frequently occurring special disturbance, such as a start-up effect or an untrained operator.
>
> It could signal improvement, but only in the bottom of R, s and attribute charts.
>
> The probability of each of the results in Figure 10.23 occurring by chance for any 5 consecutive plots is $\leqslant 0.0026$ ($= 0.1574 \times 0.1574 \times 0.1574 \times 0.1574 \times 0.840 \times 5$, where 0.1574 is the probability $1 \leqslant z \leqslant 3$ and 0.840 is the probability $-3 \leqslant z \leqslant 1$) or less than once in every 388 plots.

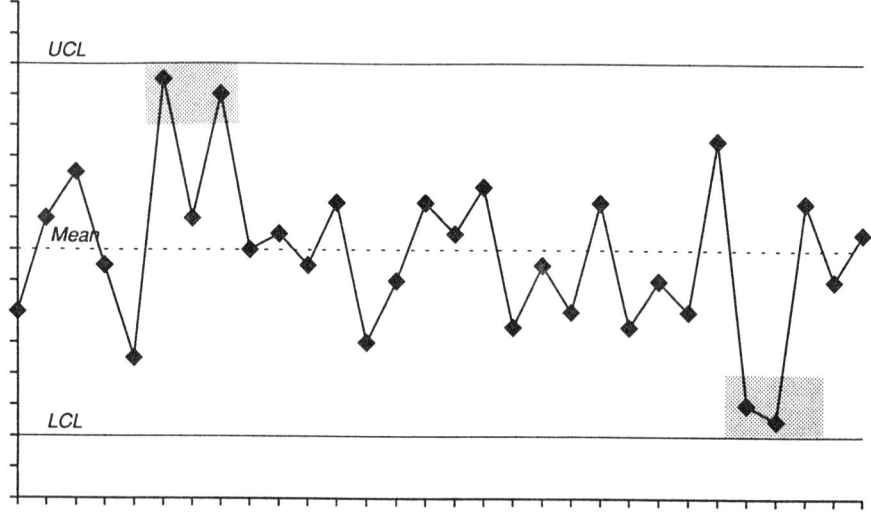

Figure 10.22 Two of three consecutive plots in top or bottom sixth of limits

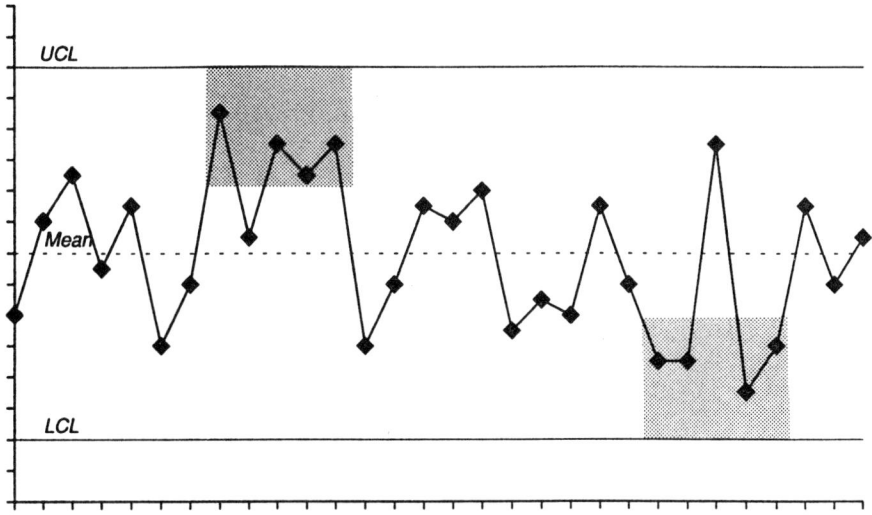

Figure 10.23 Four of five consecutive plots in top or bottom third of limits

Sometimes an occurrence indicates a more persistent special disturbance similar to those described under Figure 10.22.

The probability of each of the results in Figure 10.24 occurring by chance for any 7 consecutive plots is 0.0077 $(=0.4987^7$, multiplication rule for seven independent events) or about once in every 130 plots.

The International Standard ISO 8258 (Shewhart control charts) illustrates 9 consecutive plots (instead of 7) above or below the mean. This changes the probability to 0.0019 or about once in every 524 plots.

More usually, this pattern indicates some step change in the process, such as resetting, parts replacement, method change or management intervention.

It could signal improvement, but only in the bottom of R, s and attribute charts.

The probability of each of the results in Figure 10.25 occurring by chance for 8 consecutive plots is 0.000025 (=1/8!, which is independent of the distribution) or about once in every 40 320 plots.

The International Standard ISO 8258 illustrates only 6 (instead of 8) consecutive plots rising or falling. This changes the probability to 0.0014 or about once in every 720 plots.

The downward run only in R, s and attribute charts might signal improvement.

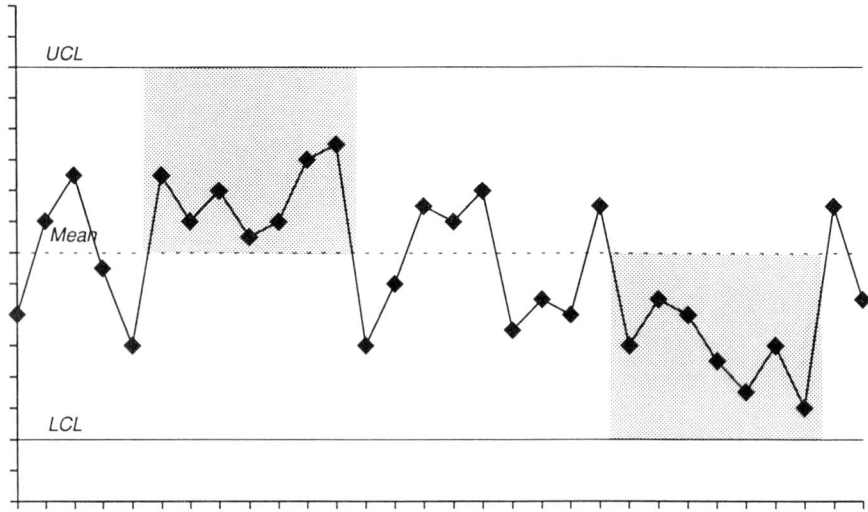

Figure 10.24 Seven consecutive plots in top or bottom half of limits

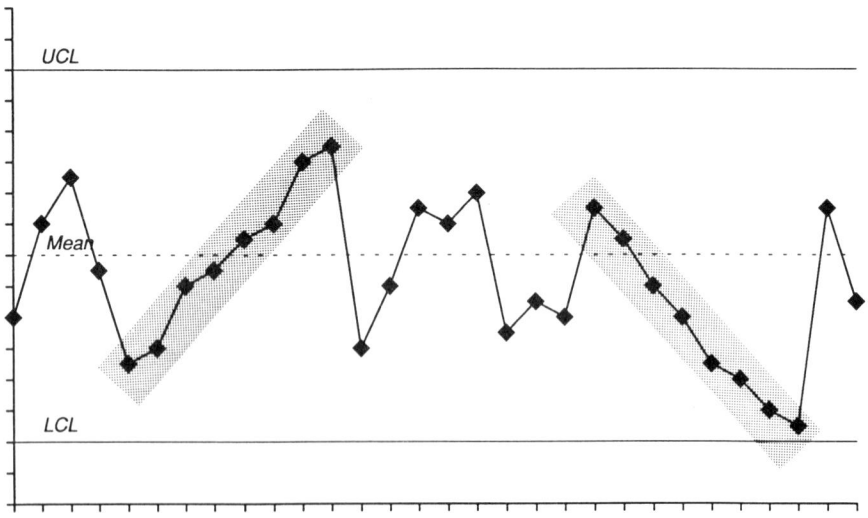

Figure 10.25 Eight consecutive plots rising or falling (seven intervals)

More usually, it indicates some progressive but inherent process change, such as equipment wear, a market shift or a seasonal change of environment.

The probability of the result in Figure 10.26 occurring by chance for any 8 consecutive plots is 0.000096 ($= 0.3148^8$, where 0.3148 is the probability $1 \leqslant z \leqslant 3$ or $-3 \leqslant z \leqslant -1$) or less than once in every 10 370 plots.

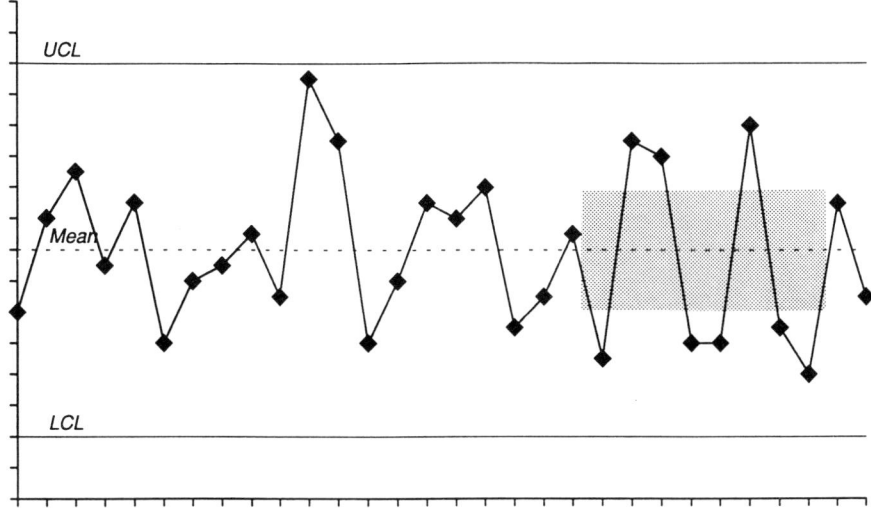

Figure 10.26 Eight consecutive plots outside centre third of limits

Because this result has a very small chance of occurring as a false signal, it might be better to take note of fewer consecutive plots outside the centre third. For example, 5 such plots have a probability of 0.0031 or about once in every 323 plots.

This pattern almost always indicates that some part of the process is out of control. It usually reflects mixed distributions that arise from samples which are drawn from different people, areas or machines.

Often, the cause is in procedures for measurement, reporting, machine setting, etc.

The probability of the result in Figure 10.27 occurring by chance for any 15 consecutive plots is 0.00325 ($=0.6826^{15}$, where 0.6826 is the probability $-1 \leqslant z \leqslant 1$) or less than once in every 307 plots.

This pattern can indicate either process improvement or that something is wrong.

Sometimes on charts for variables, especially when both charts of a pair exhibit the same pattern, it indicates a more consistent and therefore an improved process.

The reasons for something being wrong are similar to those described for Figure 10.26, in particular when, by design or by mistake, false data is used.

The pattern in Figure 10.28 is an extreme example of the effects described for Figure 10.26. Often it reflects an inappropriate charting method (see Section 10.15 for an alternative method and Chapter 9 for another approach).

229

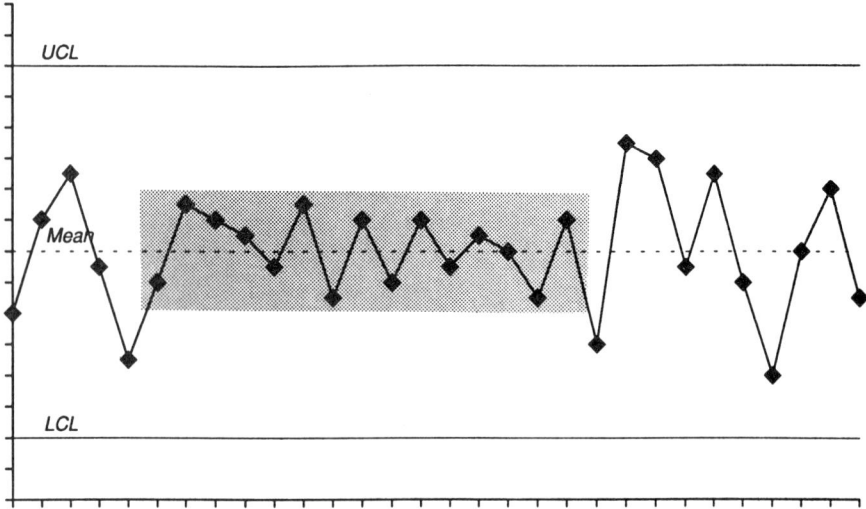

Figure 10.27 Fifteen consecutive plots inside centre third of limits

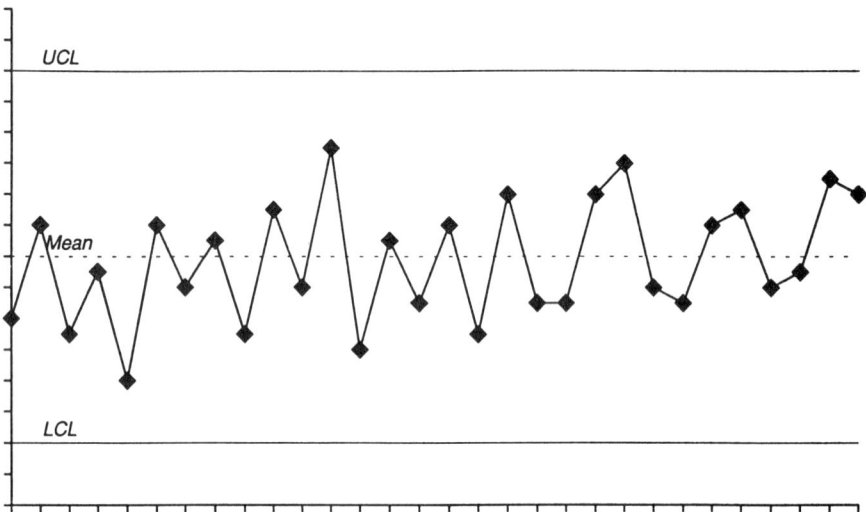

Figure 10.28 Alternating high/low plots

The pattern in Figure 10.29 reflects the effect of action to control inherently unstable processes. It illustrates several process cycles of the type illustrated in Figure 10.14.

The pattern in Figure 10.30 is a combination of the two patterns in **Figure 10.25.** It usually illustrates the long-term effects of some environmental

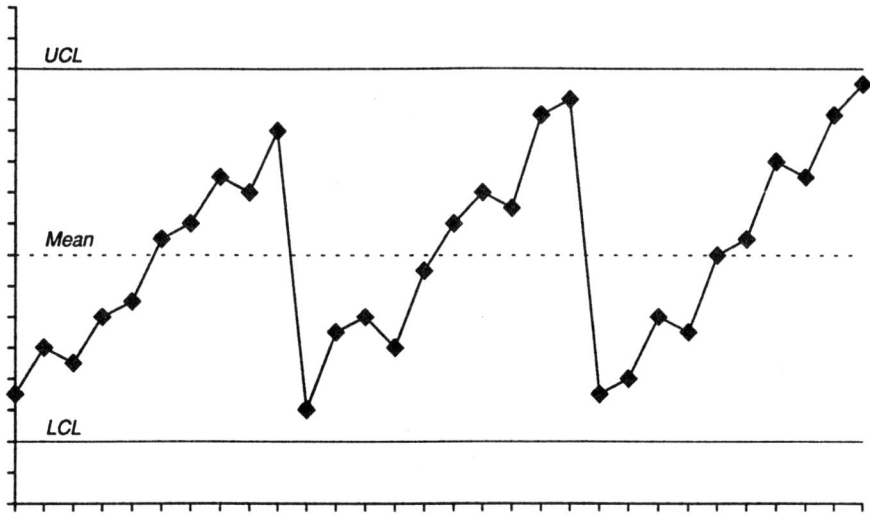

Figure 10.29 Moving means

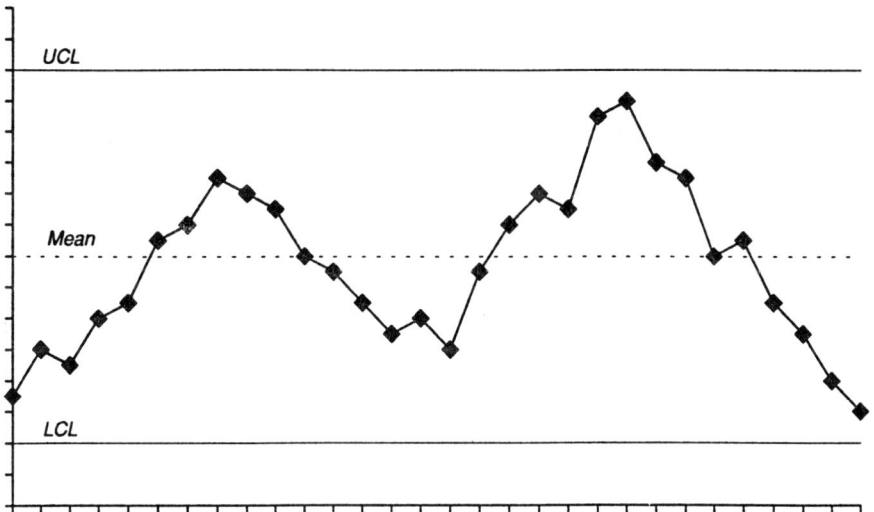

Figure 10.30 Cyclic patterns

factor such as ambient temperature, and suggests the need for compensatory action.

The various patterns described above are not intended to represent a mandatory checklist for everybody who uses a control chart. In practice, local knowledge will point to a few patterns that have day-to-day usefulness under particular process conditions.

■ 10.19 British control charts

The Shewart control charts described in this chapter are sometimes referred to as American control charts as opposed to British control charts. The latter were widely used by inspection departments in British manufacturing industry during and after the Second World War, and are still used by quality control departments in some companies.

Instead of the Shewhart convention of two control lines (one upper and one lower), there can be four lines in the British convention. These are upper and lower *action lines* (similar to Shewhart control lines) and optional upper and lower *warning lines*.

The lines for variables control charts are based on probabilities given by the normal distribution. The warning lines (*UWL* and *LWL*) are at $\mu \pm 1.96\sigma/\sqrt{n}$, the action lines (*UAL* and *LAL*) are at $\mu \pm 3.09\sigma/\sqrt{n}$, as shown in Figure 10.31.

Since for the normal distribution there would be a 1 in 40 chance of a value occurring above the *UWL*, with the same probability below the *LWL*, the warning lines are sometimes referred to as 1 in 40 lines.

Similarly, the action lines are sometimes called 1 in 1000 lines, the probability being 1 in 1000 of a value outside each of them. Note how close these action lines are to the control lines on Shewhart charts.

Attribute charts have similar lines to those used for variables charts: that is, for both types of chart, the lines are based on 1 in 40 and 1 in 1000 values outside. Attribute chart lines are calculated from binomial or Poisson distributions as appropriate (rather than from normal distributions). Constants used for calculating the lines are given in British Standards.

The rules for chart interpretation depend upon the purpose of the chart.

■ If they are being used for process control (in other words, monitoring the process setting and variability regardless of product specification), they are little different from those for Shewhart charts. The guidelines in Section

Upper action line	UAL	$= + 3.09\sigma_e$	$(1/1000)$
Upper warning line	UWL	$= + 1.96\sigma_e$	$(1/40)$
Mean	*Nominal*		
Lower warning line	LWL	$= - 1.96\sigma_e$	$(1/40)$
Lower action line	LAL	$= - 3.09\sigma_e$	$(1/1000)$

Figure 10.31 British control chart control lines for sample means

10.18 are supplemented by one other rule when warning lines are used: two consecutive plots between *UWL* and *UAL* or between *LWL* and *LAL* indicate an out-of-control situation.

■ If product quality control is the purpose (in other words, monitoring product conformance with specification), the rules are more simple: there should be an investigation either if 1 plot is above *UAL* or below *LAL* or if 2 consecutive plots are between *UWL* and *UAL* or between *LWL* and *LAL*.

In some texts the rules suggest also that the process should be stopped when these results occur. This is appropriate for processes that have a marginal capability and where product conformance with specification is controlled only by charting.

It is usual for quality control purposes to draw action and warning lines equidistant from the specification nominal rather than equidistant from the estimated process mean (which is the process control and Shewhart convention).

Conformance with a specification from upstream is sacrosanct in quality control environments; processes are often designed to achieve tolerance and no more. British charts are appropriate because they can focus on product conformance.

In process control environments, processes are designed to achieve nominal. There is some freedom for production areas to manage variability in response to downstream customers and the focus of charting is process behaviour. Although British charts can meet this purpose, the more simplistic Shewhart charts tend to find greater use because they are more readily accepted. The bulk of charting is carried out by production personnel in industries whose aim is economically to satisfy customers rather than to meet precise objective specifications.

■ 10.20 Shewhart conventions

The use of the mean ± 3 standard deviations for Shewhart chart control lines has been described as heuristic: in other words, it seems to work well in practice, even though it is not firmly rooted in probability theory. British control chart lines are based on probability theory and are described as probabilistic.

For a normal distribution (of variables), where almost all values fall within the mean ± 3 standard deviations, the justification of the Shewhart convention is that any significant change in the process mean or variability is readily identified after control line calculation. In other words, the convention does not detract from pattern interpretation.

The Shewhart argument for using these lines on attribute charts is again that, although they are simplistic, they have been found to work well in practice. Theoretically, the control lines should be calculated using the binomial distribution for p and np charts and the Poisson distribution for the c chart and in some cases for the u chart. However, the binomial distribution is approximately normal for large np and nq and the Poisson distribution is approximately normal for large λ, the fit being particularly good in the tails. Hence, control lines calculated from process mean ± 3 standard deviations are usually sufficiently close to lines calculated from the exact distributions.

However, when mean values are low numbers, it should be noted that probabilities usually associated with the charts may differ fairly substantially from the real situation. For examples:

- In c charts (where the Poisson distribution applies) with a mean of 2 per sample, 0.45% of values will occur outside the upper control line when the system is in control. This compares to 0.135% based on a normal distribution.
- If the mean is only 0.6, then 2.31% occur outside the line (again instead of 0.135%), so that the risk of a false alarm is now about 1 in 43 rather than about 1 in 740.

Those who favour probabilistic charts argue that the symmetrical control lines of Shewhart charts are illogical, in particular because they often lead to negative and therefore unusable lower control lines. In the Shewhart convention, a negative lower control line on range, standard deviation and some attribute charts is taken as zero. Once again, the simplistic justification is that this has been found to be satisfactory for practical purposes.

■ 10.21 Average run length

Average run length (ARL) is the term given to the average number of plots on a chart between occasions when a decision is required because plots are outside the control lines.

Calculation of ARL

Calculation of ARL is straightforward for all Shewhart charts and for British charts that have only action lines. The control/action lines of these charts are set, respectively, at values of $\mu \pm 3\sigma_e$ and $\pm 3.09\sigma_e$ (or z values of ± 3 and ± 3.09). From standard normal probability tables (Appendix B) the probability of a plot above an upper or below a lower line is 0.00135 and 0.001 for the respective charts. When a chart has both upper and lower lines, the probability of a plot outside either of the lines is twice the probability of a plot outside one of the lines: in other words,

P(outside control lines) $= P_c = 2 \times 0.00135 = 0.0027$
P(outside action lines) $= P_a = 2 \times 0.001$ $= 0.002$.

The ARL is the reciprocal of this probability, so that:

ARL (Shewhart chart) $= 1/P_c =$ about 370
ARL (British chart with action line only) $= 1/P_a = 500$ precisely.

When a chart has action and warning lines, ARL is calculated from the formula

$$ARL \text{ (chart with action and warning lines)} = \frac{1 + P_w - P_a}{P_a + P_w(P_w - P_a)}$$

where P_w is the probability of a plot outside the warning lines. In the case of a British chart, where action lines are set at $\mu \pm 3.09\sigma_e$ and warning lines are set at $\mu \pm 1.96\sigma_e$, P_a is as above and $P_w = 2 \times z_{1.96} = 2 \times 0.025 = 0.05$, so that:

ARL (British chart with action and warning lines)

$$= \frac{1 + 0.05 - 0.002}{0.002 + 0.05(0.05 - 0.002)} = 238.$$

Control chart sensitivity

The concept of ARL was developed originally to measure control chart sensitivity to process changes, which is especially important where small changes from optimum could be costly. It has particular application in some large-scale continuous process industries, such as pharmaceutical production and oil refining.

The above calculations of ARL assume an in-control process where the control, action or warning lines are equidistant from the process mean. Given that the lines are not recalculated, a chart that is sensitive to change in the process mean will show a significant drop in ARL for small changes in the mean. Less sensitive charts will show a smaller drop in ARL for the same change in mean.

In general, control chart sensitivity is affected by sample size (larger sample sizes usually reduce sensitivity) and operating rules (British charts used with warning lines are more sensitive than Shewhart charts).

Process management

The concept of ARL also finds use in situations where industrial engineers or resource planners need to allow time for process stoppage to investigate and make a decision on an outside lines plot. The application is not widespread because such occurrences have been found to be insignificant alongside other reasons for stoppages. However, the concept allows control lines to be set to optimize stoppages so that they suit required production rates (sometimes at the expense of quality control) or, conversely, to suit quality control (when the limits of product fitness-for-purpose are not so well defined).

For example, where stoppages could be allowed, on average, only every 1000 plots

$ARL = 1000 = 1/P_a$, $P_a = 1/1000 = 0.001 = P$(outside both action lines)

Therefore, P(outside one action line) $= 0.001/2 = 0.0005$ and from tables $z = 3.2906$, hence action lines are set at $\mu \pm 3.3\sigma_e$.

Alternatively, where stoppages were thought to be necessary, on average, every 250 plots

$ARL = 250 = 1/P_a$, $P_a = 1/250 = 0.004 = P$(outside both action lines)

Therefore, P(outside one action line) $= 0.004/2 = 0.002$ and from tables $z = 2.88$, hence action lines are set at $\mu \pm 2.9\sigma_e$.

■ References to further reading on the topics of Chapter 10

Oakland (1986) and Walpole and Myers (1993) give brief accounts of the main types of control chart, including some not described in this book. Grant and Leavenworth (1988) and Montgomery (1985) give more technical details.
See Bibliography for details of titles and publishers.

■ Self-assessment questions on Chapter 10

1. The thickness of metal sheets cut from a single rolling (specified as 0.30 ± 0.01 millimetres) was measured for 25 samples. Each sample was measured 5 times as follows:

Sample

1	2	3	4	5	6	7	8	9	10	11	12	13
0.301	0.296	0.306	0.302	0.300	0.296	0.303	0.298	0.301	0.296	0.296	0.303	0.300
0.306	0.295	0.304	0.296	0.299	0.298	0.305	0.299	0.303	0.297	0.295	0.301	0.293
0.302	0.300	0.298	0.298	0.296	0.292	0.299	0.299	0.304	0.300	0.301	0.306	0.295
0.297	0.299	0.305	0.308	0.295	0.301	0.301	0.300	0.298	0.303	0.299	0.299	0.297
0.299	0.301	0.299	0.304	0.302	0.297	0.307	0.303	0.301	0.293	0.302	0.301	0.298

Sample

14	15	16	17	18	19	20	21	22	23	24	25
0.297	0.305	0.301	0.302	0.297	0.304	0.297	0.308	0.300	0.299	0.301	0.306
0.301	0.302	0.295	0.300	0.302	0.301	0.298	0.302	0.293	0.295	0.294	0.305
0.306	0.295	0.295	0.306	0.305	0.302	0.302	0.307	0.297	0.304	0.299	0.298
0.304	0.297	0.297	0.299	0.306	0.299	0.291	0.298	0.303	0.301	0.298	0.299
0.298	0.306	0.303	0.302	0.299	0.297	0.297	0.304	0.298	0.300	0.297	0.304

(a) Plot a histogram for the 125 values and comment on the capability of the rolling process.

(b) Calculate control lines for mean and range charts. Draw the charts, plot the sample means and ranges, and say whether or not the process appears to be in control.

2. Construct a median control chart for the data in question 1. Plot the data and sample medians on the chart and comment.

3. Nominal 500 ml containers of lemonade are filled automatically. The actual amount of liquid in 25 samples (each of 6 containers) was measured and recorded. The resulting sample means and ranges were as follows, where values are in millilitres:

Sample	1	2	3	4	5	6	7	8	9	10	11	12	13
Mean	505	510	505	502	500	502	509	504	507	509	504	498	503
Range	11	18	4	16	8	12	13	24	9	17	4	9	10

Sample	14	15	16	17	18	19	20	21	22	23	24	25
Mean	506	504	500	499	503	505	505	502	503	504	502	503
Range	19	12	18	9	11	7	13	11	9	10	15	12

Construct mean and range control charts for this data, and plot the data on the charts. Comment on the state of control of the filling process.

4. The measurements described in question 3 were followed by measurements of a further 20 samples. The results were as follows:

Sample	26	27	28	29	30	31	32	33	34	35
Mean	507	502	506	499	501	503	501	503	505	502
Range	14	20	8	10	11	3	5	7	2	5

Sample	36	37	38	39	40	41	42	43	44	45
Mean	502	504	505	503	503	504	506	502	501	505
Range	6	4	8	2	4	5	2	1	3	4

(a) Plot these results on the same chart.

(b) Attempts were being made to improve the dispensing system. State whether you think the attempts were successful and if so where and in what way. Recalculate the control chart lines from the point where you think the process has settled down.

5. All new tyres are tested before dispatch, and only those passing stringent tests are dispatched. The test results are also used to provide an hourly monitor of the manufacturing process. The number of tyres tested in each hour varies slightly, and the following are results for 20 consecutive hours of production:

Hour	1	2	3	4	5	6	7	8	9	10
Volume	110	105	95	103	100	102	102	99	106	98
Rejects	4	7	6	6	9	10	8	7	7	8

Hour	11	12	13	14	15	16	17	18	19	20
Volume	107	103	101	98	107	106	104	109	103	109
Rejects	11	3	8	5	9	4	7	4	3	6

Use this data to calculate control lines for a p chart. Construct the chart, plot the data on the chart, and say whether or not you think the manufacturing process is in control.

6. Tins of food delivered to a supermarket are checked for damage occurring in transit. In 25 samples (each of 80 tins) the number of damaged tins were

4, 7, 5, 12, 3, 5, 4, 20, 6, 4, 3, 5, 3, 10, 3, 8, 22, 25, 6, 9, 5, 2, 8, 3, 6

Calculate control lines for an np chart, construct the chart and plot the data. State whether you think the transit process is in control, so far as damage is concerned.

7. The number of people injured as a result of accidents in a particular factory area was recorded for a 30-month period as follows:

Month	1	2	3	4	5	6	7	8	9	10	11	12	13	14	15
Number	9	8	15	11	6	9	12	8	4	8	10	7	8	5	4

Month	16	17	18	19	20	21	22	23	24	25	26	27	28	29	30
Number	8	6	9	3	10	7	6	2	5	8	4	6	3	5	3

Construct a c chart for this data, plot the data on the chart, and comment.

8. As part of continuous market surveys, new car buyers are asked to list any faults they find with their purchase. During the first 20 weeks that a new model was on sale, the total number of buyers and the total number of faults found were as follows:

Week	1	2	3	4	5	6	7	8	9	10
Buyers	300	321	317	298	331	345	317	328	317	309
Faults	189	197	176	183	207	193	215	208	196	184

Week	11	12	13	14	15	16	17	18	19	20
Buyers	341	327	324	318	332	317	333	321	328	339
Faults	201	192	199	186	197	189	184	176	192	173

For each week, calculate the number of complaints per buyer and construct a *u* chart. Plot the points on the chart and comment.

9. The number of videos rented out per week over a period of 25 weeks was as follows:

Week	1	2	3	4	5	6	7	8	9	10	11	12	13
Videos	243	263	213	242	207	203	189	215	163	268	203	155	164

Week	14	15	16	17	18	19	20	21	22	23	24	25
Videos	206	225	249	212	212	237	233	294	205	239	275	251

Use this data to construct a control chart for sample size 1, plot the values on the chart and comment.

11 | Significance Testing

■ 11.1 Introduction

Previous chapters have explained and illustrated applications of theories that enable statistics to be derived from data. This chapter outlines methods which compare and allow conclusions to be drawn from different sets of data and their statistics.

Within industry, questions often arise concerning the difference between sets of numbers. For example:

■ Is there a difference between the setting of one machine and that of another?

■ Is there a difference between the lifetimes of two (or more) components?

■ What is the effect of some particular action? The action might have been a process change intended to reduce variability. The question asked would be: 'is there a difference between the situation now and the situation before the change was made?'

One or other of the techniques known as *significance tests* (sometimes called *hypothesis tests* and summarized in Table 11.1) can usually provide an answer to the above type of question.

■ 11.2 Basic principles

The general approach in significance testing is described below and is outlined in Figure 11.1. The description introduces the terminology and basic principles of the technique.

Table 11.1 Summary of significance tests

Test	H_0	Formula	Degrees of freedom
Attributes	Proportions are as stated	$\chi^2 = \sum \dfrac{(O-E)^2}{E}$	classes -1
Contingency table	Independence of rows and columns	$\chi^2 = \sum \dfrac{(O-E)^2}{E}$	(rows -1) (multiplied by columns -1)
Goodness of fit to Poisson	Data from Poisson distribution	$\chi^2 = \sum \dfrac{(O-E)^2}{E}$	classes -2
Goodness of fit to normal	Data from normal distribution	$\chi^2 = \sum \dfrac{(O-E)^2}{E}$	classes -3
Variation 2 samples	$\sigma_x^2 = \sigma_y^2$	$F = \dfrac{s_x^2}{s_y^2}$ if $s_x^2 > s_y^2$	$\nu_1 = n_x - 1$ $\nu_2 = n_y - 1$
Means 1 sample	mean $= \mu$	$t = \dfrac{\bar{x} - \mu}{s/\sqrt{n}}$	$n - 1$
Means 2 samples paired	$\mu_d = 0$ or $\mu_x = \mu_y$	$t = \dfrac{\bar{d} - 0}{s/\sqrt{n}}$	$n - 1$
Means 2 samples not paired $\sigma_x^2 = \sigma_y^2$ or $n \simeq m$	$\mu_x = \mu_y$	$t = \dfrac{\bar{x} - \bar{y}}{\sqrt{\left(\dfrac{(n-1)s_x^2 + (m-1)s_y^2}{n+m-2}\right)\left(\dfrac{1}{n} + \dfrac{1}{m}\right)}}$	$n + m - 2$
Means 2 samples not paired $\sigma_x^2 \neq \sigma_y^2$ and $n \neq m$	$\mu_x = \mu_y$	$t = \dfrac{\bar{x} - \bar{y}}{\sqrt{(s_x^2/n) + (s_y^2/m)}}$	$\dfrac{(s_x^2/n + s_y^2/m)^2}{\dfrac{(s_x^2/n)^2}{(n-1)} + \dfrac{(s_y^2/m)^2}{(m-1)}}$
Kolmogorov–Smirnov small sample goodness of fit	Data from a given distribution	$D = \max \lvert S_n(x) - F(x) \rvert$	

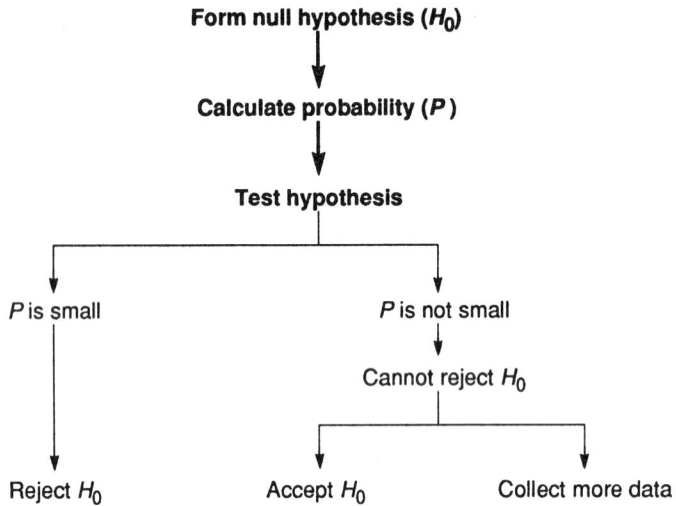

Figure 11.1 Summary of significance testing

Step 1: Form null hypothesis

The first step in a significance test is to make a statement which represents an assumption concerning the matter being considered. Referring to the examples above, such statements might be expressed as follows:

'No difference exists in the setting of the two machines.'

'No difference exists between the lifetimes of two (or more) components.'

'The action to reduce variability has not made any difference.'

Notice that all of the above statements contain the idea of 'no difference'. Problems involving this idea led to the original theory of significance tests; the required statement is referred to as the *null hypothesis* and is usually designated H_0. However, the theory now extends to cases where the null hypothesis may be that there is some specified non-zero difference. For example:

'Make A cars average 5 miles per gallon more than make B cars.'

If H_0 is true, a particular set of results is to be expected. In other words, based on H_0 there exist the *expected results*. Also, if H_0 is true, then any difference between these expected results and the actual or *observed results* is insignificant and arises purely by chance.

Step 2: Probability calculation

Once the null hypothesis has been made, the next step is to calculate the probability that the observed result, or results more unlikely than the observed result, will occur. This probability is designated P.

The probability calculation is carried out by evaluating some appropriate *test statistic* for which the probability distribution can be found in standard tables.

Step 3: Test the null hypothesis

The method of testing the null hypothesis amounts to attempts to reject it.

■ If the calculated probability (that the observed or more extreme results will occur) is small, it indicates that the observed results are unlikely to occur if H_0 is true. Therefore, H_0 is rejected because it is likely to be wrong.
■ If the calculated probability is not small, there is insufficient evidence to reject H_0. Note that this does not mean that H_0 is correct, merely that it cannot be rejected. It might be that the sample is too small to show up any difference between the observed and expected results. In practical situations, this gives the options either to assume that H_0 is correct or to collect more data.

The key issue, in testing the null hypothesis, is what constitutes a small probability? In other words, what is the value of the calculated probability which can be regarded as small enough to warrant rejection of H_0? This value is known as the *significance level*.

The significance level to be used should be decided before the data is analysed, otherwise there is a danger that the analyst might be influenced by the results. Much depends upon the problem under scrutiny.

> For example, if it is important to make sure a new drug does not have the damaging side-effects of an existing drug, then a small value of P is required. The influencing factors are that it might cost more and that it might tie up more scarce experts than for a larger P.

The value most often used is the 5% significance level, which is written as $P = 5\%$. At the 5% level, there is a 1 in 20 (or 0.05) chance of wrongly rejecting H_0. Alternatively, the significance level represents the risk that is taken in order to reach a conclusion.

Although 5% is the most frequently used significance level, it is possible to use any value: 1% and 0.1% are other levels in common use. The 1% and 0.1% significance levels give risks of wrongly rejecting H_0 of 1 in 100 and 1 in 1000 respectively.

In most practical situations the 5% level is adequate. For example:

■ Where research is being carried out and indications will be followed up if encouraging.
■ Where a system is being checked to see if there have been any changes.

243

■ Where, as is often the case, the statistical analysis is to be used to support or refute some suspicion based on existing evidence.

The 5% level will be used as the required significance level for examples in this chapter. In considering calculated probabilities (P), the following commonly used terms will be applied as defined below for working in the context of a 5% significance level.

Not significant	$P > 5\%$
Significant	$P < 5\%$
Highly significant	$P < 1\%$
Very highly significant	$P < 0.1\%$

■ 11.3 Chi-squared tests (χ^2 tests)

The three steps in significance testing are set out below using chi-squared for the calculation in step 2. Chi-squared is the appropriate statistic when the data under consideration is in the form of frequencies: in other words, counts of events or numbers of occurrences.

The test involves calculating

$$\chi^2 = \sum \frac{(O_i - E_i)^2}{E_i}$$

where O_i is the observed frequency in the ith class or group of results and
E_i is the expected frequency (based on H_0) for the same class.

Any discrepancy between O_i and E_i will tend to increase χ^2, so a large value of χ^2 indicates a substantial difference between what was observed and what should have been expected if H_0 were true. The probability of χ^2 being at least as big as the calculated value is found from chi-squared tables (Appendix G).

> For example, a machine has been producing 5% defective components over an extended period of time. Attempts have been made to improve the process and reduce the number of defectives. In order to check whether there has been an improvement, 200 components were run off and tested, 6 of them (only 3%) were found to be defective. The question arises as to whether this indicates that there has been an improvement.

Step 1: Form null hypothesis

For the example, the null hypothesis (H_0) can be expressed as the assumption that there has been no improvement: in other words, the machine is still producing 5% defectives overall.

Step 2: Probability calculations

There are three stages in the calculations of the probability.

1. Calculate χ^2

The data and calculations are laid out in the form of a table as shown below. The counted numbers are arranged in categories (observed and expected) and show the results of each possible class (defective and satisfactory). In the table O is the observed frequency and E is the expected frequency based on H_0. In this example, the expected number of defectives is 5% of 200, which is 10.

Class	O	E	$O - E$	$(O - E)^2$	$(O - E)^2/E$
Defective	6	10	-4	16	1.60
Satisfactory	194	190	4	16	0.08
	200	200			1.68

The sum of the final column (1.68) is χ^2.

2. Find the number of degrees of freedom (ν)

The value of χ^2 depends upon two factors: first, the difference between observed and expected frequencies (the larger the difference, the larger will be the contribution to χ^2 and the smaller will be P); second, the number of classes (the more classes there are, the larger χ^2 is likely to be).

Tables of χ^2 (from which probabilities are determined) need to reflect the number of classes. For convenience in the tables, classes are taken into account by the quantity called *degrees of freedom*, which was introduced in Section 7.5 in connection with the t distribution. For the χ^2 test, it is defined as follows: *the number of degrees of freedom (ν) is the number of classes with frequencies that can be assigned independently of the frequencies of other classes.* Alternatively,

ν = number of classes − number of restrictions imposed.

In the example above, only one class can be assigned independently (the value of the second class is determined by 200 minus the frequency assigned to the first), giving $\nu = 1$. Alternatively, ν = number of classes − 1 because there is just one restriction on the data, namely that the frequencies must sum to 200. Thus $\nu = 2 - 1 = 1$.

The one restriction, that frequencies sum to a fixed total, always applies to classes and categories in cases similar to this. Hence a general expression for the situation when there are a number of categories:

ν = number of classes − 1

Estimate the probability (P)

The required probability is read off chi-squared tables (Appendix G) for the calculated values of χ^2 and ν.

For the values of $\chi^2 = 1.68$ and $\nu = 1$, the tables give $25\% > P > 10\%$. In other words, the probability of the observed results (or results that differ even more from those which would be expected) occurring by chance lies between 25% and 10% if H_0 is true.

Step 3: Test the null hypothesis

It has been determined already for this chapter that 5% is to be used as the value on which to base the conclusions. In other words, the tables reading ($25\% > P > 10\%$) should be expressed in terms of the 5% level, i.e. $P > 5\%$. In line with the definitions, this result is *not significant*. Therefore H_0 cannot be rejected, which is to say that there is no (statistical) evidence of an improvement.

This result may well be due to the small number of defectives occurring. A larger sample may be required to show an improvement. For example:

Suppose that in the last example a sample of 600 components had been tested but still 3% defectives were found. The χ^2 test would be as follows.

H_0: assume no improvement.

Class	O	E	O − E	$(O-E)^2$	$(O-E)^2/E$
Defective	18	30	−12	144	4.80
Satisfactory	582	570	12	144	0.25
	600	600			5.05 = χ^2

Here $\nu = 2 - 1 = 1$ and hence from tables, $2.5\% > P > 1\%$, so $P < 5\%$, which is *significant* and therefore H_0 can be rejected.

In other words, there is fairly strong evidence of an improvement.

Often there are more than two classes, as in the following example.

The number of vehicles produced in different periods of a shift were as follows:

Hours of shift	0–2	2–4	4–6	6–8
Number of vehicles	125	108	87	80

Do the figures indicate a significant production rate variation during the shift?

H_0: assume no difference between periods (the production rate is constant during the shift)

Class	O	E	$O-E$	$(O-E)^2$	$(O-E)^2/E$
0–2	125	100	25	625	6.25
2–4	108	100	8	64	0.64
4–6	87	100	–13	169	1.69
6–8	80	100	– 20	400	4.00
	400	400			12.58 $= \chi^2$

Here $\nu = 4 - 1 = 3$, and hence from tables $0.1\% < P < 1.0\%$, so $P < 1\%$, which is *highly significant* and therefore H_0 can be rejected.

In other words, there is strong evidence that there is a difference over the shift and a look at the data verifies that there is a slowing down as the shift progresses.

■ 11.4 Contingency tables (significance of tabular data)

In the language of statistics, the simple table of data below is called a *contingency table*. It can have many more rows and columns than the example.

A contingency table can be analysed, using the χ^2 test, to check the independence of its rows and columns. The example is of the numbers of incomplete assemblies on day shifts and night shifts at a factory.

	Day	Night
Incomplete	20	12
Complete	190	178

It looks as though the percentage of incomplete assemblies is smaller on the night shift (6.3%) than on the day shift (9.5%). Before looking for reasons for this, the production team wants to make sure that the figures indicate that there is a real difference between shifts. It may be that the apparent difference could easily occur by chance and that there is no real difference. The statistical analysis is as follows.

H_0: assume no real difference between shifts.

Sum the rows and columns:

	Day	Night	Total
Incomplete	20	12	32
Complete	190	178	368
Total	210	190	400

Calculate the expected frequency within each class, which is

$$\frac{\text{row total} \times \text{column total}}{\text{grand total}}$$

Thus for 'incomplete on day shift' the expected frequency

$$= \frac{32 \times 210}{400} = 16.8$$

Similar calculations for the other classes give the following table of expected frequencies. The calculations can be checked, since totals should be the same as in the original table.

	Day	Night	Total
Incomplete	16.8	15.2	32
Complete	193.2	174.8	368
Total	210	190	400

Observed and expected frequencies are now tested as in the previous section.

Class	O	E	$O - E$	$(O - E)^2$	$(O - E)^2/E$
Day incomplete	20	16.8	3.2	10.24	0.61
Night incomplete	12	15.2	-3.2	10.24	0.67
Day complete	190	193.2	-3.2	10.24	0.05
Night complete	178	174.8	3.2	10.24	0.06
					1.39 $= \chi^2$

Here $\nu =$ (number of rows $-$ 1) $\times$ (number of columns $-$ 1)
$\qquad = (2 - 1) \times (2 - 1) = 1$
From tables: $10\% < P < 25\%$, thus $P > 5\%$ which is *not significant*. H_0 cannot be rejected: in other words, there is no evidence of a difference between shifts.

Large contingency tables

Contingency tables with any number of rows and columns are analysed in a similar way.

Null hypothesis for contingency tables

The null hypothesis in general is to assume that rows and columns are independent.

In the example, the implication is that the occurrence of incomplete and complete assemblies is independent of day and night shifts. Also, the expected values tie in with H_0 (the proportion of day shift incomplete assemblies is $16.8/210 = 0.08$, and for the night shift, the proportion is $15.2/190 = 0.08$).

Expected class frequencies in contingency tables

For any class in a contingency table,

$$\text{expected frequency} = \frac{\text{row total} \times \text{column total}}{\text{grand total}}.$$

Degrees of freedom in contingency tables

$\nu = (\text{number of rows} - 1) \times (\text{number of columns} - 1).$

The reason for this statement is explained below with reference to Table 11.2, which contains 3 rows and 4 columns.

In this type of analysis, the totals are fixed. In Table 11.2, the classes containing an X can take any value, but the values taken by the remaining empty cells are fixed because they have to satisfy the totals and therefore are not independent. Similarly, for any table when one row and one column are crossed out, the number of cells remaining is the number of degrees of freedom, giving the rule set out above.

Calculation precision

Calculate expected values to one decimal place and χ^2 values to one or two decimal places.

Limitation on expected frequency

The theory behind the χ^2 test is based on an approximation which assumes that expected frequencies are not small. There is some debate about how small the expected frequencies can be for the test still to be valid, but it is usually understood as meaning that every class should have an expected frequency of at least 5. It is generally accepted that the lower limit of 5 is a conservative value. If there are likely to be smaller classes, convenient rows and columns should be grouped together before starting the analysis. Note that the limit of 5 refers to *expected* frequencies only. This matter is discussed further in Section 11.5.

Table 11.2   Degrees of freedom in a contingency table

	Column 1	Column 2	Column 3	Column 4	Row totals
Row 1	X	X	X		T_{R1}
Row 2	X	X	X		T_{R2}
Row 3					T_{R3}
Column totals	T_{C1}	T_{C2}	T_{C3}	T_{C4}	

For example, in the following table which lists the frequency of car replacement by car size, there will be small expected frequencies in the first two columns and the bottom row.

Size of car (cc)		1000	1500	2000	2500
	1 year	3	3	9	16
Time	2 years	2	5	7	11
before	3 years	7	2	5	6
replacement	4 years	9	3	2	1

Therefore group together columns 1 and 2, also rows 3 and 4, giving

	1000/1500	2000	2500
1 year	6	9	16
2 years	7	7	11
3/4 years	21	7	7

This grouped table can be analysed using χ^2.

■ 11.5 Goodness-of-fit tests (identification of distributions)

Sometimes it is necessary to know if data has arisen from a particular distribution: for example, to confirm assumptions made for purposes of other calculations (in this book there are several instances where data is assumed to have come from a normal distribution). The *goodness-of-fit tests* described below can be used to confirm any distribution.

In earlier chapters it has been pointed out that a sample taken from a distribution (or population) often appears to differ quite markedly from that population, particularly if it is a small sample. The differences gradually disappear as the sample size increases, although it may require very large samples for these differences to become negligible.

For some samples it may be necessary to see whether any differences are significant. The use of χ^2 is a particularly useful and powerful test if there is a fairly large sample. This book describes the following other methods:

■ Draw a histogram. This quick and easy method may clearly verify the underlying distribution, particularly if it is a normal distribution. However, it often requires a large amount of data (about 100 values or more).

■ Use probability paper. This is another simple method and can be used for quite small samples (about 20 values or more). With appropriate papers (linear, logarithmic, binomial, Weibull, etc.) it can be applied usefully to most distributions, but it is particularly helpful with normal distributions.

■ Kolmogorov–Smirnov test. This is useful for small samples.

The following example illustrates the χ^2 test. It uses data on the number of faults found on 150 cars at an electrical test station. The question arises as to whether the distribution of faults is random or not. If it is random, it is probably typical of the process. If it is not random, there might be some particular cause that only affected this batch of cars. A Poisson distribution is to be expected if the faults occur at random.

Number of faults found	(x)	0	1	2	3	4	5
Number of cars affected	(f)	35	45	40	23	4	3

H_0: assume the data arises from a Poisson distribution.

To find expected frequencies the mean (λ) must be estimated. The sample mean is used as the estimate. From the table below,

$$\lambda = \bar{x} = \frac{\Sigma fx}{\Sigma f} = \frac{225}{150} = 1.5.$$

Expected frequencies (E) are set out in the table below. They are given by $(\Sigma f) \times f(x) = 150 \times f(x)$ and have been calculated using the Poisson probability function (Section 6.4). For example:

$$f(0) = e^{-1.5} = 0.223 \quad \text{and} \quad 150 \times f(0) = 33.5$$
$$f(1) = 1.5 \times f(0) = 0.335 \quad \text{and} \quad 150 \times f(1) = 50.2$$

The value of χ^2 is calculated, with the rule for small frequencies applying.

x	$f(=O)$	fx	E	$O-E$	$(O-E)^2$	$(O-E)^2/E$
0	35	0	33.5	1.5	2.25	0.07
1	45	45	50.2	-5.2	27.04	0.54
2	40	80	37.7	2.3	5.29	0.14
3	23	69	18.8	4.2	17.64	0.94
4	4	16	7.1			
5	3	15	2.1	9.8 -2.8	7.84	0.80
6	0	0	0.5			
7	0	0	0.1			
Total	150	225	150.0		2.45	$= \chi^2$

In this example, $\nu =$ number of classes after grouping -2. This has been derived using the second definition of ν (number of classes $-$ number of restrictions imposed). There are two restrictions, namely that observed frequencies must sum to 150 and that they must combine with the x values to give the observed sample mean ($\bar{x}$).

Thus $\chi^2 = 2.49$ and $\nu = 5 - 2 = 3$.

From tables $25\% < P < 50\%$, so $P > 5\%$, which is *not significant* and therefore H_0 cannot be rejected. In other words, the data is a *good fit* to a Poisson distribution.

In practical terms, the analysis indicates that faults occurred at random and there is unlikely to be any special cause for some cars having 4 or 5 faults; they are to be expected as part of the general distribution.

Expected frequency in the tails of distribution

Although the greatest number of faults on a car in the example was 5, x was taken to 7 in the analysis (with zero frequencies at 6 and 7). The reason for this is that the expected frequencies should be calculated until they are negligible, as in theory the distribution goes on to infinity. Summing expected values is a check on the calculations, since the sum should be about the same as that for the observed frequencies. It is sometimes marginally less due to rounding errors and the fact that the rest of the tail has been ignored.

Analysis for other distributions

Similar analyses can be carried out for other distributions. The following example is an analysis for a normal distribution.

> Data of a sample of 100 items was taken and grouped into a frequency table. The sample mean ($\bar{x}$) is 4.70 and the sample standard deviation (s) is 1.81. The analysis is to determine whether or not the data can be assumed to be a random sample from a normal distribution.
>
> H_0: assume the distribution is normal.
>   Take $\mu = 4.70$ and $\sigma = 1.81$, then use $z = (x - \mu)/\sigma$ to determine the proportion of the normal distribution lying below x_i for $x_i = 1, 2, \ldots 10$ and hence the proportions between x_i and $x_{(i-1)}$ as shown in Table 11.3.
>   Multiplying by the total frequency of 100 gives expected frequencies which are then compared to observed frequencies using the χ^2 test.

Table 11.3   Goodness-of-fit test of normal distribution

Upper group limit x_i	Z $\dfrac{x_i - \mu}{\sigma}$	Probability of value below x_i $P(x < x_i)$	Probability of value between x_i and x_{i-1} $P(x_i < x < x_{i-1})$	Expected value E	Observed value O	$O - E$	$\dfrac{(O-E)^2}{E}$
1	-2.04	0.0207	0.0207	2.07 ⎫ 6.81	1 ⎫ 7	0.19	0.01
2	-1.49	0.0681	0.0474	4.74 ⎭	6 ⎭		
3	-0.94	0.1736	0.1055	10.55	10	0.55	0.03
4	-0.39	0.3483	0.1747	17.47	19	1.53	0.13
5	0.17	0.5675	0.2192	21.92	22	0.08	0.00
6	0.72	0.7642	0.1967	19.67	21	1.33	0.09
7	1.27	0.8980	0.1338	13.38	9	4.38	1.43
8	1.82	0.9656	0.0676	6.76 ⎫	7 ⎫		
9	2.38	0.9913	0.0257	2.57 ⎬ 10.20	4 ⎬ 12	1.80	0.32
10	2.93	0.9983	0.0070	0.70	1		
>10			0.0017	0.17 ⎭	0 ⎭		
Totals				100	100		$\chi^2 = 2.01$

The value of $\chi^2 = 2.01$ and $\nu = $ number of classes $- 3$ since three degrees of freedom are lost in imposing three restrictions on the data (mean, standard deviation and total). Thus $\nu = 4$. From tables $75\% > P > 50\%$, so $P > 5\%$, which is *not significant*. In other words, the data is a good fit to a normal distribution.

Limitation on expected frequency

It must be remembered that the theory used to calculate the values in χ^2 tables assumes that expected values are not small. Since tests for most distributions involve small frequencies in the tails of the distribution, classes are grouped together (as in the example) until expected frequencies are 5 or more. It is, of course, necessary to group the corresponding observed frequencies before calculating χ^2. Note that in the example the grouping is done after calculating expected values.

■ 11.6 *F*-tests for comparison of variability

The ultimate objective of statistical process control and therefore of the techniques described in this book is to reduce variability in products or processes. The *F*-test is one way of assessing whether or not this has been achieved. Since variability is most reliably measured by variances (or by standard deviations), the test is based on variances.

Suppose that the distribution of a variable x has variance σ_x^2 and the distribution of a similar variable y has variance σ_y^2. For example, as illustrated in Figure 11.2, the x values may be the weights of packets before attempts are made to reduce their variability, the y values occur after a number of modifications have been made to the process that produces the packets. The true situation regarding the variability of the distributions may be as in the upper part or as in the lower part of Figure 11.2. The test attempts to decide between the two.

The weights of six packets before process changes (x values) were

310, 300, 308, 317, 320 and 305 g.

The weights of five packets after process changes (y values) were

323, 325, 324, 327 and 329 g.

Whether or not the later packets are heavier does not matter, since the average weight could be adjusted easily. The requirement is to test to see whether there has been an improvement in variability. The null hypothesis as usual will be to assume no real difference, i.e.

H_0: assume no difference in variability (or written mathematically, $\sigma_x^2 = \sigma_y^2$).

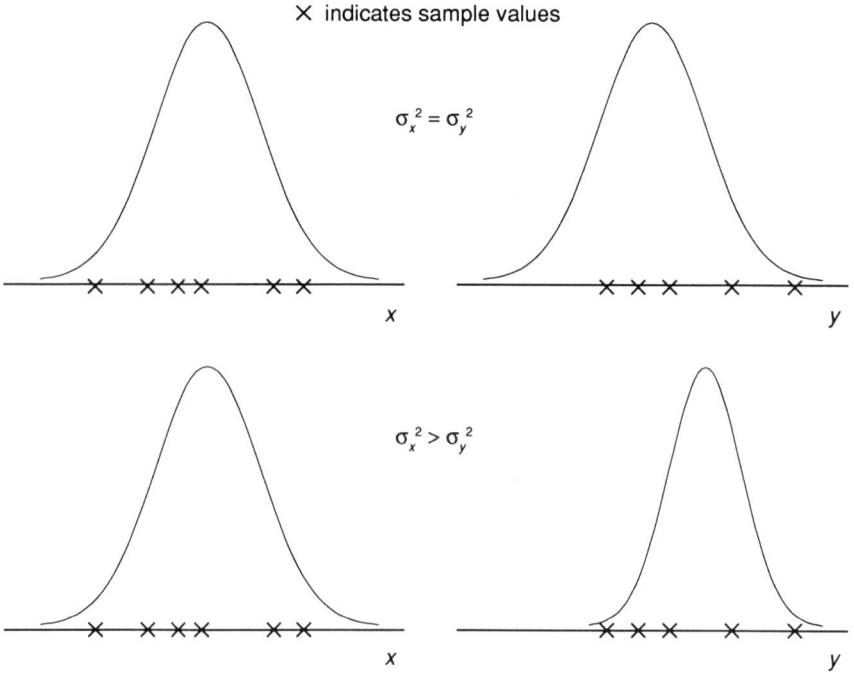

Figure 11.2 Variability of populations

Since only sample values are known, the population variance is estimated as the sample variance: that is, s_x^2 estimates σ_x^2 and s_y^2 estimates σ_y^2. Next, the quantity F is calculated, which is the result of dividing the larger of s_x^2 and s_y^2 by the smaller of the two, so that $F > 1$. Expressed mathematically, that is

$$F = \frac{s_x^2}{s_y^2} \text{ if } s_x^2 > s_y^2 \text{ or } F = \frac{s_y^2}{s_x^2} \text{ if } s_y^2 > s_x^2.$$

Calculating from x and y values, $s_x^2 = (7.46)^2$ and $s_y^2 = (2.41)^2$. Therefore $F = s_x^2/s_y^2 = 9.6$.

Tables of F are given in Appendix H. There are degrees of freedom for each of the two samples equal to the size of the sample minus one. If there are n values in a sample, there would appear to be n degrees of freedom; but in fact one is lost in the calculation of s^2, giving $n - 1$. The degrees of freedom for the samples are

$\nu_1 = $ (size of sample with larger s^2) $- 1$
$\nu_2 = $ (size of sample with smaller s^2) $- 1$

Hence, $\nu_1 = 6 - 1 = 5$ and $\nu_2 = 5 - 1 = 4$.

From tables $P \approx 2.5\%$, so $P < 5\%$, which is *significant* and therefore H_0 can be rejected. In other words, the figures indicate an improvement.

Note that, once the probability of F is obtained, the analysis is similar to χ^2 tests.

Alternative hypotheses

Strictly speaking, the procedure for any significance test should include specifying a null hypothesis and an *alternative hypothesis*, the alternative hypothesis being accepted if the null hypothesis is rejected. In the last example, where H_0 $(\sigma_x^2 = \sigma_y^2)$ was rejected, an alternative H_A $(\sigma_x^2 > \sigma_y^2)$ could have been accepted.

One- and two-sided tests

The last example focused on whether or not there had been an improvement in variability. In cases such as this, where there is interest only in a one-way change, the test is called *one-sided* or *one-tailed* and the tables are used as they stand.

Sometimes it is only possible to get a one-way change. However, there are occasions when the interest is in a change that can be either way. In such cases, the test is called *two-sided* or *two-tailed* and with H_0 $(\sigma_x^2 = \sigma_y^2)$, the alternative would be H_A $(\sigma_x^2 \neq \sigma_y^2)$. For two-sided tests the probabilities in the table are doubled: in other words, the 10% level of significance becomes the 20% level, the 5% becomes 10% and so on.

Because of the danger of being influenced by the data itself, it is important to decide before data is examined not only the significance level to be used, but also whether to carry out a one-sided or a two-sided test.

Often the decision is not obvious, and for this reason two-sided tests are usually performed because they err on the side of caution. In particular, when new medical techniques are tested, a two-sided test is usually selected. For example:

A method in common use (x), to find the density of a liquid, gives results which have a standard deviation (based on a very large number of values) of $s_x = 0.04$ units. Data from six independent evaluations of a new method (y) have $s_y = 0.03$ units. It would be an important advance if real difference in variability between the two methods could be confirmed.

H_0: assume no difference in variability: that is, $\sigma_x^2 = \sigma_y^2$.
H_A: $\sigma_x^2 \neq \sigma_y^2$ for a two-sided test.
Then $s_x^2 = (0.04)^2 = 0.0016$ and $s_y^2 = (0.03)^2 = 0.0009$
So that $F = s_x^2/s_y^2 = 0.0016/0.0009 = 1.78$.

Because of the large data base, the variance of the commonly used method is known almost exactly. This is indicated by taking its degrees of freedom as $\nu_x = \infty$ and $\nu_y = 6 - 1 = 5$.

The following are abstracted from the group of values in Appendix H (a table of one-sided values) given for ν_1 $(= \nu_x) = \infty$ and ν_2 $(= \nu_y) = 5$.

F	P(1-sided test)	P(2-sided test)
3.10	10%	20%
4.36	5%	10%
9.02	1%	2%

The calculated F (= 1.78) is less than the tabulated F of 3.10 for which $P > 20\%$ and therefore $P > 5\%$ which is *not significant* and H_0 cannot be rejected. In other words, there is no evidence of a difference in variability between the commonly used and the new methods. However, more data might reveal a difference.

Assumption of normality for the *F*-test

The theory assumes that the samples are taken from normal distributions. If they are almost normal then the table will be approximately true. If they are non-normal then what are called non-parametric tests should be used.

Assumption of random sample selection

The theory of all significance tests in this chapter assumes that samples are taken at random. Samples should be random, but sometimes that is almost impossible.

The difficulty in obtaining a good representative sample is one reason why opinion polls vary so much. It was recognized years ago that the same difficulty affected inspectors choosing components: they were more likely to unconsciously choose defective ones than good ones, unless the choice was carefully controlled.

■ 11.7 *t*-tests for comparison of means

Frequently, questions are asked as to whether there is a difference between two quantities, whether one make of engine performs more efficiently than another, whether one make of tyre lasts longer than another, whether one machine is producing components with different dimensions from another, and so on.

These questions can be answered by using a *t*-test to compare means of observations which (as for the *F*-test) are measurements, not frequencies as in the χ^2 test.

■ 11.8 *t*-test of single sample mean

Sometimes, there is only one sample. For example:

> A firm making tyres claims that, on a particular car model under normal driving conditions, the tyres should last for 50 000 miles. A customer tests 8 tyres which last for the following mileages:
>
> 42 800 47 600 51 500 48 400 43 800 46 100 50 200 47 900

This data is used with the *t*-test to check the claim. A calculation is made of *t*. The value is looked up in tables (Appendix E) against degrees of freedom ($v = n - 1$). The table reading is interpreted in a way similar to that used in χ^2 and *F* tests. As in Section 7.5:

$$t = \frac{\bar{x} - \mu}{s/\sqrt{n}}$$

where

$\bar{x}$ = sample mean
s = sample standard deviation
n = sample size.

> The null hypothesis will assume no real difference between the mean of the observations and the claim except a difference due to chance. In other words, the claim will be rejected if the sample mean is significantly less than 50 000. H_0: real mean (μ) = 50 000 (miles).
>
>   In this example, the *t*-test is one-sided because the alternative hypothesis (that the sample mean is less than 50 000, expressed mathematically as H_A: $\mu <$ 50 000) would be accepted automatically if H_0 were rejected. Note that H_0 is accepted if $\bar{x} \geqslant$ 50 000.
>
>   Calculations give $\bar{x} = 47\ 287.5$, $s = 2967.6$, $n = 8$ and
>
> $$t = \frac{47\ 287.5 - 50\ 000}{2967.6/\sqrt{8}} = \frac{2712.5\sqrt{8}}{2967.6} = -2.59.$$

The negative sign is ignored: it indicates only that $\bar{x}$ is less than μ, and only the size of *t* is important. $v = n - 1 = 7$, and from tables $1\% < P < 2.5\%$ so $P < 5\%$ which is *significant*. Therefore H_0 is rejected (and H_A is accepted), which is evidence that the claim is for too high a mileage.

Assumption of normal distribution

The theory assumes that the distribution from which the sample is taken is normal. This is not a critical assumption because the test is based on sample means and, as a consequence of the central limit theorem, sample means will be almost normal even if the original data is not. In other words, unless the sample size (n) is very small (3 or 4), the derived probabilities will not differ very much from the true values and the test is reliable.

Using pocket calculators

Beware of very large or very small numbers (with many digits) when using a pocket calculator to determine s. Errors can arise due to digits being lost if the calculator cannot hold squares of numbers (most calculators can cope with up to 6-digit original numbers). Any problem is overcome easily by simplifying the data by adding, subtracting, multiplying or dividing all numbers (including μ) by a constant.

> In the last example, subtracting 40 000 and dividing by 100 gives new data as 28, 76, 115, 84, 38, 61, 102 and 79. Making the same change to μ gives a new μ of 100. The t-test on this new data gives the value of t as -2.59 as before.

Two-sided t-tests

Similar to two-sided F-tests, the probabilities in the table are doubled.

Using σ instead of s

Calculations commonly use s because σ is not known. However, there are occasions when the standard deviation, σ, is known accurately from previous data; then σ can be used in the formula for t instead of the sample standard deviation, s. In such cases the degrees of freedom are taken as $\nu = \infty$. For example:

> A machine has been producing components with a mean length (μ) of 10 cm and a standard deviation (σ) of 0.2 cm for a long time. Then there is a suspicion that the setting of the machine (and hence the component mean length) has changed, perhaps because of wear in a part of the machine. However, it is believed that variability has not changed and therefore σ can be assumed still to be 0.2 cm. In this case, 0.2 can be used for s in the calculation of t with degrees of freedom $\nu = \infty$ which is found in the bottom row of the t-tables.

Significance tests and confidence limits

Note that the theory of confidence limits and significance tests are interrelated. If a 95% confidence limit is found for μ, then a significant test on $\bar{x}$ within the interval will be not significant at the 5% level, but significant if $\bar{x}$ is outside the level. The same applies at all other levels of significance.

This means that another way of testing for a significant result is to find the 95% confidence interval for μ and if $\bar{x}$ is outside this, then it is significant. However, it does not indicate how significant the value is. Significance tests in general indicate whether or not there is a difference, and if there is, confidence intervals indicate how large that difference might be.

■ 11.9 *t*-test of means of two samples with data in pairs

Sometimes there are two samples and the data arises in pairs, one half of each pair being influenced by one factor, the other half by a different factor. For example:

It is thought that increasing tiredness (a changing factor) causes work rate to reduce during a shift. Six workmen performing the same routine job are timed near the beginning of a shift (the first sample) and also near the end (the second sample). The data is in pairs (an early time and a later time for each workman).

Worker	1	2	3	4	5	6
Early time	2m12s	2m31s	1m51s	2m05s	1m45s	2m10s
Later time	2m23s	2m35s	2m01s	2m02s	1m47s	2m16s
Difference	11	4	10	-3	2	6

The data is used with the *t*-test to check if there is a significant difference between early and late times. In general terms, the first step is to calculate the mean difference of the pairs. In other words, if the first sample (x) gives results $x_1, x_2, \ldots x_n$ and the second sample (y) gives $y_1, y_2, \ldots y_n$, then the differences (d) are $d_1 = x_1 - y_1$, $d_2 = x_2 - y_2$, etc. and the mean difference is $(d_1 + d_2 + \cdots + d_n)/n = \bar{d}$.

If there is no real difference between x and y, the ds should have a real mean of zero. In other words, the test is only of ds and this two-sample case has been reduced to a single-sample case where the null hypothesis is H_0: $\mu_d = 0$ and the test is the same as for the single sample in Section 11.7.

$$t = \frac{\bar{x} - \mu}{s/\sqrt{n}} \text{ becomes } t = \frac{\bar{d} - 0}{s/\sqrt{n}}$$

where s is standard deviation of the d values and, as before, $\nu = n - 1$.

Continuing the example, H_0: assume no real difference ($\mu_d = 0$).
$\bar{d} = 5$, $s = 5.22$ hence $t = (5\sqrt{6})/5.22 = 2.35$ and $\nu = 6 - 1 = 5$.
The test is two-sided (μ_d could be higher or lower than 0). Therefore the probabilities in the table are doubled. From the table $10\% > P > 5\%$ with $H_A: \mu_d \neq 0$ which is *not significant*.

H_0 cannot be rejected, so there is no evidence of a difference. However, P is close to 5% and the not significant result might be due to the very small sample. Also, 5 of the 6 workmen had longer times later in the shift, and it is quite likely that more data would show a significant slowing in times.

It could be argued that a one-sided test should be used here, since workmen are not going to speed up as they tire. In this case, $5\% < P < 2.5\%$ and the result is getting close to being significant.

This particular paired test (carried out only on individual time differences) has focused on the factor of changing tiredness rather than variation in skill among individuals. From the original data, it is clear that there are differences in speed between the workmen, but by using differences, the variability in speed has been virtually eliminated from the analysis. Hence, the t-test applied to pairs is more sensitive than the general t-test described in the next section. It is preferable to pair, if at all possible, although it cannot always be done.

■ 11.10 t-test of means of two samples with data not in pairs

Sometimes there are two samples that may or may not be of the same size, and in either case the data is not paired. For example:

Two types of tyre are being assessed for durability on a particular motor vehicle. Five tyres of one specification lasted for 42 600, 51 200, 39 300 44 600 and 48 700 miles before having to be replaced. Eight tyres of another specification were tested under similar conditions and lasted for 55 300, 47 200, 54 600, 59 500, 44 300, 57 800, 59 200 and 55 300 miles. The sample sizes are different and so the results cannot be paired.

The theory in this case assumes that the populations of the two samples have equal variances. This assumption is in addition to the assumption that data arises from random samples of normal distributions. It is usual to test that the new assumption is reasonable, by first carrying out an F-test on the standard deviations of the two samples. If the result is not significant, then the t-test can be used to check if there is a difference between the two specifications.

The null hypothesis is H_0: assume no real difference between x and y, or rather between the mean of x and the mean of y: that is, $\mu_x = \mu_y$. Then, test the sample means $\bar{x}$ against $\bar{y}$ using

$$t = \frac{\bar{x} - \bar{y}}{\sqrt{\left(\dfrac{(n-1)s_x^2 + (m-1)s_y^2}{n+m-2}\right)\left(\dfrac{1}{n} + \dfrac{1}{m}\right)}}$$

where s_x and s_y are the standard deviations of x and y respectively, $\nu = n + m - 2$, n is the sample size of x and m is the sample size of y.

If there is a significant difference between variances, the t-test can still be used, provided the sample sizes are almost equal. Otherwise there are two possible ways to proceed.

■ *If n and m are both greater than or equal to 30*, then calculate z, the standard normal deviate:

$$z = \frac{\bar{x} - \bar{y}}{\sqrt{(s_x^2/n) + (s_y^2/m)}}.$$

The probability of z can be found from standard normal tables (see Chapter 5) and although approximate is close enough to the correct probability.

■ *If n or m is less than 30*, calculate

$$t = \frac{\bar{x} - \bar{y}}{\sqrt{(s_x^2/n) + (s_y^2/m)}}$$

with degrees of freedom $\nu = \dfrac{(s_x^2/n + s_y^2/m)^2}{\dfrac{(s_x^2/n)^2}{(n-1)} + \dfrac{(s_y^2/m)^2}{(m-1)}}.$

The probability is found from t-tables and is again an approximate value.

Continuing the example, where $\bar{x} = 45\,280$, $s_x = 4747$, $\bar{y} = 54\,150$ and $s_y = 5550$. First, an F-test: $F = s_y^2/s_x^2 = (5550/4747)^2 = 1.37$, with $\nu_y = m - 1 = 7$ and $\nu_x = n - 1 = 4$. The test is two sided so $P > 20\%$, which is clearly greater than 5% so the result is *not significant*. Therefore continue with a t-test.

$$t = \frac{45\,280 - 54\,150}{\sqrt{\left(\dfrac{4(4747)^2 + 7(5550)^2}{5 + 8 - 2}\right)\left(\dfrac{1}{5} + \dfrac{1}{8}\right)}} = -2.95$$

Here $\nu = n + m - 2 = 11$, for two-sided test from tables $1\% < P < 5\%$ with $H_A: \mu_x \neq \mu_y$, which is *significant* and H_0 can be rejected. This is evidence of a difference in lifetimes between the tyres and therefore they should not be recognized as equal in durability.

■ 11.11 Kolmogorov–Smirnov test

The Kolmogorov–Smirnov test should be used, rather than χ^2, to test if observed data comes from a particular distribution when the sample size is small. In this context, a small sample is one that is not large enough to form a reasonable number of classes with frequencies greater than 5.

The theory of the Kolmogorov–Smirnov test assumes that the data comes from a continuous distribution. However, it can be used on a discrete distribution provided it is recognized that probabilities will be approximate. The test is based on the fact that the difference between the cumulative frequencies of a sample and its population is independent of the actual distribution.

In addition, the probabilities (in the table which is used) have been calculated assuming that the parameters of the distribution are known: that is, the distribution is known completely. However, in most practical cases, one or more parameters will have to be estimated from the sample. In the example that follows, the mean μ and the standard deviation σ are both estimated. The probabilities then are only approximately true, but the error is of a conservative nature. If D in the table is exceeded by the observed value, then the probability is less than indicated and H_0 may be rejected with considerable confidence.

To illustrate the test, consider the data of crankshaft flange separations in Figure 3.12 and suppose that, instead of data of 125 crankshafts, the only available data is that of the 10 crankshafts in the first two columns (73, 53, 50, 42, 55, 63, 54, 66, 82 and 48). The mean ($\bar{x}$) and standard deviation (s) of this sample of 10 are 58.6 and 12.28 respectively ($\bar{x}$ and σ_{n-1} on most calculators).

At first sight, the way to compare the sample distribution with a normal distribution may seem to be a plot of the sample on a line together with a sketch of a normal distribution having mean $\mu = 58.6$ and standard deviation $\sigma = 12.28$, as shown in Figure 11.3(a). In fact, it does not help very much because a majority of points are close together at the lower end.

A better comparison is obtained when plots of the sample cumulative frequency distribution $S_n(x)$ and the normal distribution cumulative distribution function $F(x)$ are made on the same graph, as shown in Figure 11.3(b).

The value of $S_n(x)$ is the number of observations less than or equal to x divided by the sample size (n) and $F(x)$ is the proportion of the normal population less than x.

Comparing the plots in Figure 11.3 of $S_n(x)$ and $F(x)$ by eye would seem to show that there is little difference between them. This indicates that the sample arises from a normal distribution. In practice, the diagrams in Figure 11.3 are not plotted unless a picture is required and the test is made through calculations in the following way.

Note that the use of normal probability paper, described in Section 8.3, is just another way of checking $S_n(x)$ against $F(x)$. In that method, $F(x)$ is converted to a straight line to make the comparison easier.

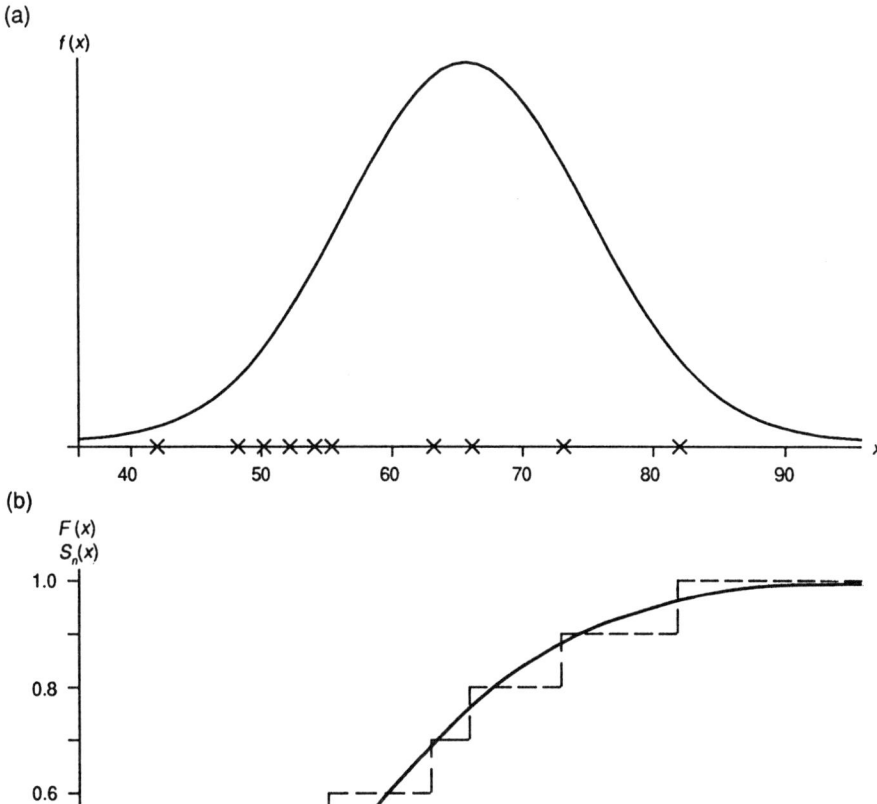

Figure 11.3 Kolmogorov–Smirnov test

■   The sample values (x) are set out in ascending order.
■   The standardized deviate (z) is calculated for each x value.

$$z = \frac{x - \mu}{s} = \frac{x - 58.6}{12.28}.$$

263

■ $F(x)$ is found, using the methods described in Chapter 5, from the standard normal tables in Appendix B. It is equal to the probability of a standard normal variable being equal to z or less.

■ Values of $S_n(x)$ are calculated; they increase by $1/n$ at each x.

■ The difference (d) between $S_n(x)$ and $F(x)$ is calculated for each x:

d is equal to $S_n(x) - F(x)$ if $S_n(x) > F(x)$
and $F(x) - S_n(x) + 1/n$ if $F(x) > S_n(x)$ (see Figure 11.4)

■ The factor D is determined as the maximum d value and is expressed mathematically as $D = \text{maximum} \ | S_n(x) - F(x) |$. The vertical straight lines in this expression are *modulus signs* which indicate that all differences are taken as positive.

■ The probability of D or greater is given in Appendix I and is tested in a way similar to that used in other significance tests.

x	42	48	50	53	54	55	63	66	73	82
z	−1.35	−0.86	−0.70	−0.46	−0.37	−0.29	0.36	0.60	1.17	1.91
$F(x)$	0.089	0.195	0.242	0.323	0.356	0.386	0.641	0.726	0.879	0.972
$S_n(x)$	0.1	0.2	0.3	0.4	0.5	0.6	0.7	0.8	0.9	1.0
d	0.011	0.005	0.058	0.077	0.144	0.214	0.059	0.074	0.021	0.028

H_0: assume data arises from a normal distribution.
Scanning the d values shows that $D = \text{maximum } d = 0.214$ and $n = 10$.
From the table in Appendix I, $P > 20\%$ which is greater than 5% and is *not significant* (D would have to be greater than 0.410 to be significant). H_0 cannot be rejected. The data are a good fit to a normal distribution.

Of course, the result does not mean that the distribution from which the sample comes is definitely normal. It has shown that the data could quite easily come from a normal distribution, but with such a small sample there are other distributions from which it could also come. A good fit implies that it has not been possible to show it is not a normal distribution.

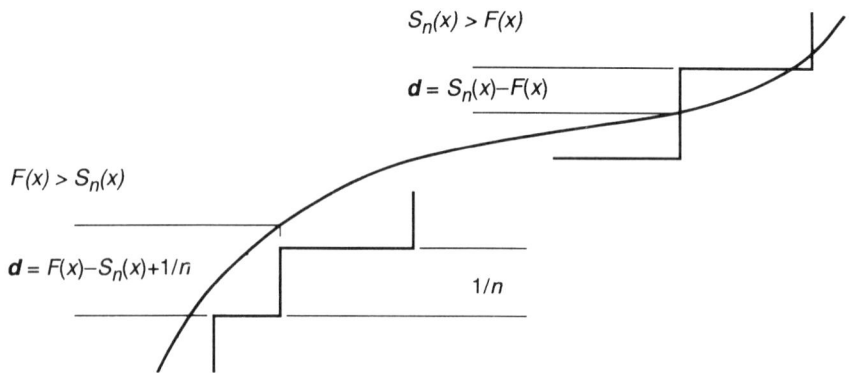

Figure 11.4 Calculation of d

■ 11.12 Tests for outliers

Sometimes one or more values in a data set appear to be unduly large or small compared to the rest and yet may not be so extreme as clearly not to belong to the distribution. A test is required to see whether these values are outliers (that is, they do not belong to this distribution) or whether they are just extreme values of the parent distribution. Outliers were introduced in Section 2.4, where it was pointed out that such values should not be discarded because they appear different to the rest; they may in fact be more important than the rest.

In many industrial situations, such as those where control charts are used to monitor a process mean, the bulk of the sample data can be considered as coming from a basic distribution representing 'common disturbances'. However, some of the data may come from one or more other distributions that represent 'special disturbances'.

Slippage and contamination in normal distributions

A normal distribution that represents special disturbances may differ from the basic distribution by having its mean shifted or its variance altered (usually increased). These effects are called *slippage* in the mean or in the variance, as appropriate. Observations of data from distributions affected by slippage can be thought of as *contaminating* the data from the basic distribution. The practical effect of this contamination will depend upon two factors:

■ The proportion of contaminated data.
■ The extent of slippage and whether it affects the mean or the variance.

Of course, this does not mean that there has necessarily been a real change to the mean or variance. An apparent extreme value may simply be a misreading or one recorded wrongly, but its effect is that of contaminating the data. The same arguments that apply to outliers also apply to mixed distributions and therefore they can be analysed by the same methods. Figure 11.5 and the discussion below illustrates these ideas.

Figure 11.5 shows two cases of slippage in the mean. In each case there are six observations from the basic distribution (A) and two from a contaminating distribution (B). The proportion of contaminated data is 2/8 (25%) in each case.

In the first case, Figure 11.5(a), large slippage in the mean ensures that observations from B are clearly identified as outliers from A. In the second case, Figure 11.5(b), the distributions overlap, so that contamination from B may be masked. If there is more than one contaminating distribution, both situations may occur.

Although the primary reason for investigation may be to identify outliers, the presence of other contaminated data may give a distorted view of the basic

(a) Large slippage in the mean

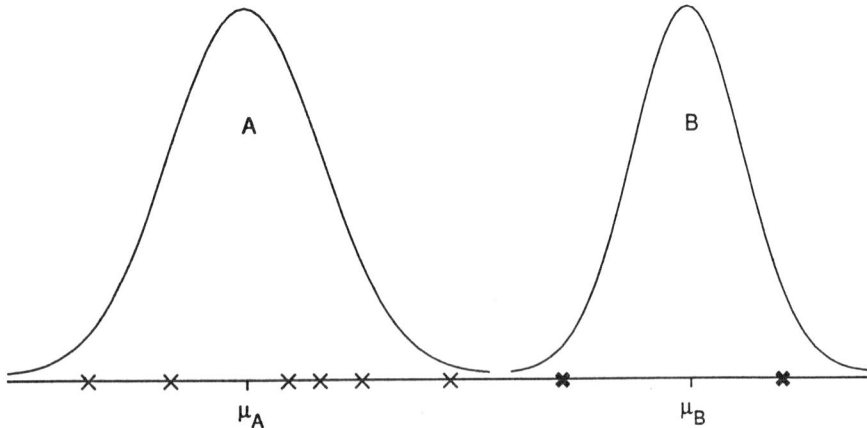

(b) Smaller slippage in the mean

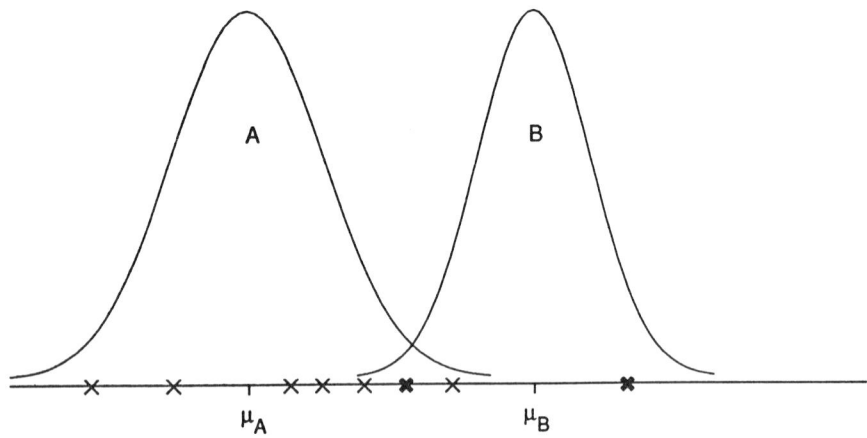

Figure 11.5 Basic distribution A contaminated by distribution B

distribution. In order to detect its presence, some specific null hypotheses must be made about the shape of the distributions concerned; this can be tested against various alternative hypotheses about slippage.

Tests for slippage with normal distributions

When there is no slippage the expected values of the measures of shape for a complete normal population are as follows:

■   Zero for skewness.
■   3 for kurtosis.

If the sample coefficients of skewness (C_s) and kurtosis (C_k) are very different from these expected values, it is an indication that the sample is unlikely to have been drawn from a single normal distribution. Perhaps it might have been drawn from a single distribution belonging to some other family, such as the Weibull. But, if the variable being observed is something like a sample mean (which is expected to be normally distributed because of the central limit theorem), it is more reasonable to assume that some contamination has occurred because of slippage in mean or variance.

The test procedure is as follows:

■ Set H_0: the entire sample of n observations comes from a single normal distribution.
■ Set H_A: at least one observation comes from some other normal distribution.
■ Calculate the sample coefficients of skewness and kurtosis:

$$C_s = \frac{\sqrt{n}\Sigma(x - \bar{x})^3}{[\Sigma(x - \bar{x})^2]^{3/2}} \qquad C_k = \frac{n\Sigma(x - \bar{x})^4}{[\Sigma(x - \bar{x})^2]^2}$$

■ Refer to Table 11.4 which shows values of C_s and C_k which would be exceeded with probability $P = 0.05(5\%)$ if the null hypothesis were true. The table shows these values for various sample sizes.
■ If the calculated numerical value of C_s is greater than the value in the tables for a particular P, then the sample value of P is smaller than this and therefore the null hypothesis should be rejected. A large positive C_s is interpreted as indicating that there is at least one outlier to the right of the distribution. A large negative C_s indicates an outlier to the left.
■ If the calculated value of C_k (which must be positive) exceeds the tabulated value, the null hypothesis is rejected.
■ The reason for using the two test statistics, C_s and C_k, is to detect either slippage in the mean or slippage in variance. The statistic C_s is sensitive to slippage in the mean when up to half the sample may be contaminated: C_k is less sensitive to slippage in the mean for heavy contamination, but is sensitive to slippage in variance for any amount of contamination.

Table 11.4 Values of C_s and C_k for slippage tests on normally distributed data

Sample size (n)	Skewness (C_s)		Kurtosis (C_k)	
	5%	1%	5%	1%
5	1.0	1.3	2.9	3.1
10	0.9	1.3	3.95	5.0
15	0.8	1.2	4.13	5.30
20	0.8	1.1	4.17	5.36
25	0.71	1.06	4.16	5.30
30	0.66	0.99	4.11	5.21

■ The most extreme value (the highest if C_s is positive, the lowest if C_s is negative) should be removed from the sample and the test repeated.

Example of a test for slippage

In the following sample of 20 recorded times (in minutes) for carrying out a routine maintenance task, one figure looks rather higher than the rest and is suspected of being an outlier. The test is to see whether this is so and whether there is any other contamination in the data.

75 79 **95** 83 81 74 85 78 80 78 80 77 76 74 74 77 81 76 74 83

The calculations for C_s and C_k are shown in Table 11.5, where the very large contributions made to each of these by the third observation (95) are obvious.

The calculated $C_s = 1.64$ is high when compared to the value of 1.1 at the 1% level in Table 11.4 for $n = 20$. Also $C_k = 6.16$ is greater than the tabulated value of 5.36 at the 1% level. Therefore in both cases $P < 1\%$ which is a *highly significant* result. This is strong indication of slippage and the calculations are repeated after removing the most extreme value (95) from the sample.

Now $\bar{x} = 78.16$, $\Sigma(x - \bar{x})^2 = 208.53$, $\Sigma(x - \bar{x})^3 = 264.15$
and $\Sigma(x - \bar{x})^4 = 4786.96$ to give $C_s = 0.38$ and $C_k = 2.09$ and hence $P > 5\%$ for both C_s and C_k.

Table 11.5   Calculations of C_s and C_k

x	$x - \bar{x}$	$(x - \bar{x})^2$	$(x - \bar{x})^3$	$(x - \bar{x})^4$	Statistics
75	−4	16	−64	256	
79	0	0	0	0	
95	16	256	4096	65 536	
83	4	16	64	256	
81	2	4	8	16	$\bar{x} = 1580/20 \quad = 79.0$
74	−5	25	−125	625	
85	6	36	216	1296	
78	−1	1	−1	1	$C_s = \dfrac{\sqrt{20(3822)}}{(478)^{3/2}} = 1.64$
80	1	1	1	1	
78	−1	1	−1	1	
80	1	1	1	1	
77	−2	4	−8	16	
76	−3	9	−27	81	$C_k = \dfrac{20(70\,330)}{(478)^2} = 6.16$
74	−5	25	−125	625	
74	−5	25	−125	625	
77	−2	4	−8	16	
81	2	4	8	16	
76	−3	9	−27	81	
74	−5	25	−125	625	
83	4	16	64	256	
1580	0	478	3822	70 330	

In other words, there is no longer any indication of slippage and the third observation (95) should be treated as an outlier (Section 2.3).

■ References to further reading on the topics of Chapter 11

Cass (1973) and Chatfield (1983) give simple introductions to significance testing. More details are given by Walpole and Myers (1993). Barnett and Lewis (1984) give details of many different tests for outliers.
See Bibliography for details of titles and publishers.

■ Self-assessment questions on Chapter 11

1. Five sections of a factory employ almost the same number of workers. The absenteeism, in lost working hours, was recorded for the 5 sections over a given period with results as shown. Is there a significant difference between the sections?

Section	A	B	C	D	E
Lost working hours	98	72	105	85	112

2. An automatic machine is used to make components of which 20% have to be scrapped because they are faulty. A modification to the machine results in a trial run of 14 defective components in 100. Does this signify an improvement?

3. Two types of spray were used to apply paint to 200 car panels. The panels were then inspected and assessed either as satisfactory or as requiring a respray. Use the following recorded results to test the hypothesis that there is no difference between the sprays.

	Spray 1	Spray 2
Satisfactory	94	67
Requiring respray	14	25

4. Three factories each produce 4 grades of a particular article. From the data given in the following table, is there sufficient evidence to assert that there is a difference between proportions produced by each factory in the 4 grades?

Grade	A	B	C	D
Factory a	11	25	17	13
Factory b	7	14	7	12
Factory c	13	14	15	14

5. From the following record of the number of items taken from stock over 100 days, is it reasonable to assume that demand follows a Poisson distribution?

Number of items	0	1	2	3	4	5	6
Number of days	9	15	27	24	15	8	2

6. The actual amounts of oil in six 2-litre cans were measured as

   2.15, 2.09, 1.97, 2.03, 1.91, 1.99

   After introduction of statistical process control the amounts in another eight cans were

   2.06, 2.03, 2.05, 2.05, 2.01, 2.04, 1.99, 2.06

   Do the results indicate that there has been a significant improvement in the variability?

7. The nominal diameter of a metal cylindrical bolt is 1.5 cm. Are the following measured values, of a sample of 7 bolts, consistent with a true mean of 1.50 cm?

   1.51, 1.53, 1.49, 1.55, 1.52, 1.52, 1.50

8. A standard cell, the voltage of which is known to be 2.50 volts, was used to test the accuracy of 2 voltmeters. Eight independent readings of the voltage of the cell were taken with each voltmeter, with results as follows.

1st voltmeter	2.51, 2.57, 2.50, 2.61, 2.55, 2.53, 2.56, 2.52
2nd voltmeter	2.54, 2.47, 2.51, 2.53, 2.52, 2.57, 2.48, 2.51

   (a) Is there evidence of bias in either voltmeter?
   (b) Is there evidence that one voltmeter is more consistent than the other?

9. The daily scrap from a certain manufacturing process varies over the days of the week. The amount of scrap in kilograms was recorded over a week and then again one year later after several modifications had been made to reduce the amount of scrap. Do the results indicate that the modifications had been successful?

	Mon.	Tue.	Wed.	Thu.	Fri
1st period	20.7	18.8	16.2	14.1	25.7
2nd period	19.3	18.9	14.2	13.7	24.8

10. Do the following results, of strength tests on 2 different types of windscreen, indicate that one type is significantly better than the other?

Type A	37.6, 34.2, 39.1, 36.3, 35.1
Type B	34.7, 35.1, 33.3, 35.6, 36.2, 34.9, 32.3

11. The skill of 10 workers was assessed and scored for a particular job as follows:

    53, 43, 78, 66, 64, 73, 65, 67, 34, 77

    After attending a training course they were reassessed and, in the same order, scored

    48, 52, 87, 78, 77, 66, 68, 68, 48, 80

    (a) Do the figures show a significant improvement?
    (b) What would the *t*-test indicate if the workers were not in the same order for both assessments and the data could not be paired?

12. Ten fuel pumps are run to failure, the times in hours when they fail being as follows:

    20, 65, 80, 125, 170, 210, 420, 450, 1175, 2085

    Use the Kolmogorov–Smirnov test to see whether the data is a good fit to the exponential distribution, $f(t) = \alpha e^{-\alpha t}$. (For this distribution $F(t) = 1 - e^{-\alpha t}$, mean $= 1/\alpha$, and let this equal the sample mean to estimate α. Failures of many electronic devices have this distribution.)

13. The following coded values were obtained for measurements of tappet clearances:

    12, 14, 11, 10, 15, 26, 10, 14, 13, 24, 14, 17

    The data is expected to follow a normal distribution, but appears to have one or two outliers. Carry out a test for outliers.

12 | Advanced Techniques

■ 12.1 Introduction

In the next chapter, various qualitative methods are discussed for subjectively determining factors that might influence processes. This chapter introduces some techniques that help to quantify such factors. These are some of the more advanced statistical techniques which reflect the immense potential for statistical work in process improvement.

The techniques considered here are particularly useful for identifying sources of variability and for establishing the kinds of statistical relationship which may exist between various measured quantities. In practice, most of the calculations will be done by special computerized packages. However, the techniques described can be used with a scientific calculator and hand calculations will certainly help in the understanding of factors which have an effect upon processes.

■ 12.2 Single-factor analysis of variance (ANOVA)

In Section 11.7, the *t*-test was described as a method for testing the difference between two sample means. When there are more than two samples, *single-factor ANOVA* is used. It is a test for difference in means of samples from different populations.

When control charts are used for variables (Section 10.4), they are constructed from sample data. The samples may be taken at various time intervals or possibly from several different machines. If they are taken from the same machine or process then hopefully it is in control: that is, the setting remains constant. This may need to be tested. If the data comes from several different machines, the requirement may be to test that there is no difference between machines as far as that variable is concerned. In such cases, the single-factor ANOVA tests the null hypothesis of no real difference in means. This is exactly what the *t*-test does for two samples, so this technique can be thought of as an extension of the *t*-test.

Theory

Suppose that data from k samples, each of size n, is laid out in a way similar to that used for charting (Section 10.5), but is expressed mathematically as shown in Table 12.1. Note that the grand mean, $\bar{\bar{x}}$, is the mean of all the x values as well as the mean of the $\bar{x}$ values.

In general, the means will differ, but this is to be expected even if they are all drawn from the same population. However, the differences might be such as to suggest that the samples are not drawn from the same population.

This will be so if the variation in means is greater than would be expected when considering the variation of the whole set of results: or, to be more specific, if the *variation between samples* is significantly greater than the *variation within samples*.

If it is, then there is evidence of a difference between samples which might reflect the existence of different populations. If not, then there is no evidence of a difference between samples: in other words, the results could easily have arisen by chance from the same population.

The null hypothesis (H_0) as usual is to assume no real difference between populations from which samples are taken. To be more precise, if μ_i is the mean of the population from which sample i is taken, then they are assumed to all have the same value, i.e.

$H_0 : \mu_i = \mu$ for $i = 1$ to k

with H_A: $\mu_i \neq \mu$ for at least one value of i.

The basis of the ANOVA technique is that it separates the *total variation* of measurements into the *variation between samples* and the *variation within samples*. To understand the terminology consider a single sample x_1, x_2,

Table 12.1 Notation for single-factor ANOVA data

Sample number i	\multicolumn{5}{c}{Observation number, j}	Sample mean $\bar{x}_i$
	1	2	3	$\cdots$	n	
1	x_{11}	x_{12}	x_{13}	$\cdots$	x_{1n}	$\bar{x}_1$
2	x_{21}	x_{22}	x_{23}	$\cdots$	x_{2n}	$\bar{x}_2$
$\vdots$	$\vdots$	$\vdots$	$\vdots$	$\ddots$	$\vdots$	$\vdots$
k	x_{k1}	x_{k2}	x_{k3}	$\cdots$	x_{kn}	$\bar{x}_k$
				Grand mean		$\bar{\bar{x}}$

$x_3, \ldots x_n$, from a population with variance σ^2. In Section 3.19 it was shown that one way of measuring the amount of variation was by calculating s^2, where

$$s^2 = \frac{\Sigma(x_i - \bar{x})^2}{n - 1}.$$

This sample variance estimates the population variance, σ^2, and may be written in more general terms as

$$\frac{\text{sum of squares}}{\text{degrees of freedom}} = \text{mean square or } \frac{SS}{DF} = MS.$$

Within a given sample i in the k samples under consideration in this section,

$$\frac{SS}{DF} = \frac{\Sigma_j(x_{ij} - \bar{x}_i)^2}{(n - 1)}.$$

This formula expresses variation within samples. A similar formula can be written for the variation between samples, based on the way the sample means vary.

The calculations leading to rejection or not of the null hypothesis are set out in an ANOVA table as shown in Table 12.2. Each summation is over i and j because the total sum of squares is actually partitioned into the individual sums of squares. Two estimates of variance are obtained and these are tested for a significant difference by the F-test (Section 11.6), so the final calculation is of F.

The F value is tested for significance using one-sided F-tables (Appendix H) with $\nu_1 = n - 1$ and $\nu_2 = n(k - 1)$.

The only way that E (within samples mean square) can be greater than D (between samples mean square) is by chance. So, if F turns out to be less than 1, then the result is immediately declared *not significant* and the null hypothesis cannot be rejected.

The theory of ANOVA assumes that data is sampled from a normal distribution, but fortunately the result is not affected much by non-normality.

Table 12.2 Single-factor ANOVA table

Variation	Sum of squares	Degrees of freedom	Mean square	F
Between samples	$\Sigma_i\Sigma_j(\bar{x}_i - \bar{\bar{x}})^2$	$n - 1$	$\dfrac{\Sigma(\bar{x}_j - \bar{\bar{x}})^2}{(n - 1)} = D$	
				D/E
Within samples	$\Sigma_i\Sigma_j(x_{ij} - \bar{x}_i)^2$	$n(k - 1)$	$\dfrac{\Sigma(x_{ij} - \bar{x}_i)^2}{n(k - 1)} = E$	
Total	$\Sigma_i\Sigma_j(x_{ij} - \bar{\bar{x}})^2$	$nk - 1$		

The following points help to simplify the calculations.

- Between samples DF + within samples DF = total DF.
- Between samples SS + within samples SS = total SS.
- Calculations are best carried out with sums of rows rather than means, using the following two formulae which can be derived by algebra:

$$\text{total } SS = \Sigma_i \Sigma_j (x_{ij} - \bar{\bar{x}})^2 = \Sigma_i \Sigma_j x^2 - T^2/N$$
$$\text{between samples } SS = \Sigma_i \Sigma_j (\bar{x} - \bar{\bar{x}})^2 = \Sigma T_i{}^2/n - T^2/N$$

where $N\,(=k \times n)$ is the total number of observations, T_i is the total for sample i and T is the sum of the T_is, which is therefore the total of all N observations.

Bearing in mind all these points, the new table of calculations is as shown in Table 12.3.

Practical method

Users generally find it easier to remember the calculations as a series of steps rather than trying to remember chunks of formula. These steps are set out in Table 12.4 where

$$A = \text{total } SS \qquad = \Sigma_i \Sigma_j x_{ij}{}^2 - T^2/N$$
$$B = \text{between samples } SS = \Sigma_i T_i{}^2/n - T^2/N.$$

It may help to note that the divisor in a sum of squares formula is equal to the number of observations which are added to make the total. For example, n values are added for T_i, so $T_i{}^2$ is divided by n; N values are added for T, so T^2 is divided by N; and so on.

Since the calculations leading to an F value are based on the ratio of variances, the location and scale of the distribution of original observations can be changed without affecting the value of F. In practical terms, this means

Table 12.3 Simplified single-factor ANOVA calculations

Sample number i	Observation number, j				Total T_i	(Total)^2 $T_i{}^2$	Total squares $\Sigma_j x_{ij}{}^2$
	1	2	$\cdots$	n			
1	x_{11}	x_{12}	$\cdots$	x_{1n}	T_1	$T_1{}^2$	$\Sigma_j x_{1j}{}^2$
2	x_{12}	x_{22}	$\cdots$	x_{2n}	T_2	$T_2{}^2$	$\Sigma_n x_{2j}{}^2$
$\vdots$	$\vdots$	$\vdots$	$\ddots$	$\vdots$	$\vdots$	$\vdots$	$\vdots$
k	x_{1k}	x_{2k}	$\cdots$	x_{kn}	T_k	$T_k{}^2$	$\Sigma_j x_{kj}{}^2$
			Grand total		T	$\Sigma_i T_i{}^2$	$\Sigma_i \Sigma_j x_{ij}{}^2$

Table 12.4 Simplified single-factor ANOVA table

Source of variation	SS	DF	MS	F
Between samples	B	$k - 1 = \nu_1$	$B/\nu_1 = D$	
				D/E
Within samples	$A - B = C$	$\nu - \nu_1 = \nu_2$	$C/\nu_2 = E$	
Total	A	$N - 1 = \nu$		

obtaining more convenient numbers. A constant can be added to or subtracted from the original values and they can be multiplied or divided by a constant as shown in the following example.

Interpretation of a control chart suggested that the diameter of a standard ball-bearing had been changing over a period of time. The measurements of samples of the ball-bearings taken at four times during the period were as follows:

Sample	Diameters (cm)				
1	0.028	0.030	0.034	0.032	0.035
2	0.025	0.026	0.030	0.029	0.030
3	0.028	0.024	0.031	0.037	0.027
4	0.030	0.024	0.027	0.028	0.028

The null hypothesis was taken as H_0: assume no real difference between populations so that $\mu_1 = \mu_2 = \mu_3 = \mu_4$ where these are the population means at the four times.

The data was simplified (it was multiplied by 1000 and then 25 was subtracted) and retabulated as follows:

Sample, i	Observation, j					T_i	T_i^2	$\Sigma_j x_{ij}^2$
1	3	5	9	7	10	34	1156	264
2	0	1	5	4	5	15	225	67
3	3	−1	6	12	2	22	484	194
4	5	−1	2	3	3	12	144	48
				Totals		83	2009	573

Note the calculation of $\Sigma_j x_{ij}^2$. For example, 264 is $3^2 + 5^2 + 9^2 + 7^2 + 10^2$.

An ANOVA table was constructed from the retabulated data using the following:

■ Total $SS = \Sigma_i \Sigma_j x_{ij}^2 - T^2/20 = 573 - (83)^2/20$
= $573 - 344.5 = 228.5$

■ SS between samples $= \Sigma T_i^2/5 - T^2/20 = 2009/5 - (83)^2/20$
= $401.8 - 344.5 = 57.3$

Source of variation	SS	DF	MS	D
Between samples	57.3	$\nu_1 = 3$	19.1 (57.3/3)	1.79 (19.1/10.7)
Within samples	171.2 (228.5 − 57.3)	$\nu_2 = 16$ (19 − 3)	10.7 (171.2/16)	
Total	228.5	19		

From Appendix H for $\nu_1 = 3$ and $\nu_2 = 16$, F is 3.24 at $P = 5\%$, which is higher than F in the ANOVA table. Hence the result is *not significant*, H_0 cannot be rejected and there is no evidence of a difference between samples. Therefore, the diameter was not shown to have changed over the period of time.

■ 12.3 Two-factor analysis of variance

The importance of the ANOVA technique lies in the fact that it can cover situations with several sources of variation. In two-factor ANOVA, the analysis is extended to cover two sources of possible differences.

Theory

To illustrate the method, suppose that just one observation is taken from each of n machines for each of k shifts. The results with the required totals will be as shown in Table 12.5. Note that T_j is the sum of column j, T_i is the sum of row i and T is the sum of all the xs.

Variation may be due to a difference between shifts, to a difference between machines and to chance. The effects of chance are often called *residuals* because they represent what is left unexplained by other effects. The

Table 12.5 Two-factor ANOVA calculations

Shift	Machine j				Total	(Total)^2	Total squares
i	1	2	$\cdots$	n	T_i	T_i^2	$\Sigma_j x_{ij}^2$
1	x_{11}	x_{12}	$\cdots$	x_{1n}	T_1	T_1^2	$\Sigma_j x_{1j}^2$
2	x_{21}	x_{22}	$\cdots$	x_{2n}	T_2	T_2^2	$\Sigma_j x_{2j}^2$
$\vdots$	$\vdots$	$\vdots$	$\ddots$	$\vdots$	$\vdots$	$\vdots$	$\vdots$
k	x_{k1}	x_{k2}	$\cdots$	x_{kn}	T_k	T_k^2	$\Sigma_j x_{kj}^2$
Total T_j	T_1	T_2	$\cdots$	T_n	T	$\Sigma_i T_i^2$	$\Sigma_i \Sigma_j x_{ij}^2$
T_j^2	T_1^2	T_2^2	$\cdots$	T_n^2	$\Sigma_j T_j^2$		

purpose of the analysis is to test for a difference between shifts and for a difference between machines, the null hypothesis being that there is no difference in either case.

Practical method

The ANOVA table is written as shown in Table 12.6, where

$$N = k \times n$$
$$B = \Sigma \, T_i^2/n - T^2/N \quad \text{(the SS between shifts)}$$
$$C = \Sigma \, T_j^2/k - T^2/N \quad \text{(the SS between machines)}$$
$$A = \Sigma_i\Sigma_j x_{ij}^2 - T^2/N \quad \text{(the total SS)}$$

Also

total $SS =$ (between shifts SS) + (between machines SS) + (residual SS)

and

total $DF =$ (between shifts DF) + (between machines DF)
    + (residual DF)

As in the single-factor case, if E is less than H, because that could only occur by chance, the result for shifts is declared *not significant*. Similarly for machines if G is less than H, the result for machines is declared *not significant*.

The following is an example of a test for difference between machines and between shifts. The raw data is of metal bar diameters (after subtraction of 2.5 and multiplication by 10) for four machines and five shifts.

Shift, i	Machine, j				T_i	T_i^2	$\Sigma_j x_{ij}^2$
	A	B	C	D			
a	3	0	3	5	11	121	43
b	5	1	0	-1	5	25	27
c	9	5	6	1	21	441	143
d	10	7	13	5	35	1225	343
e	14	9	5	8	36	1296	366
T_j	41	22	27	18	108	3108	922
T_j^2	1681	484	729	324	3218		

H_0: assume no difference between shifts or between machines. In other words,

$$\mu_1 = \mu_2 = \mu_3 = \mu_4 = \mu_5 \text{ and } \mu'_1 = \mu'_2 = \mu'_3 = \mu'_4$$

where μ_i is the mean for shift i and μ'_j is the mean for machine j.

Table 12.6 Two-factor ANOVA table

Variation	SS	DF	MS	F
Between shifts	B	$k-1=\nu_1$	$A/\nu_1=E$	E/H
Between machines	C	$\nu-1=\nu_2$	$C/\nu_2=G$	G/H
Residual	$A-B-C=D$	$\nu-\nu_1-\nu_2=\nu_3$	$D/\nu_3=H$	
Total	A	$N-1=\nu$		

Now $k=5$, $n=4$, $N=20$, $T=108$, $\Sigma\,T_i^2=3108$, $\Sigma\,T_j^2=3218$, $\Sigma_i\Sigma_j x_{ij}^2=922$.

between machines $SS = \Sigma\,T_j^2/5 - T^2/20$
$$= 3218/5 - 108^2/20 = 60.4$$

between shifts SS $= \Sigma\,T_i^2/4 - T^2/20 = 3108/4 - 108^2/20$
$$= 193.8$$

total SS $= \Sigma x_{ij}^2 - T^2/20 = 922 - 108^2/20 = 338.8$

Variation	SS	DF	MS	F
Between machines	60.4	3	20.1 (60.4/3)	2.85 (20.1/7.05)
Between shifts	193.8	4	48.45 (193.8/4)	6.87 (48.45/7.05)
Residual	84.6 (338.8 − 60.4 − 193.8)	12 (19 − 3 − 4)	7.05 (84.6/12)	
Total	338.8	19		

Machines: $F=2.85$, $\nu_1=3$, $\nu_2=12$, $P>5\%$. The result is *not significant*, H_0, cannot be rejected and there is no evidence of a difference between machines.

Shifts: $F=6.87$, $\nu_1=4$, $\nu_2=12$, $P<1\%$. The result is *highly significant*, H_0 can be rejected and there is strong evidence of a difference between shifts.

■ 12.4 Other aspects of ANOVA

Interaction

In the last example, the diameters produced by each machine changed from one shift to another. There was a moderate increase in diameters in shift *c* compared to *a* and *b*, and shifts *d* and *e* recorded even larger increases. With a few exceptions a similar sort of pattern occurred for all four machines.

A similar situation is illustrated in Figure 12.1(a) for three machines, where production from all machines clearly changes in a similar way from one shift to another.

A different situation is shown in Figure 12.1(b), where production changes from only two machines in a similar way, diameter values increasing from one shift to another. Production from the third machine behaves in a completely different way; in this case diameter values reduce when the other machines increase them. This effect is called an *interaction*. In general, given two possible sources of variability (say, machines and shifts), an interaction occurs if the way the results differ for one factor (say, machines) depends upon the level of the other factor (the shift).

Replication

In the two-factor numerical example above, only one reading for each machine/shift combination was obtained and for that case it is not possible to separate chance (residual) effects from the interaction. If several results are obtained for each machine/shift combination (in other words, *replication* of data), then it is possible to separate the interaction and chance (residual) effects in the ANOVA.

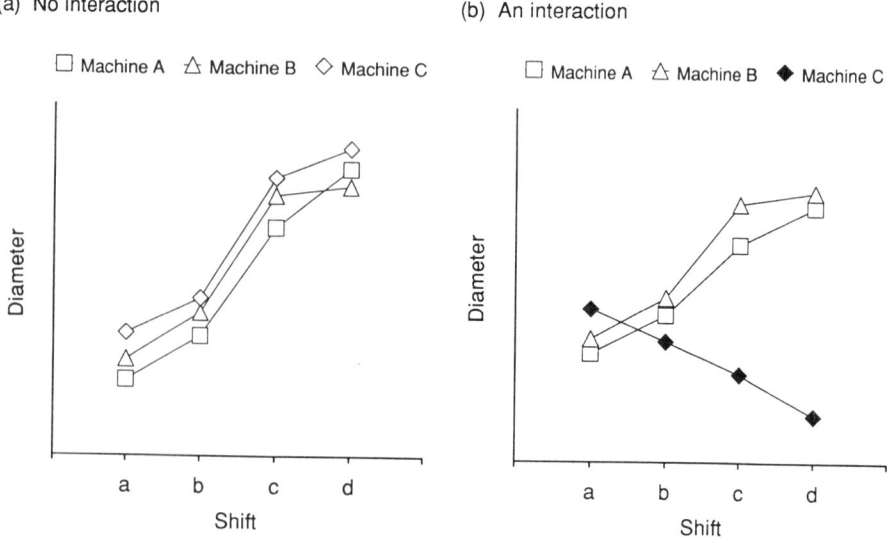

Figure 12.1 Patterns of change

ANOVA with more than two factors

With computer packages, the ANOVA technique can be used to analyse experiments involving a large number of possible sources of error and various possible interactions between them. Sources of more information on these techniques are indicated in the Bibliography.

■ 12.5. Regression analysis

Regression analysis is the term used to describe techniques that help in deciding whether there is a relationship between particular variables. The most simple illustration of an analysis is when only two variables are considered.

Figure 12.2(a) is a scatter diagram for two variables: it shows the speed of a cutting tool (x) against the wear on that tool (y). The diagram indicates that a relationship exists between the variables, and regression analysis can be used to find that relationship.

One approach is to try to find a line which gives the best representation of the relationship between x and y. Mathematically, this means finding the equation of the *best-fit line* through the set of points. The easiest case is that where the line is straight and the relationship between variables is said to be *linear*. Although there is a fair amount of scatter in Figure 12.2(a), clearly the best relationship between x and y is given by a straight line.

The linear relationship may seem somewhat limited, but in practice the majority of cases are satisfied by a straight line over the region of interest. In cases where this is not so, sometimes a linear relationship can be obtained by transforming one or more of the variables. For example, instead of using x, use some function of x such as $\ln x$ or $\sqrt{x}$ or $1/x$ or x^2. This idea is used in the probability plotting methods described in Chapter 8.

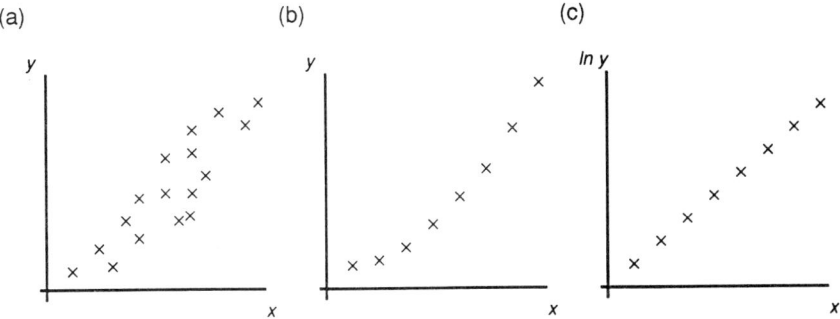

Figure 12.2 Regression

The type of diagram illustrated in Figure 12.2(b) is often obtained when plots are made of the size (y) of populations or investments against the time (x) in years or months. Obviously it is not a straight line, but there is a mathematical relationship between x and y which is of the form $y = a \exp(bx)$ where a and b are constants. If ln y is plotted against x, then a linear relationship is obtained as in Figure 12.2(c).

Extensions to the case of a linear relationship between two variables are described later, but the basic principles are the same as stated here.

The method used to find a linear best-fit line is called the *method of least squares*, where x is the *independent variable* and y is the *dependent variable*, i.e. it depends on x.

If, as in Figure 12.2(a), the points represent tool speed and wear, x is chosen to be tool speed and y to be tool wear because the wear depends on the speed. If the points show company profits against year, x is taken as the year and y as the company profits. It is important to label x and y correctly.

The principle used to find the best straight line is shown in Figure 12.3, where the line is such that $e_1^2 + e_2^2 + e_3^2 + e_4^2 + e_5^2$ is a minimum. More generally, where there are n points, the line is chosen so that $e_1^2 + e_2^2 + e_3^2 + \cdots e_n^2$ is a minimum, where e (called the *residual*) is the vertical distance from each point to the line.

The best-fit line is called the *regression line of y on x* and that shown in Figure 12.3 has the mathematical expression $\hat{y} = a + bx$ where a is the *intercept* and b is the *slope* or *gradient*.

The line is calculated from the sample of points using the criterion that $e_1^2 + e_2^2 + e_3^2 + \cdots$ is a minimum. Some standard mathematics gives the following formulae for a and b.

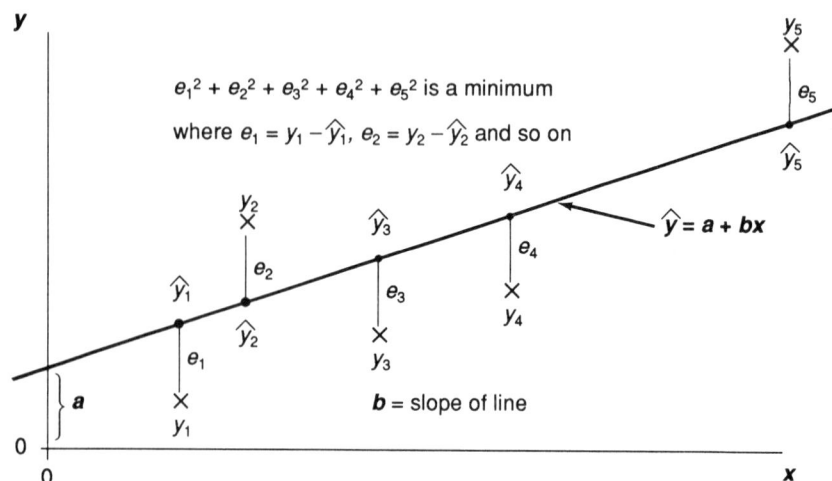

Figure 12.3 Principle of the regression line

$$b = \frac{n\Sigma\,xy - (\Sigma\,x)(\Sigma\,y)}{n\,\Sigma\,x^2 - (\Sigma\,x)^2} \quad \text{and} \quad a = \Sigma\,y/n - b\,\Sigma\,x/n.$$

Substituting into the equation $\hat{y} = a + bx$ gives the line which is used to predict y from x.

The line calculated from sample data is an estimate of the true relationship between x and y. The mathematical model describing the true relationship is assumed to be of the form

$$y = \alpha + \beta x + \varepsilon.$$

The calculated values a and b are estimates of α and β respectively. The residual e is an estimate of ε which gives the deviation of y from the line $\mu_y = \alpha + \beta x$. In this case, μ_y is the mean of all y values which might be observed at a particular x: that is, the expected value of y for that x.

Statistical assumptions

The theory of linear regression with two variables, set out above, makes the following statistical assumptions which are illustrated in Figure 12.4:

■ The means of all distributions of y for a given x fall on a straight regression line.

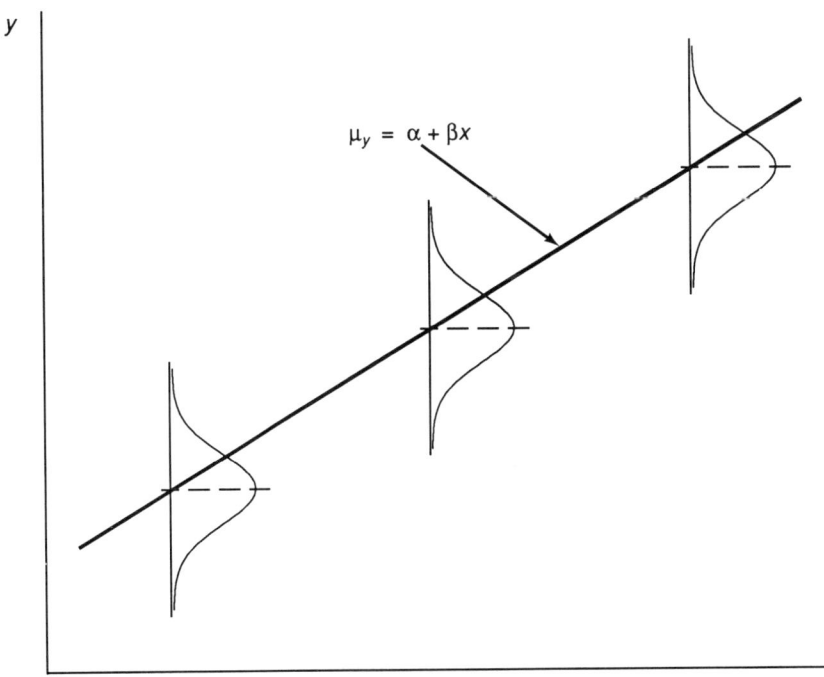

Figure 12.4   Distribution of y at different values of x

- For all x, the errors (ε) in y are normally distributed with zero mean.
- The variances of the distributions of y are the same for all x.
- The errors in y are independent of one another.

Calculations to estimate the regression line are as in the following example.

> In a chemical process, it was known generally that the temperature affects the amount of a given impurity A, but a need arose to estimate A when the temperature is $89°C$ from the following data.
>
> | Temperature ($°C$) | 81 | 84 | 83 | 88 | 85 | 90 | 87 |
> | Amount of A (%) | 0.1 | 0.3 | 0.2 | 0.4 | 0.3 | 0.4 | 0.3 |
>
> The method used was to draw a scatter diagram using the available data and then to find the regression line for predicting the amount of A from temperature.
>
> Since the aim is to predict amount from temperature, temperature is the independent variable and is designated x on the scatter diagram, so the amount of A is y. The scatter diagram is shown in Figure 12.5 and verifies that a straight line is appropriate.

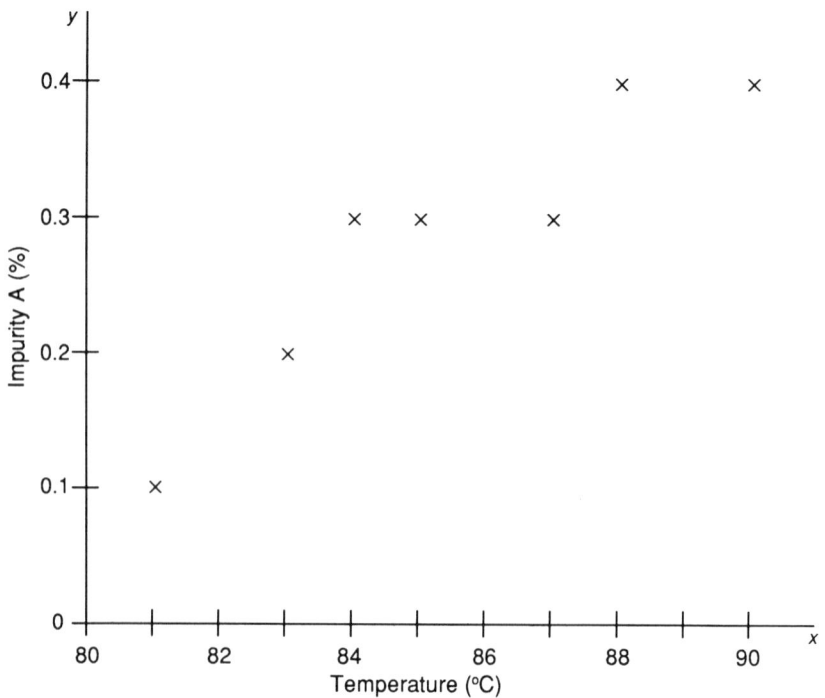

Figure 12.5 Scatter diagram for impurity and temperature

To simplify the calculations, let $X = x - 85$ and $Y = 10A$.

x	y	X	Y	X^2	XY	Y^2	
81	0.1	-4	1	16	-4	1	
84	0.3	-1	3	1	-3	9	
83	0.2	-2	2	4	-4	4	(Y^2 is required later)
88	0.4	3	4	9	12	16	
85	0.3	0	3	0	0	9	
90	0.4	5	4	25	20	16	
87	0.3	2	3	4	6	9	
Total			3	20	59	27	64

$$b = \frac{n \, \Sigma \, XY - (\Sigma \, X)(\Sigma \, Y)}{n \, \Sigma \, X^2 - (\Sigma \, X)^2} = \frac{7 \times 27 - 3 \times 20}{7 \times 59 - (3)^2} = 0.319$$

$$a = \Sigma \, Y/n - b \, \Sigma \, X/n \qquad = 20/7 - 0.319 \times 3/7 = 2.72$$

Therefore the line in terms of X and Y is $Y = 2.72 + 0.319X$. To find the line in terms of x and y, reverse the earlier simplification, i.e. replace X by $x - 85$ and Y by $10y$. In other words, the line becomes

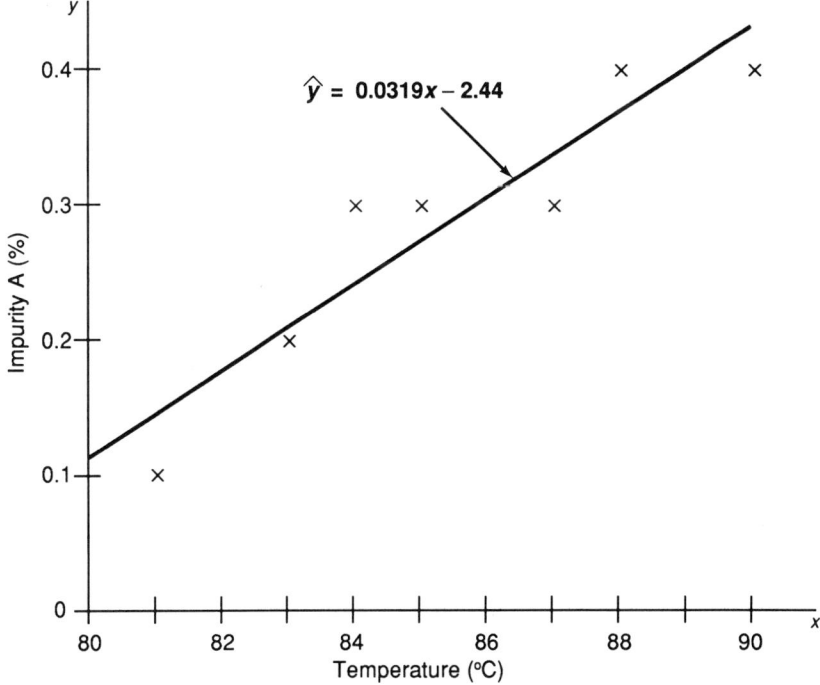

Figure 12.6 Regression line for predicting impurity from temperature

$10y = 2.72 + 0.319(x - 85)$, which gives the regression line of y on x and drawn on Figure 12.6 as

$$y = -2.44 + 0.0319x.$$

The estimated amount of impurity at a temperature of $89°C$ could be read off Figure 12.6 or calculated as $-2.44 + 0.0319 \times 89 = 0.40$.

Residuals

In order to check that the assumptions for the mathematical model are correct, the residuals are calculated and plotted against the x values. The residuals are calculated by reading off y values on the line ($\hat{y}$, sometimes called the fitted values) at given x values and then taking the difference from the observed y values as shown in Table 12.7 from Figure 12.6. The residuals (e) are plotted in Figure 12.7. They should show no observable pattern and the variation about the zero should show no definite trend. This follows from the assumptions of independence of errors about the line and a constant variance of y for all x.

The other assumption of normality about the line can be tested by plotting the residuals on normal probability paper. Of course, in this case with only 7 values it is difficult to come to any definite conclusion about the model, but the plot of residuals is much as expected.

In practice, the calculations required to find the line and the analysis of the residuals is most easily carried out on a computer using statistical packages such as MINITAB.

Table 12.7 Calculation of residuals for data plotted in Figure 12.6

x	y	$\hat{y}$	$y - \hat{y} = e$
81	0.1	0.14	-0.04
84	0.3	0.24	0.06
83	0.2	0.21	-0.01
88	0.4	0.37	0.03
85	0.3	0.27	0.03
90	0.4	0.43	-0.03
87	0.3	0.34	-0.04

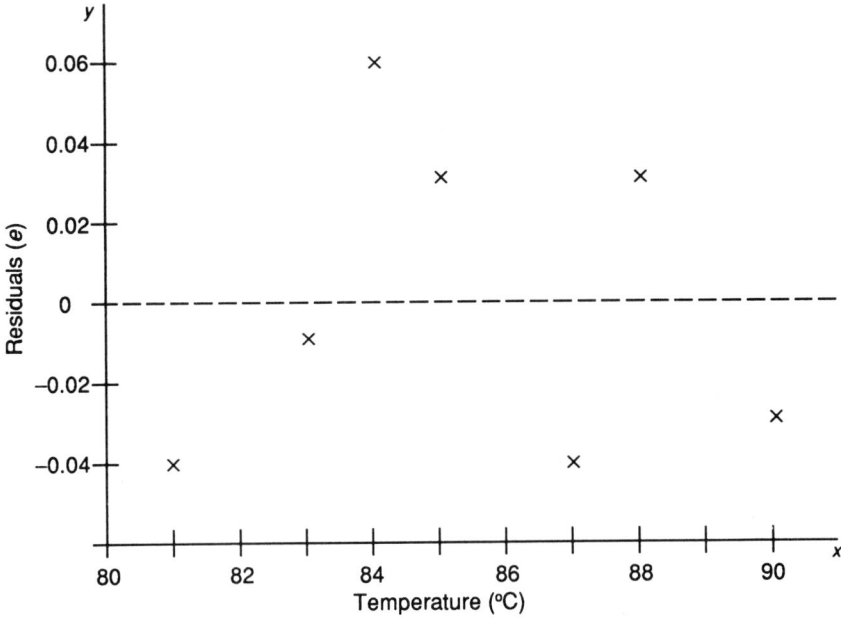

Figure 12.7 Scatter diagram for residuals and temperature

■ 12.6 Correlation

Correlation is a measure of relationships; more specifically, the *correlation coefficient r* is a measure of the *linear relationship* between x and y.

In Figure 12.8(a) there is a good relationship or good correlation between x and y. In 12.8(b) there is little or no relationship and in 12.8(c) there is a fairly weak relationship.

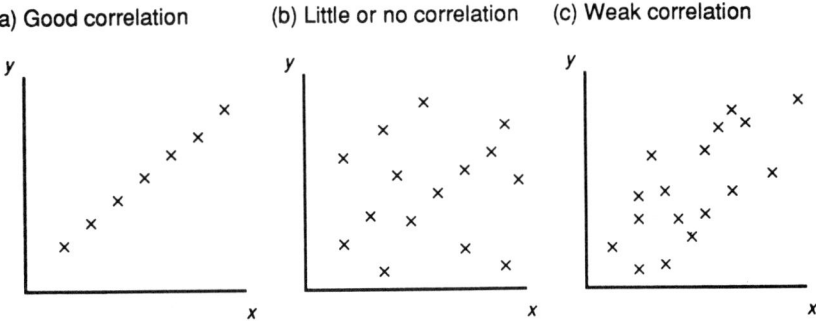

Figure 12.8 The meaning of correlation

It is possible to find a regression line in all three cases, but if there is no correlation then the line is meaningless. The correlation coefficient enables assessment of a relationship between the variables and hence whether it is worth finding the regression line. The coefficient r is given by

$$r = \frac{n \, \Sigma \, xy - (\Sigma \, x)(\Sigma \, y)}{\sqrt{(n \, \Sigma \, x^2 - (\Sigma \, x)^2).(n \, \Sigma \, y^2 - (\Sigma \, y)^2)}}.$$

The correlation coefficient always has a value between minus one and plus one, which is expressed mathematically as $-1 \leqslant r \leqslant +1$.

- When $r = +1$, there is *perfect positive correlation* and all the points lie on a straight line with positive slope, as shown in Figure 12.9(a).
- When $r = -1$, there is *perfect negative correlation*, all the points lying on a straight line with negative slope, as in Figure 12.9(b).
- When $r = 0$, there is no correlation, as in Figure 12.9(c).

When r is positive, y increases as x increases; and when r is negative, y decreases as x increases. For a given number of points, the larger the absolute value of r, the greater is the correlation. This is illustrated by Figures 12.9(d) and (e).

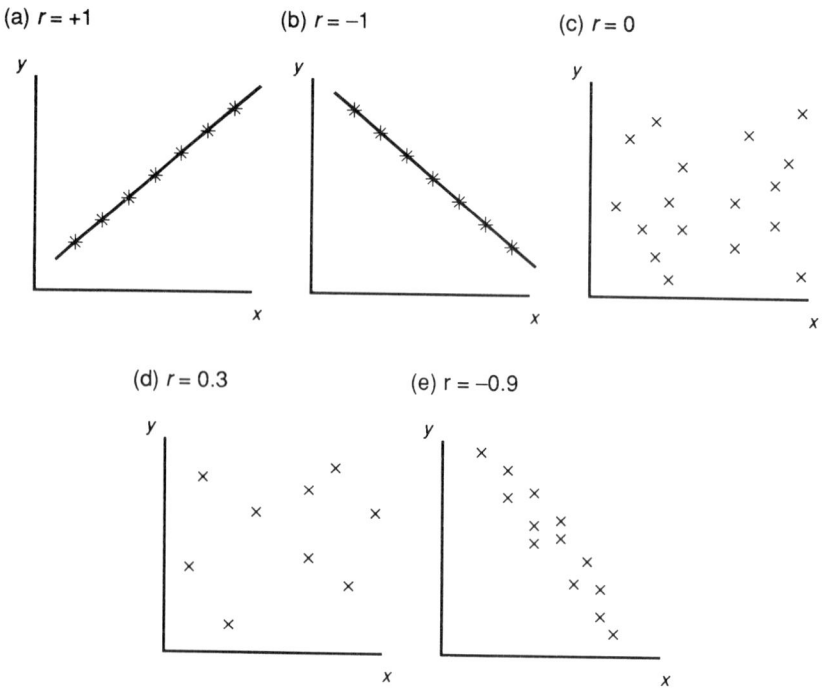

Figure 12.9 Different values of r

Significance of *r*

When there is no relationship at all between x and y, the value of r might not be zero. Figure 12.9(d) could quite easily arise by chance with 9 points when the real correlation is actually zero. For data of a given number of points (n) to indicate a real relationship, r has to be greater than the value shown in the 5% row of Table 12.8. This means that there is a probability of 0.05 of obtaining r as large as the tabulated value purely by chance if there is no real correlation.

Thus for 5 points, r needs to be greater than 0.878 (or less than -0.878), but for 20 points it need only to be greater than 0.444.

In fact, with the null hypothesis H_0: assume no real correlation, Table 12.8 enables a significance test to be carried out on r at the 5% or 1% significance level.

Alternatively, using $t = r\sqrt{[(n-2)/(1-r^2)]}$ with $\nu = n-2$, a t-test can be performed in the normal way.

The following illustration of the calculations involved in finding the correlation coefficient and testing its significance uses data from the example in Section 12.5.

Because r is unchanged by a linear transformation to x or y, the example values relating to X and Y can be used. They are $\Sigma\, X = 3$, $\Sigma\, Y = 20$, $\Sigma\, X^2 = 59$, $\Sigma\, XY = 27$ and $\Sigma\, Y^2 = 64$.

$$r = \frac{7 \times 27 - 3 \times 20}{\sqrt{(7 \times 59 - (3)^2).(7 \times 64 - (20)^2)}} = 0.926.$$

This calculated value of r is greater than the value of 0.875 tabulated for the 1% level in Table 12.8. Hence there is a strong indication of a relationship.

Many of the calculations involved in finding the correlation coefficient are the same as those used in the calculation of the regression line. Therefore, if both are to be found, it is possible to save time on some of the calculations; in fact, $r = b(s_x/s_y)$ where s_x and s_y are the sample standard deviations of x and y.

Table 12.8 Values of *r*

	Sample size (*n*)											
	3	4	5	6	7	8	9	10	12	20	40	60	100
5%	0.997	0.950	0.878	0.811	0.755	0.707	0.666	0.632	0.576	0.444	0.312	0.254	0.197
1%	0.9999	0.990	0.959	0.917	0.875	0.834	0.798	0.765	0.708	0.561	0.403	0.329	0.254

The table shows critical values of *r* which need to be exceeded to indicate a real linear relationship. Ignore the sign of *r* when using the table.

Use of analysis of variance

Analysis of variance can be used as an alternative method of testing the relationship between x and y. It has particular use in multiple regression analyses (Section 12.7).

In this approach the total variation in y is divided into the *variation explained by regression* and the *variation about the regression line* as in Table 12.9, where

y = actual sample value at x
$\hat{y}$ = value of y given by the regression equation at x
$\bar{y}$ = sample mean of the y values.

Thus

$$F = \frac{MS \text{ explained by regression}}{MS \text{ about regression}} \text{ with } \nu_1 = 1 \text{ and } \nu_2 = n - 2.$$

An assumption of no real correlation is equivalent to stating that the slope of the regression line is zero. So in this case the null hypothesis is H_0: assume no real correlation (that is, $\beta = 0$) and a one-sided test is performed.

The ANOVA table for this test on the chemical process data in Section 12.5 is as follows:

Variation	SS	DF	MS	F
Explained by regression	5.88	1	5.88	
				29.4
About regression	0.98	5	0.20	
Total	6.86	6		

$F = 29.4$, $\nu_1 = 1$, $\nu_2 = 5$, $P < 1\%$. The result is *highly significant* and H_0 can be rejected. $\beta \neq 0$ so there is strong evidence of correlation.

Note that the result of this test will be the same as in the t-test for correlation. If one is significant, so is the other.

Table 12.9 Calculations in ANOVA applied to regression

Variation	SS	DF	MS	F
Explained by regression	$\Sigma(\hat{y} - \bar{y})^2$	1	$\dfrac{\Sigma(\hat{y} - \bar{y})^2}{1} = A$	
				A/B
About regression	$\Sigma(y - \hat{y})^2$	$n - 2$	$\dfrac{\Sigma(y - \hat{y})^2}{n - 2} = B$	
Total	$\Sigma(y - \hat{y})^2$	$n - 1$		

Coefficient of determination

A value which is sometimes useful is $r^2 \times 100\%$. It is called the *coefficient of determination* and is the percentage of the total variation in y which is explained by regression.

In Figure 12.10(a) most of the variation in y is explained by regression. In Figure 12.10(b) very little is, most variation being caused by factors other than x.

For the chemical process in Section 12.5, the coefficient is $(0.926)^2 \times 100\% = 85.7\%$. In other words, 85.7% of the variation in impurity can be explained by the temperature, leaving 14.3% due to other causes.

Interpretation of *r*

Great care has to be taken when interpreting the correlation coefficient. The following points should be borne in mind:

■ The correlation coefficient is a measure of the *linear relationship*: that is, when a straight line is the best line through the points. It is therefore not appropriate in cases when a linear relationship does not exist.

The correlation coefficient for the points in Figure 12.11 is zero even though there is a perfect relationship in that x defines y exactly.

■ If there are only a few points, r can be quite large even when there is no real relationship between the variables. In fact, as explained above, r can be almost 1 in this situation if n is very small. It is not difficult to find examples in published work where conclusions are drawn on the value of r and no mention is made of the number of points, even though only a few values were obtained.

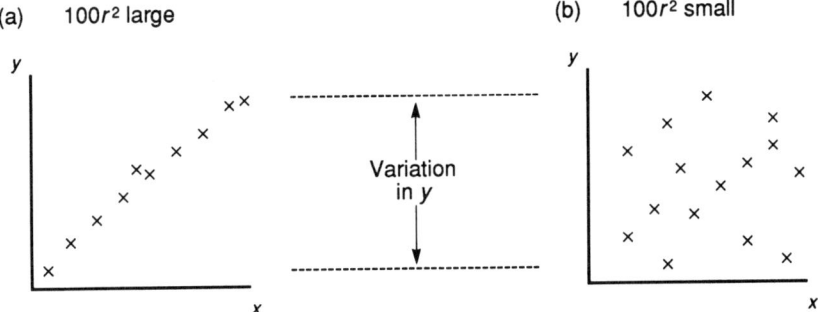

Figure 12.10 Illustrations of the coefficient of determination

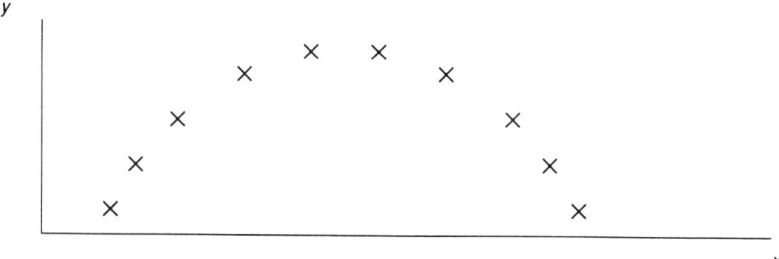

Figure 12.11 Data giving a zero correlation coefficient

■ One of the most common mistakes in interpretation is to assume that a significantly large value of *r* automatically indicates a *causal relationship* between the two variables. There are many cases where each variable depends on a third variable and the apparent relationship is a *spurious correlation*. The correlation coefficient is a particular statistic that needs supporting evidence provided by other techniques, including common sense.

There often seems a lack of common sense when spurious correlations lead to some authoritative claims based on 'proven research'. There are a number of amusing instances of real data to illustrate this point. In one example, a graph of 'the number of brooding storks' against 'the number of births of babies' recorded over the years in Germany apparently shows a strong linear relationship; it is said that this verifies what every youngster knows anyway. Could it point to another method of birth control?

■ Finally, great care must be taken with extrapolations when trying to predict *y*. Predictions are fairly safe within the range of known data and distributions, but the same conditions may not apply outside their range.

Curvilinear regression

If it is believed that the line of best fit is a curved line and it is not possible to transform in order to linearize, then it may be possible to fit a curve of the mathematical form

$$\hat{y} = a + bx + cx^2 + dx^3 + \cdots$$

A study of the plot of residuals will often show when this is likely. The calculations become more tedious as the number of terms is increased, and it is usual to use a statistical package on a computer for the analysis.

■ 12.7 Multiple regression

The case of two variables has been looked at in some detail in this chapter, but quite often the dependent variable is influenced by several others.

This multiple situation can be analysed by considering equations of the form

$$y = \beta_0 + \beta_1 x_1 + \beta_2 x_2 + \beta_3 x_3 + \cdots + \varepsilon$$

where x_1, x_2, x_3 and so on are the independent variables and ε is the error term.

> For example, y might be the colour depth of a stoved enamel paint finish, which depends upon the stoving temperature (x_1), humidity (x_2), paint viscosity (x_3), tint dispersion and so on. Balancing these variables is particularly important when matching replacement parts are expected.

The ideas and techniques developed for simple regression with just one independent variable can be extended to two or more independent variables, the regression equation that is derived being of the form

$$\hat{y} = b_0 + b_1 x_1 + b_2 x_2 + b_3 x_3 + \cdots$$

where b_0 is an estimate of β_0, b_1 is an estimate of β_1 and so on.

Consider the case of two independent variables, x_1 and x_2, where the relationship is assumed to be linear and therefore has a regression equation of the form $y = b_0 + b_1 x_1 + b_2 x_2$. Note that x_1 and x_2 may not be completely independent and in practice they are often found to be highly correlated which usually indicates dependence.

This equation can be represented by a plane in 3-dimensional space, called the *regression plane*, where the sample values are points (x_1, x_2, y) in that space, as shown in Figure 12.12.

Just as e was the vertical distance from the point to the line in the 2-dimensional case, so now e is the vertical distance from the point to the plane, and again Σe^2 is minimized in order to find the values of b_0, b_1 and b_2.

It follows that b_0 is the value of y when $x_1 = 0$ and $x_2 = 0$, b_1 is the slope of x_1 and b_2 is the slope of x_2.

The assumptions concerning the distribution of residuals about the correct regression line in the case of one independent variable (normally distributed, independent and so on) also apply here, except that the residuals this time will be the distances from the correct plane.

The technique can be extended to any number (m) of independent variables where the equation represents a hyperplane in $m + 1$ dimensional space.

As in the curvilinear case, because of the many calculations required, analysis is almost always carried out using a computer package. One of the important aspects of these analyses is to discard any of the variables which do not affect the dependent variable y. In these cases the b_i will not differ

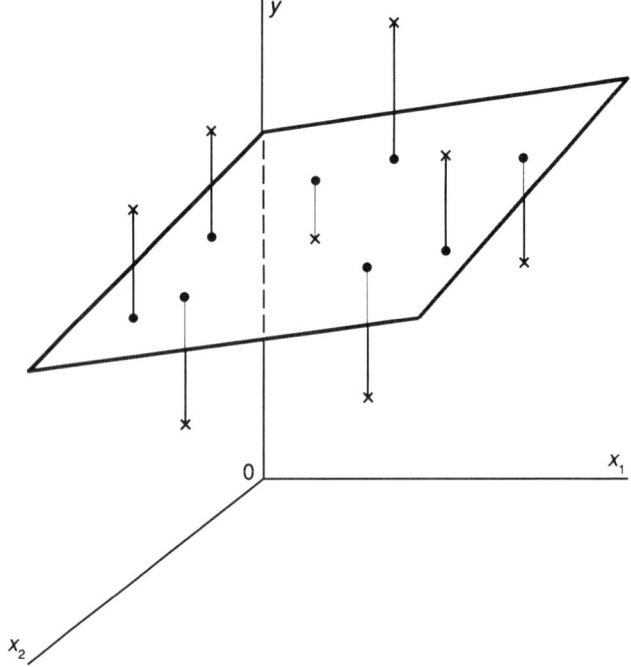

Figure 12.12 Regression plane $\hat{y} = b_0 + b_1x_1 + b_2x_2$

significantly from zero for those discarded. Most packages have methods of finding just which variables should be included, by adding variables one at a time, by starting with all variables and discarding one at a time or, if there are only a few variables, by looking at all possible subsets.

Multicollinearity

One of the things that a researcher needs to check is that there is no large correlation between so-called independent variables. These variables are often slightly dependent on one another, but this does not matter if the dependence is not large; in fact, the multiple regression analysis is ideal for handling such data. However, if there is a large dependence between two or more variables, then there are often problems in that the resulting equation is unreliable. This is illustrated by the tube effect shown in Figure 12.13.

In this case, because x_1 and x_2 are highly correlated, all the results lie close to a line – in fact, within a tube. In estimating the plane, the addition or deletion of a point can cause quite large swings to the plane. Thus the resulting equation gives very unreliable relationships between the variables.

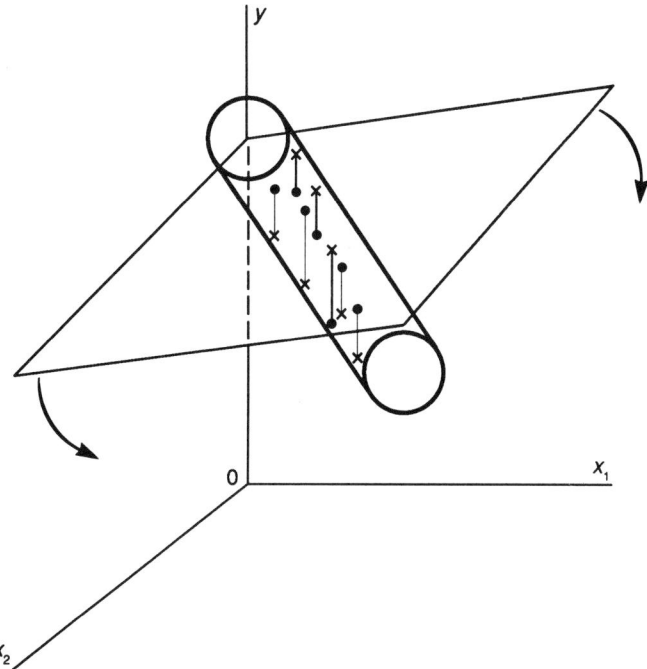

Figure 12.13 The tube effect

The dependence between any pair of so-called independent variables can be checked by calculating the correlation coefficient for them. It is usual to do this for all possible pairs. In practice there is a potential problem if the absolute value is greater than 0.7 for a pair, the solution being to leave one of them out of the analysis. After all, if there is such a large dependence between two, then only one is needed.

■ 12.8. Design of experiments

The word *experiment* in this context means any practical trial or test that is intended to obtain information about a product or a process. Before carrying out an experiment a great deal of thought needs to go into the design of that experiment. What is the purpose of the experiment? How many factors should be involved in the experiment? What variation in the factors needs to be taken into account?

A *factor* is any feature of the product or process which may affect the required information. The petrol consumption of a car depends upon the size of engine, weight of car, speed, type of petrol and many other factors. It will almost certainly not be practicable to accommodate every possible variation of every factor involved. Some features will not be varied at all and they need to be kept constant throughout the experiment. If that is impossible, then the design of the experiment should be such as to eliminate their effect as much as possible.

There are other questions: for example, how much data should be collected? Should a pilot experiment be performed first? In all, there is a great deal to consider before embarking upon the practical part of the exercise and sometimes this can take up more time than the experiment itself. The aim of most experiments is to obtain as much information as possible within available or allocated time and resources.

It has to be said that in many cases not enough thought is given to the design of experiments. Sometimes an experiment is set up, data is collected and then the researcher starts thinking about how it should be analysed, when it is too late to change anything. The design of experiments is a specialist topic which has given rise to much literature, and some references are given in the Bibliography. In this section, an attempt is made to give the reader an introduction to the subject and to cover a few basic principles.

One small aspect of design is touched on in Section 11.9 which covered paired *t*-tests. The example given in that section, of times for completing a task by six workers early and later in a shift, concluded that there was no significant difference between early and later times. Suppose that the later times of workers 4 and 5 had been slightly different, as shown below.

Worker	1	2	3	4	5	6
Early time	2m12s	2m31s	1m51s	2m05s	1m45s	2m10s
Later time	2m23s	2m35s	2m01s	2m08s	1m57s	2m16s
Difference	11	4	10	3	12	6

A *t*-test on the differences now gives $t = 4.9$, $v = 5$, $P < 1\%$, which is highly significant. Different from the earlier test, this shows clearly that the workmen had slowed down.

But suppose that the experiment had been carried out by recording times for six workmen early in a shift and a different six later. Or suppose that the same six were recorded early and later but that no information was kept on which workman recorded which time: just six figures for the early time, another six for the later time. In both cases there is no way the data can be paired, and it is necessary to perform the unpaired *t*-test explained in Section 11.10. For the same data the test result is $t = 0.86$, $v = 10$, $P > 20\%$, *not significant*. In other words, there is no evidence of a difference in time.

In this last test, the actual time difference from one period to the next has been completely masked by the large variation between workers. By designing the experiment as a paired one, the test has become very much more sensitive. This illustrates the importance of design, although of course there are times when such a design is impossible and there is no option but to do the general two-sample *t*-test. The same reasoning would be applied by a psychologist who preferred to work with identical twins when measuring human performance under different conditions.

Randomization

Often there is considerable uncontrolled variation in an experiment. If this is a function of time or correlates with other factors in the experiment, then it will most likely produce trends in the response over and above those due to the factors being tested. One way to deal with this difficulty is by *randomization*.

> Three racing cars, A, B and C are to be compared. Each will be driven by the same driver round a circuit for three separate recorded laps. There will be nine results in all, three for each car. Clearly, it would not be advisable for the driver to drive the first three laps in car A, the next three in B and finally three in C. There are many things which can influence the results as the time goes on. The driver may improve as he goes through the trials because he is learning about the track as he drives, or the weather may deteriorate as the day progresses. This problem can be mainly overcome by randomizing the sequence in which the cars are driven.
>
> The sequence in which the cars are driven is determined from a *random number table*. Figure 12.14 contains just three rows from a typical table. Numbers 1 to 3 are assigned to car A, 4 to 6 to car B, and 7 to 9 to car C. Starting at any point in the figure and moving along a row or down a column (or, in fact, in any prearranged pattern), the runs are arranged in the order that the assigned numbers occur. In this example, zeros are ignored and also any number which has come up before. Suppose that the decision is made to work across row 3 starting from the left. The first number is 7 so the first trial is on car C, then 6 (car B), then 5 (car B), then 9 (car C), then 1 (car A), then 8 (car C), then 3 (car A), then 7 and 5 which have already occurred so they are ignored, then 0 which is also ignored, then 4 (car B), then eventually the final number 2 (car A). In summary, the sequence will be
>
> C, B, B, C, A, C, A, B, A
>
> If four runs were to be made in each car, then it would be necessary to allocate 1 to 4 to A, 5 to 8 to B, and 9 to 12 to C. The numbers in the figure are now looked at in pairs and only numbers 1 to 12 are considered; all others are ignored. Starting at the left of the first row the numbers across the row are 90, 11, 92, 32, 06 and so on. The first car will be C (from 11), the second B (from 06) and so on.

90	11	92	32	06	75	63	60	05	39	07	24	29
14	08	26	50	20	49	95	60	13	36	41	68	50
76	59	18	37	50	46	13	77	49	89	39	93	13

Figure 12.14 Random numbers (extracted from tables)

It may be almost impossible to compensate completely for the effect of uncontrolled factors over time, but randomization will certainly help.

Factorial design

Engineers often have to determine optimum operating conditions involving a number of different factors, each of which can vary. Decisions have to be made on the number of experiments to be conducted and also on the experimental design, including the values of variable factors. The value that a factor takes in an experiment is called the *level* of the factor. Frequently in the design of experiments, a number of the variable factors are set at two different levels, although there can be more.

> Consider a chemical process where the output (or the *response variable*) is affected by changes in temperature, pressure and time of stirring. The aim is to find optimum values for each factor in combination with the others such that output is greatest. A decision is made that the effect of each can be adequately investigated by considering each factor at two levels: say, T_1 and T_2 for temperature, P_1 and P_2 for pressure, S_1 and S_2 for time of stirring. Another decision is made that only eight experiments are to be carried out in all.
>
> Within these decisions, one possibility is to carry out two experiments at $T_1P_1S_1$ and then, to find out the effect of temperature, to carry out two more at $T_2P_1S_1$. A further two at $T_1P_2S_1$ allows an assessment of the effect of pressure change, leaving the last two at $T_1P_1S_2$ to assess the effect of a change in stirring time. These experiments are illustrated by the 3-axis picture in Figure 12.15.
>
> The effect of a change in temperature is estimated from the difference between the average of the two responses at $T_2P_1S_1$ and the average of the two at $T_1P_1S_1$, i.e.
>
> $$\frac{[R_2P_1T_1(1) + T_2P_1T_1(2)]}{2} - \frac{[T_1P_1S_1(1) + T_1P_1S_1(2)]}{2}.$$

However, a much better design, also using eight experiments, is to test at each of the eight combinations of temperature, pressure and time of stirring, as illustrated in Figure 12.16.

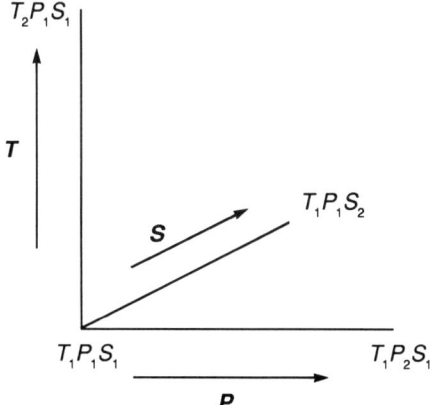

Figure 12.15 Different temperatures (*T*), pressures (*P*) and times of stirring (*S*)

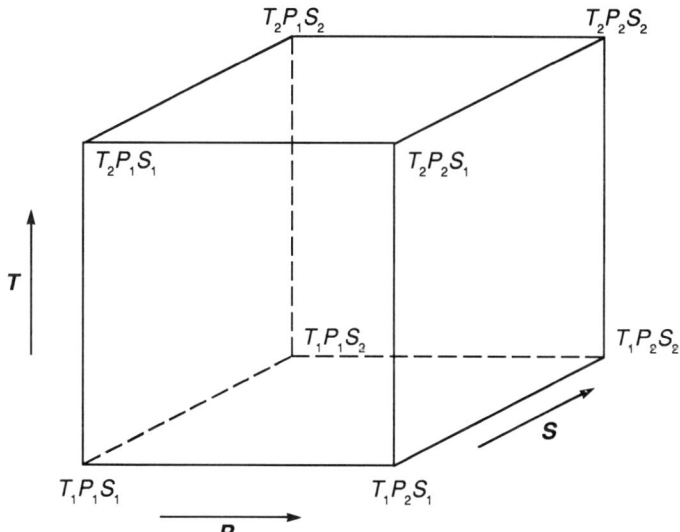

Figure 12.16 Factorial design

Now, the effect of the change in temperature is given by the average of the four responses at the upper temperature minus the average of the four at the lower temperature, i.e.

$$\frac{[T_2P_1S_1 + T_2P_2S_1 + T_2P_1S_2 + T_2P_2S_2]}{4}$$

$$- \frac{[T_1P_1S_1 + T_1P_2S_1 + T_1P_1S_2 + T_1P_2S_2]}{4}.$$

The same applies to pressure and time of stirring. This will be twice as accurate as the previous design. Moreover, it is now possible to test for possible interactions – something which could not be done in the previous case.

These factorial experiment designs (two levels of three factors, expressed as $2^3 = 8$) are special cases of *orthogonal* designs, which are generally desirable. One of the important properties of orthogonal designs is that one effect is not affected by changes in any of the other effects.

While it is true that *full factorial* designs have many advantages over other designs, they have the disadvantage that the total number of tests becomes very large when there are a large number of factors. With seven factors each at two levels, the number of tests in a full factorial design is $2^7 = 128$. There are ways of reducing the full factorial number and still testing the main factors and interactions. Readers wishing to pursue the subject further should look at one or more of the books listed in the Bibliography.

Taguchi methods

One of the so-called quality gurus (Section 1.1) of recent years, the Japanese engineer Genichi Taguchi, has had a large influence on the design of experiments as applied to quality improvement. This is of great importance to engineers who are involved in attempts to reduce variability in processes and products, ideally before they are commissioned for commercial production. Taguchi is known also for the cost function (Section 12.10), which is used to the same end. There is a reference to his work in the Bibliography.

A feature of the Taguchi approach is to make the product or process robust so that it operates just as well under different conditions. In the laboratory or factory, a unit may function perfectly well under ideal controlled conditions, but it is important that the unit works just as well in service, where conditions may be very different. In service, it is of little benefit having an engine that starts perfectly every time under normal weather conditions but causes trouble if it is very damp or very cold.

Taguchi suggests the use of two types of variable in experiments. There are *control factors*, which are variables that can be controlled both in the tests and in use. Then there are *noise factors*, which are variables that may or may not be possible to control in the tests, but which certainly cannot be controlled in use.

The aim of the product design is to choose the levels of the control factors which are most robust or insensitive to changes in the noise factors: in other words, to design in such a way that the product or process works just as well (or almost as well) under adverse conditions.

In order to achieve this, noise factors which can be controlled in the experimental stage are included as factors of the experiment. Orthogonal designs are usually employed, and often far fewer experiments are carried out than would be involved in a full factorial design. The argument in much of this work is that interactions either are unlikely to occur or, if they do, will only be weak, and so it is unnecessary to test for them.

> In the 2^3 factorial design covered previously, the changes in the main effects P, T and S would be assessed using a so-called L4 *orthogonal array*, which requires experiments at $T_1 P_1 S_1$, $T_1 P_2 S_2$, $T_2 P_1 S_2$ and $T_2 P_2 S_1$.
>
> The effect of a change in T is then calculated from the results at these values using
>
> $$\frac{[T_2 P_1 S_2 + R_2 P_2 S_1]}{2} - \frac{[T_1 P_1 S_1 + T_1 P_2 S_2]}{2}.$$
>
> Thus with an L4 orthogonal array, only four experiments are required. The effects of changes to P and S are estimated from the same data with similar calculations.

In a similar way, instead of a full factorial $2^7 = 128$ experiments for seven factors at two levels, it is possible to carry out just eight tests using an L8 orthogonal array. If interactions are suspected, the design can be modified to include responses to test these.

Taguchi argues that it is important to include a noise factor in experiments.

> In the chemical process example above (where four experiments are required to estimate the effects of changes to control factors) with just one noise factor, each of the four experiments would be carried out at one level of the noise factor and another four at another level, giving eight experiments in all. The effect of the noise factor can then be determined by calculating the mean response at each level of the noise factor and subtracting one from the other. If two noise factors were to be included, each at two levels in the experiment, then the 2^2 orthogonal design for these would be carried out at each of the four combinations in the L4 array, and so on. Orthogonal designs are thus used for control and noise factors, and combined for the experiment.

From experience it has been found that the response variable of experiments designed for use in manufacturing has to be chosen with care. In general, yields and failure rates should be avoided because they often entail interactions. If possible, the response variable should be in terms of a continuous variable which is energy related, such as pressure or velocity.

Another innovation of Genichi Taguchi is to use what are called *signal-to-noise ratios*. They enable the investigator to analyse the effects of the control factors on variability and hence to assess the robustness of the process or product. The ratios are calculated for each combination of control variables that are being tested and then analysed by ANOVA techniques which use them as the response variable. Different formulae are used for the signal-to-noise ratios depending on the objective, but they are designed in such a way that the desired combination of control variables is the one that maximizes the signal-to-noise ratio.

Designs based on orthogonal arrays work well in practice: they have been shown to have a success rate better than 9 out of 10. The risks associated with Taguchi methods, compared with full factorial designs, are generally recognized to be worth the reduction in experiment costs. Whatever experimental design is used, however, once an optimum combination has been found it is most important to make a confirmation run at this combination. It is all too easy for mistakes to have been made or for a freak result to have occurred.

■ 12.9 Reliability testing

This section summarizes the principle types of reliability test. The types are generally classified as parametric or non-parametric, which indicates the statistical treatments that are used for the evaluation of test results.

Parametric tests

Parametric tests are so called because the parameters of failure distribution can be estimated from the reliability test data. Knowledge of the parameters enables estimation of an item's reliability at any time (or distance) between the first and last test failure. The estimations require the selection of a mathematical model which fits the distribution of test results.

The Weibull distribution (Section 6.8) is the most widely applied model for treatment of test results because it can accommodate other distributions of continuous data, such as normal and exponential. Application of the model is usually through use of the probability paper described in Section 8.8 or the plotting methods previously explained in Chapter 8.

Non-parametric tests

The term 'non-parametric' is a loose description which covers situations when the distribution of test data is not expressed in terms of its parameters. This happens when the data does not suggest a particular family of distributions such as normal, exponential or Weibull.

The methods described later in this section under binomial trials and the success run theorem are examples of non-parametric tests.

Complete tests

This is a self-explanatory description of parametric tests which are terminated when all units in a sample have been tested to failure.

Time-truncated tests

Time-truncated is the description given to parametric tests which are terminated after a predetermined time. They are sometimes called *censored tests*. A worked example is given in Section 8.8.

Estimation of *MTBF* (mean time between failures) from time-truncated tests uses the values of times to failure. For example, 7 identical units were tested for 700 hours, and during that time 6 failed, one each at 220, 400, 530 and 600, and two at 650 hours. One had not failed when the test was stopped at 700 hours.

$$\text{MTBF} = \frac{\text{total test time}}{\text{number of failures}}$$

$$= \frac{220 + 400 + 530 + 600 + 650 + 650 + 700}{6} = 625 \text{ hours}$$

Time-truncated tests of repairable systems

In time-truncated tests of repairable systems, the sample size is maintained throughout the test period: in other words, units that fail before the predetermined time are replaced. At the end of a test, some units being tested will not have failed; others are replacements for units that have failed.

Estimation of MTBF from time-truncated tests of repairable systems takes account of whether or not failed units are replaced. For example, 7 identical units were tested for 700 hours, and during that time 8 failures occurred which were immediately replaced (a total of 15 units were used).

$$\text{MTBF} = \frac{\text{total test time}}{\text{number of failures}} = \frac{7(700)}{8} = 612 \text{ hours}$$

Failure-truncated tests

Failure-truncated tests are parametric tests which are terminated when a predetermined number of units have failed in the sample being tested. These tests are designed to reduce test time: for example, it usually takes less time to test 20 units and stop the test at the 10th failure than to make a complete test of 10 units.

When testing of 20 units begins simultaneously, the 10th failure occurs after only about 24% of the time for a complete test of 10 units when the Weibull shape parameter $\beta = 1$ and after about 50% of the time when $\beta = 2$.

Sudden-death tests

Sudden-death tests are failure-truncated tests where the sample is divided into sub-samples of equal size and the test of each sub-sample is truncated (usually at the first failure).

Information was required of the time by which 5% of exhaust valves would fail in a prototype six-cylinder engine. An accelerated sudden-death test was carried out using 10 engines. Each engine was tested until its first exhaust valve failure. The results were as follows:

Engine	1	2	3	4	5	6	7	8	9	10
Age at failure (hours)	7	13	20	30	36	42	49	61	72	92

The sample size (n) was 10, and each engine was taken as a class (i) and as a sub-sample of six. Plotting positions (Pi) were calculated using the Hazen formula (Section 8.2).

Class (i)	1	2	3	4	5	6	7	8	9	10
Cumulative failure % (P_i)	5	15	25	35	45	55	65	75	85	95

Age is shown plotted against cumulative failure on a Weibull probability paper in Figure 12.17 with a best-fit 'sample line' drawn through the plots. The sample line represents the failure distribution of the first out of six possible failures: in other words (again using the Hazen formula, but with $n = 6$), it represents the distribution of 8.3% of failures.

To obtain an estimate of the distribution of all times to failure, a second line is drawn. This is the 'population line', which is drawn parallel to the sample line through the point where the 8.3% cumulative failure line passes through the age at which 50% cumulative failure occurs in the sample.

The time by which 5% of all exhaust valves would fail is estimated from the population line. It is about 28 hours, as shown in Figure 12.17.

Extrapolations

Test methods aimed at reducing test time (for example, failure-truncated and sudden-death tests) amount to extrapolation of the life curve beyond the test life. These methods are valid only when the mathematical model is known from past experience of similar units. Only the known portion of the life curve

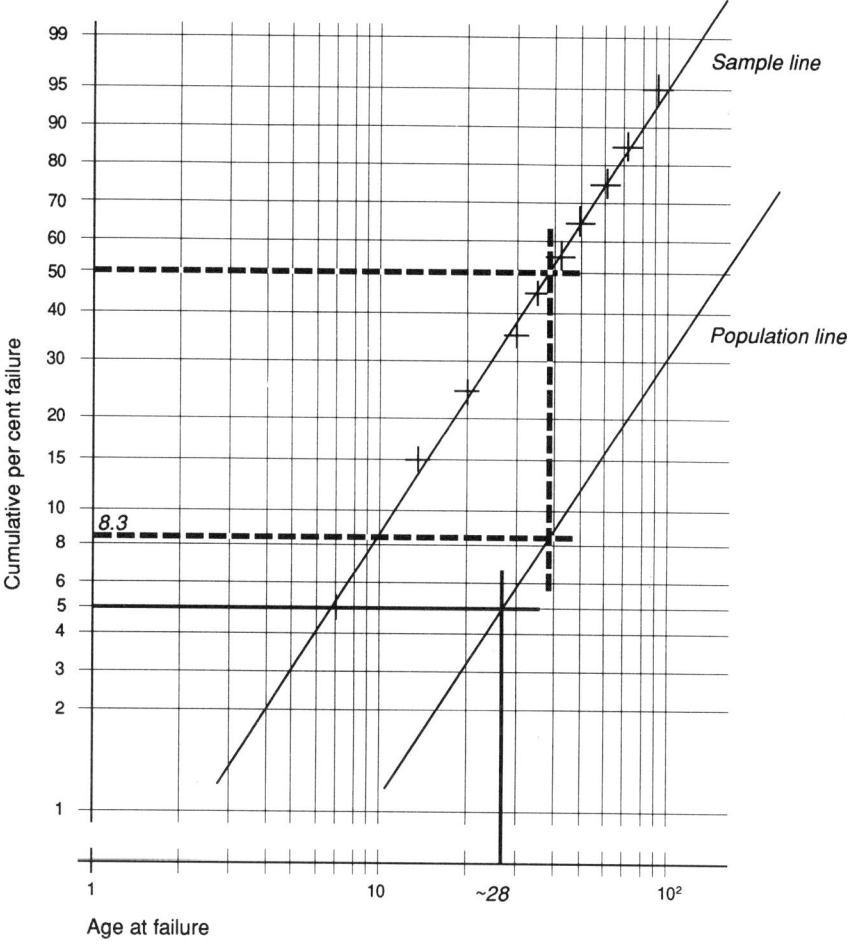

Figure 12.17 Weibull probability plot for sudden-death test data

should be used for estimations, and care should be taken to go not too far beyond the test period.

Sequential tests

Sequential testing describes a procedure where units are tested one at a time to a predetermined condition. It is used because the total test time to reach a decision is often shorter than for the previously described tests. A sequential test is stopped when the number of failures and the cumulative test time of all units tested show (at a given confidence level) that reliability is greater or less than a predetermined value.

Interrupted tests

Tests where functioning units are withdrawn from test at various times are called interrupted tests, and the withdrawn units are called *suspended units*.

Accelerated tests

Tests that involve subjecting units to conditions more severe (and hence resulting in a shorter life) than those encountered in normal use are called accelerated tests. The results from accelerated tests can be extrapolated to normal conditions provided the following apply:

■ The more severe conditions do not give rise to different failure modes.
■ The correlation between test and normal use is known.

Binomial trials

In a binomial trial, a sample of several items is tested for a fixed period of time or until failure, whichever is shorter. The items need not be tested simultaneously, but the maximum test period must be the same for each item. When the trial is completed, reliability (R) and failure rate (F) are calculated respectively as the percentages surviving and failing in the sample tested.

Ten vehicle emission systems (the sample) were tested for the equivalent of 100 000 miles under 'average use' conditions. The test was carried out to obtain data that would enable estimation of the reliability of a larger population with a confidence level of 90%. There are 4 failures: that is, $F = 40\%$ and $R = 60\%$.

The estimation of population reliability for this example is read off from a family of curves for 90% confidence level, as shown in Figure 12.18.

The lines for $F = 40$ and $R = 60$ intersect the sample size 10 upper limit line at about 15 and 28 respectively. They also intersect the sample size 10 lower limit line at about 69 and 84. In other words, for up to 100 000 miles and at 90% confidence level (or with a 10% risk of error), the probable failure rate in the larger population is between 15% and 69%. Also, the probable reliability is between 28% and 84%.

If only reliability is of interest (a one-sided confidence level), the statement can be made that, up to 100 000 miles, the larger population will have a reliability of at least 30% with a 95% confidence level (5% risk of error).

306

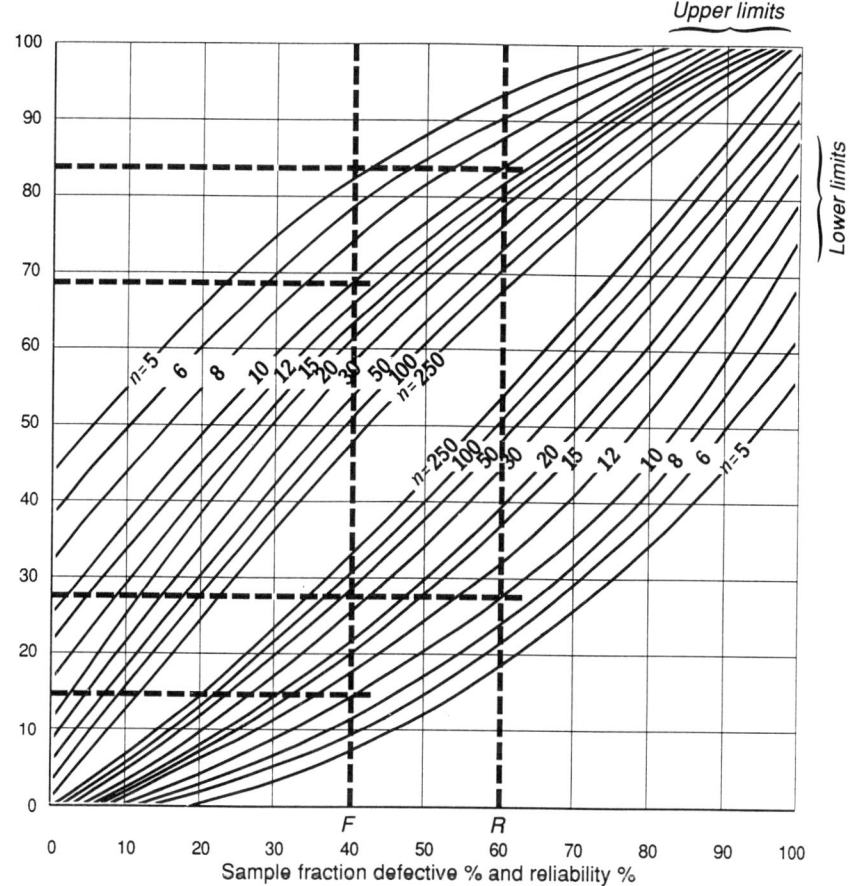

Figure 12.18 Family of curves for a binomial trial at 90% confidence level

Success run theorem

There are occasions when no failures occur during tests and the reliability estimation methods described so far are of little use. In this situation, a concept known as the success run theorem can be used. The result is a statement about reliability at a particular confidence level, and the statement is dependent upon sample size as shown in the formula

$$R_C = (1 - C)^{1/n}$$

where R_C is the reliability with confidence level C and n is the sample size.

307

Table 12.10 Success run theorem sample size table

R minimum	Confidence level (C)															
	0.500	0.550	0.600	0.650	0.700	0.750	0.800	0.850	0.900	0.950	0.960	0.970	0.980	0.990	0.995	0.999
0.500	1	2	2	2	2	2	3	3	4	5	5	6	6	7	8	10
0.550	2	2	2	2	3	3	3	4	4	6	6	6	7	8	9	12
0.600	2	2	2	3	3	3	4	4	5	6	7	7	8	10	11	14
0.650	2	2	3	3	3	4	4	5	6	7	8	9	10	11	13	17
0.700	2	3	3	3	4	4	5	6	7	9	10	10	11	13	15	20
0.750	3	3	4	4	5	5	6	7	9	11	12	13	14	17	19	25
0.800	4	4	5	5	6	7	8	9	11	14	15	16	18	21	24	31
0.850	5	5	6	7	8	9	10	12	15	19	20	22	25	29	33	43
0.900	7	8	9	10	12	14	16	19	**22**	29	31	34	38	44	51	66
0.950	14	16	19	21	24	28	32	37	45	59	63	69	77	90	104	135
0.960	17	20	23	26	30	34	40	47	57	74	79	86	96	113	130	170
0.970	23	27	31	35	40	46	53	63	76	99	106	116	129	152	174	227
0.980	35	40	46	52	60	69	80	94	114	149	160	174	194	228	263	342
0.990	69	80	92	105	120	138	161	189	230	299	321	349	390	459	528	688
0.995	139	160	183	210	241	277	322	379	460	598	643	700	781	919	1058	1379
0.999	693	799	916	1050	1204	1386	1609	1897	2302	2995	3218	3505	3911	4603	5296	6905

More usually, the success run theorem is used to determine the sample size required to demonstrate a particular minimum reliability at a given confidence level. This approach uses a different expression of the formula:

$$n = \frac{\ln(1 - C)}{\ln R_C}.$$

The same information can be obtained, without the need for calculation, by using a table such as the one shown in Table 12.10. Because the method does not relate to test time, it is important that test times are appropriate, otherwise the results will be meaningless.

> An example that uses the success run theorem is to answer the question: how many items must be tested to demonstrate that their reliability is 90% with 90% confidence?
> Referring to Table 12.10, the intersection of the $R = 0.900$ row with the $C = 0.900$ column is $n = 22$. In other words, 22 items must be tested without failure to demonstrate 90% reliability with 90% confidence.

■ 12.10 Variability costs

The basic idea is that, whenever a feature deviates from its optimum, some cost is incurred. This will include some readily quantifiable cost, perhaps representing scrap, rework or additional time needed for assembly in some subsequent stage of manufacture, and some less tangible costs, such as a measure of dissatisfaction on the part of the eventual consumer. No matter how the cost is represented mathematically, it should increase as the feature gets further from its optimum.

Genichi Taguchi has suggested that there is a quadratic relationship between the cost of a feature and its deviation from optimum value. In other words, where the feature measures x and the optimum value is T, the cost is proportional to the square of the deviation from optimum, $(x - T)^2$.

For easy comparison of costs, it is convenient to recentre measurements: that is, to work with their deviations from the mean (μ). The recentred deviate corresponding to x is $y = (x - \mu)$ and so the cost of a measured value x can be written as a quadratic function of y,

$$C(y) = k[y - (T - \mu)]^2$$

where k is an arbitrary scale factor.

The total expected cost is found by multiplying $C(y)$ by the recentred pdf, $f(y)$, to obtain the weighted cost,

$$K(y) = C(y)f(y)$$

and then finding the total area under the curve for $K(y)$. The shape of this

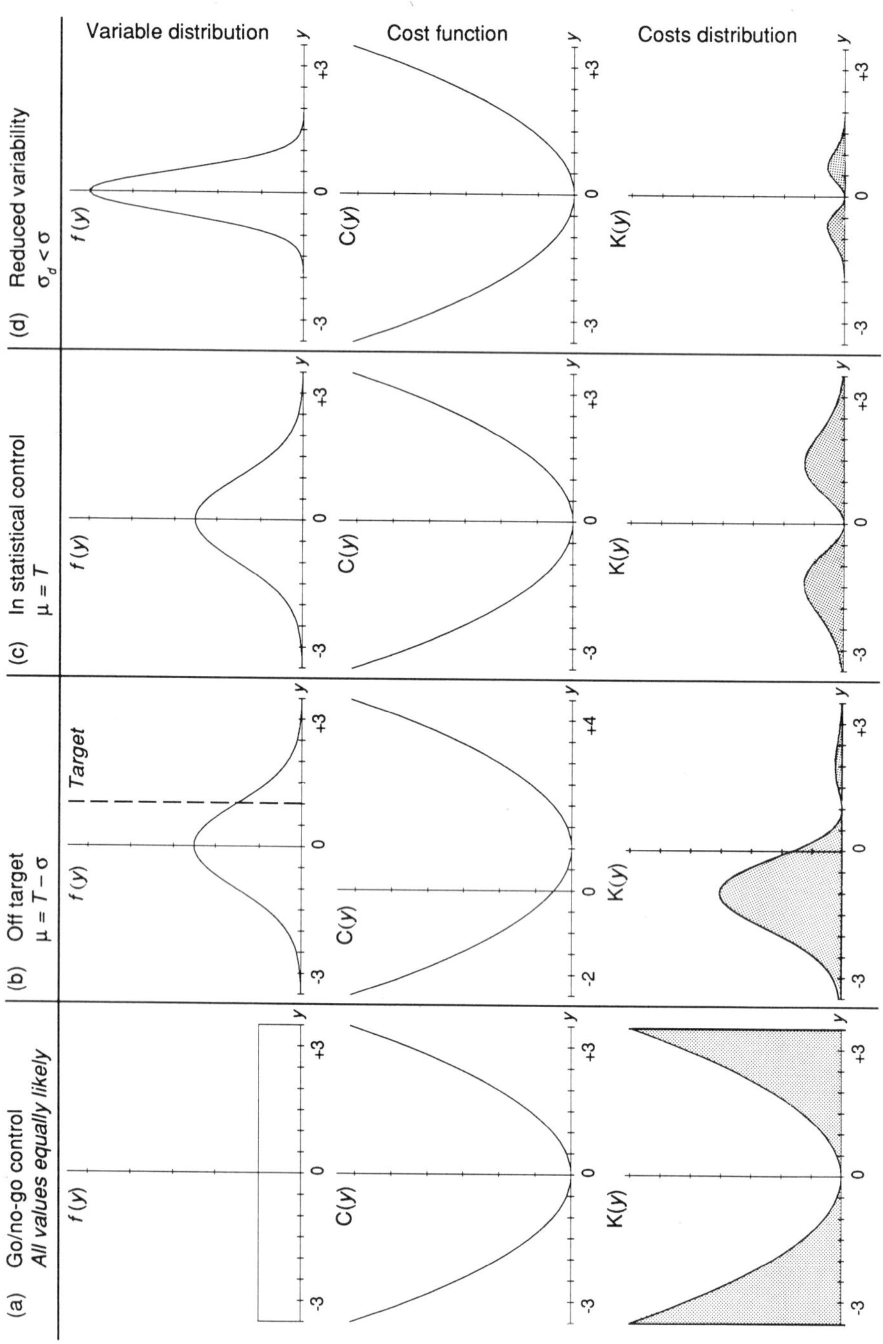

Figure 12.19 Cost to society

curve and the area beneath it vary considerably if the relative values of μ, σ and T are changed. This is illustrated by the four sets of pictures in Figure 12.19. Each set shows the relationship between $f(y)$, $C(y)$ and $K(y)$ for particular conditions.

The costs shown as shaded areas in Figure 12.19 are described as the *costs to society*: in other words waste or opportunity for economic improvement.

> The first set of pictures (a) includes a large shaded area. This set represents a go/no-go control situation where x has a uniform distribution (Section 6.2) with the mean on target. In other words, items are made anywhere within a specified band so that all values between the limits are equally likely to occur.
>
> The second set (b) represents the situation when control is applied such that the values have a normal distribution but the mean is not on target. Again, this set includes a large shaded area. In the example, μ is one standard deviation less than T.
>
> The third set (c) represents another situation when values have a normal distribution with the same standard deviation as set (b), but control is applied such that most items are made at or near to the specified target or nominal (in other words, $\mu = T$). The shaded area is significantly less than in (a) or (b).
>
> The last set (d) represents the mean on target, as in (c), but the standard deviation σ_d is smaller than σ in (b) and (c). This reduction in variability leads to a very noticeable reduction in the shaded area that represents the cost to society.

All the pictures relate to the same product, so the cost function, $C(y)$, is the same for all conditions of manufacture.

The varying height of the $K(y)$ curve shows how various values of y contribute to the total expected cost. In Figure 12.19(b) the largest contributions are made by values of y near $-\sigma$, whereas in Figure 12.19(c) the largest contributions are made by values near $\pm\sqrt{2}\sigma$ (approximately $\pm 1.4\sigma$).

The total area under the $K(y)$ curve when $\mu = T$ is half that when $\mu = T - \sigma$, so the total expected cost is halved by shifting the mean by one standard deviation in this case. It can be shown that, if the cost function is quadratic, the total expected cost will always be smallest when $\mu = T$, regardless of whether or not the underlying distribution is normal. This indicates that it makes economic sense to keep the process mean on nominal.

Another consequence of assuming that the cost function is quadratic is that, when the process mean is equal to nominal, the total expected cost will be directly proportional to the process variance. This implies, for instance, that if the standard deviation (σ) is reduced by a factor of 2, then the total expected cost (proportional to σ^2) will be reduced by a factor of 4, as illustrated in Figure 12.19(d).

Although the cost function might sometimes not be quadratic, this kind of analysis suggests two conclusions:

- Processes should be set and maintained as close as possible to the target (nominal) value. This ideal might be difficult to achieve if the process is not in statistical control (Section 9.2).
- There is economic benefit in continuous process improvement that reduces product variability, even for processes which are in statistical control and have high capability.

These points were made earlier and illustrated for a particular case in the discussion of process control (Section 9.7). Figure 9.4 in that section is an elegant representation of actual costs which is commonly but erroneously used in the context of Taguchi's theory. If it had been derived using the theory explained in this section, its shaded areas would appear like those in Figure 12.19.

■ References to further reading on the topics of Chapter 12

Walpole and Myers (1993) give a very clear account of the material covered in this chapter but at a more technical level. Draper and Smith (1981) should be read before undertaking any serious regression analysis. Cox (1958) and Davies (1956) are classic works on the design of experiments. Newer methods are described by Taguchi (1991) and Logothetis (1990). For further discussion of variability costs, see Taguchi (1991).

See Bibliography for details of titles and publishers.

■ Self-assessment questions on Chapter 12

1. Pellets are mass produced by the automatic compression of a powder. The specific gravity determined for each of a random sample of 5 pellets from each of 4 machines is given below. Is there a significant difference between machines with respect to the specific gravity of the pellets they produce?

Machine	A	B	C	E
	1.72	1.68	1.70	1.70
	1.70	1.67	1.72	1.69
	1.71	1.68	1.70	1.72
	1.69	1.70	1.71	1.69
	1.73	1.66	1.72	1.71

2. To minimize the corrosion action in aqueous antifreeze solutions, an inhibitor is incorporated in the concentrated antifreeze. The effect of 4 inhibitors is to be tested in 3 types of antifreeze. Metal samples are allocated at random to each antifreeze/inhibitor combination. All samples are immersed for an equal time and the corrosive action is assessed by the loss in weight. The results are in the following table in milligrams per square decimetre per day.

Inhibitor	A	B	C	D
Antifreeze 1	16	17	16	14
Antifreeze 2	15	16	13	17
Antifreeze 3	16	17	15	13

What conclusions can be drawn from a two-factor analysis of variance?

3. For 12 test specimens the normal stress and shear resistance were determined, with coded results as follows:

Specimen

1	2	3	4	5	6	7	8	9	10	11	12

Normal stress
30.6 27.6 32.4 31.9 33.1 29.7 28.9 32.7 28.6 33.9 30.4 31.8
Shear resistance
24.9 25.8 21.4 21.7 22.5 26.3 25.9 23.6 27.1 24.2 27.3 26.5

(a) Find the correlation coefficient and test its significance.
(b) Find the linear regression line for estimating shear resistance from normal stress, and predict the shear resistance when stress is 35.0.

4. The data given below is yearly output measured in thousands of tonnes over an 8-year period.

Year	1	2	3	4	5	6	7	8
Output	15.7	15.8	16.8	17.2	16.9	17.9	18.3	18.5

(a) Plot the data on a graph.
(b) Calculate the regression line for predicting output.
(c) Plot the line on the graph.
(d) Calculate the correlation coefficient.
(e) Test the correlation coefficient for significance and comment.
(f) Forecast output in year 9.

5. The weight (grams) of a particular powder which will dissolve in 100 millilitres of water at different temperatures ($^\circ$C) was found to be as shown in the table below.

Temperature	0	10	20	30	40	50	60	70	80	90	100
Weight		43.4	49.3	55.1	60.7	65.5	70.4	75.9	80.1	84.9	89.3	93.6

(a) Plot the data on a scatter diagram.
(b) Estimate the linear regression line to predict weight from temperature and plot on the scatter diagram.

(c) Identify the fitted values of weight at each of the given temperatures.

(d) Determine the residuals (observed y – fitted y) and plot these against the temperature.

(e) Find the correlation coefficient r and the coefficient of determination.

Note when attempting question 5

With the points close to a straight line and the coefficient of determination $= 99.7\%$, the linear regression line appears to be an adequate mathematical model for the data. However, the plot of the residuals clearly shows that this is not so. In fact, fitting a quadratic of the form $y = a + bx + cx^2$ gives an even better fit, with the coefficient of determination almost 100% and the residuals showing no recognized pattern.

13 Qualitative Methods

■ 13.1 Introduction

Statistical methods are used to arrange measured and counted numerical data; they provide quantitative information that helps conclusions to be drawn from the data. This chapter gives brief descriptions of some important methods that prompt the use or consider the results of statistical methods. They are included in this 'statistical' textbook to emphasize the point that statistics have relevance only in the context of particular situations.

Notice in this chapter that some words, such as 'expectation' and 'probability', are used occasionally in ways that are different from their statistical usage as technical terms in earlier chapters.

■ 13.2 Brainstorming

Brainstorming is a particular title that has been given to industrial practice of a commonsense activity. Whenever there is a need, perhaps a problem, it is sensible to obtain all relevant information before putting together a possible plan of action. The activity is not only commonsense, it is also commonplace and is carried out in a variety of guises. In parliamentary democracies there is political debate, in commercial companies there are board-room reviews, in legal proceedings there are arguments and pleas. All these situations are governed by strict codes of practice.

The same applies to industry, where relevant information might be held by anyone, from the experienced chief executive to the latest new employee. Brainstorming is designed to elicit that information in the context of a product or process problem. It is essentially a small group discussion with a code of practice designed to encourage contribution from all.

The ground-rules for the activity were described succinctly by the United States Federal Aviation Authority during the investigation of an air disaster:

Obtain facts
- Do not analyse.
- Do not select potential causes.
- Do not discard potential causes.
- Do not focus.

The group atmosphere should be disciplined but informal. It is important that rank or other social considerations do not inhibit discussion, and that all get the chance to participate.

The subject matter might be the positioning of a new machine or the review of a new design or an unusual control chart pattern or a source of customer dissatisfaction. All contributions are recorded. The following list is an example of brainstorming the question, 'what aspects of a case-hardening process might contribute to gear wheel failure?'

soft metal structure	metal composition	hard metal structure	surface area
gear weight	inert gas	carrier gas	positioning in furnace
conditioning gas	carburizing agent	furnace temperature	time in furnace
furnace lining	furnace conveyor	heater tubes	calibration procedure
quench composition	quench temperature	contamination	metal movement
material handling	retreatment process	scrap procedure	batch identification

The next stage is to assemble the information in a manner that allows for proper decision making. The following three sections outline commonly used techniques. Note the corollary to this, that these methods are best used when preceded by brainstorming.

■ 13.3 Cause-and-effect diagram

Cause-and-effect diagrams are associated with the Japanese engineer Ishikawa and, because of their appearance, are sometimes called 'fishbone' diagrams. Their construction starts from a list of the possible causes of an effect – this could be a list generated by brainstorming in response to a problem.

The first step is to define precisely the *effect* or problem under consideration and then to construct a diagram linking it to the five *main groups* or sources of *causes*. Figure 13.1 illustrates this conventional diagram; in practice, other terms might be used to describe the main groups and, at some stage, one or more of the groups might be discarded. However, at the start it is helpful to have headings that can accommodate every possible cause.

In the illustration, the people, methods and material groups are self-explanatory. Facilities includes tools, machines, buildings and so on. The

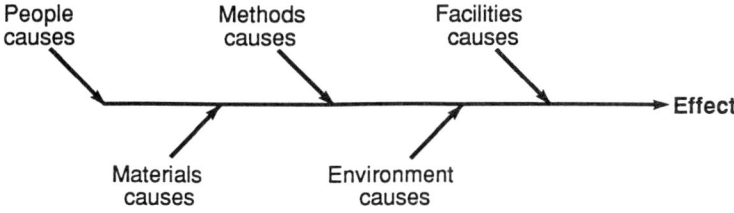

Figure 13.1   Cause groups in a cause-and-effect diagram

environment group is for those causes that are outside the direct control of the assembled company: for example, the weather, legislation and market conditions.

Especially when there is a large number of possible causes, *sub-groups* might be formed. This is done by deciding what type or aspect of the main groups seem to be involved in the effect. For example, the people involved might be technicians, managers, executives and so on. The sub-groups are represented in Figure 13.2 (which shows sources of failure or success in quality achievement) by branch arrows leading to the main group. For convenience or clarity in some presentations, the sub-group headings could replace the main group headings. This might seem a tedious procedure so far, but in using any technique it is important to have a disciplined approach.

The next step is to use the original list of possible causes and give the reason why a group or sub-group is a source of the problem. The *immediate*

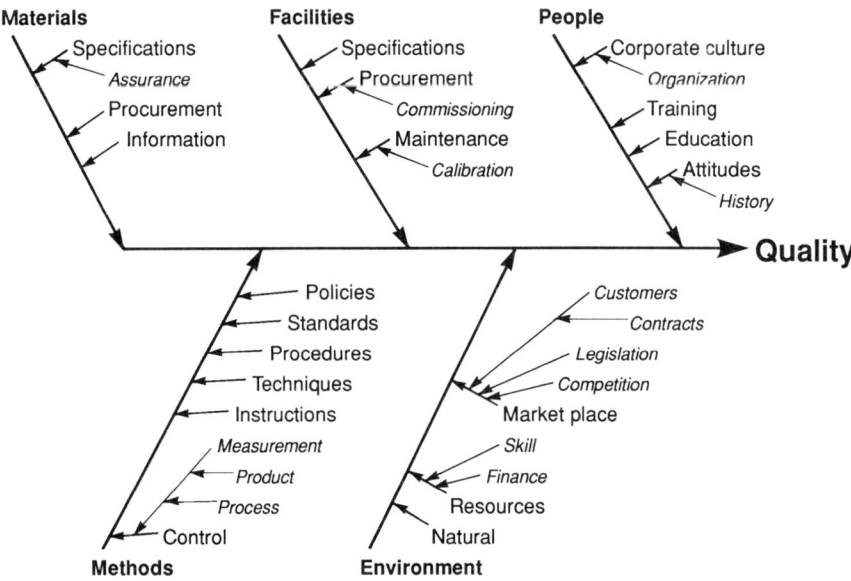

Figure 13.2   Groups and sub-groups of causes

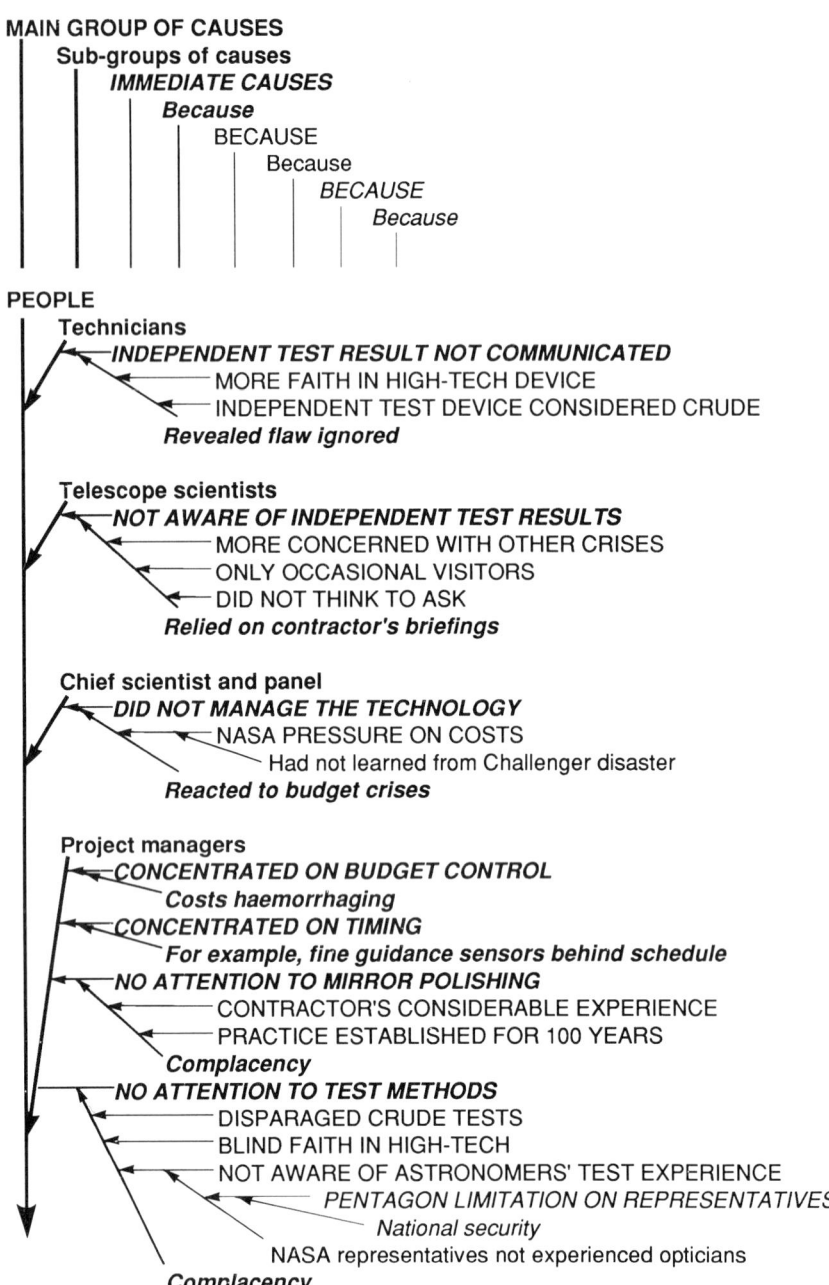

MAIN GROUP OF CAUSES
   Sub-groups of causes
      IMMEDIATE CAUSES
         Because
            BECAUSE
               Because
                  BECAUSE
                     Because

PEOPLE
   Technicians
     INDEPENDENT TEST RESULT NOT COMMUNICATED
         MORE FAITH IN HIGH-TECH DEVICE
         INDEPENDENT TEST DEVICE CONSIDERED CRUDE
     Revealed flaw ignored

   Telescope scientists
     NOT AWARE OF INDEPENDENT TEST RESULTS
         MORE CONCERNED WITH OTHER CRISES
         ONLY OCCASIONAL VISITORS
         DID NOT THINK TO ASK
     Relied on contractor's briefings

   Chief scientist and panel
     DID NOT MANAGE THE TECHNOLOGY
         NASA PRESSURE ON COSTS
         Had not learned from Challenger disaster
     Reacted to budget crises

   Project managers
     CONCENTRATED ON BUDGET CONTROL
      Costs haemorrhaging
     CONCENTRATED ON TIMING
      For example, fine guidance sensors behind schedule
     NO ATTENTION TO MIRROR POLISHING
         CONTRACTOR'S CONSIDERABLE EXPERIENCE
         PRACTICE ESTABLISHED FOR 100 YEARS
     Complacency
     NO ATTENTION TO TEST METHODS
         DISPARAGED CRUDE TESTS
         BLIND FAITH IN HIGH-TECH
         NOT AWARE OF ASTRONOMERS' TEST EXPERIENCE
           PENTAGON LIMITATION ON REPRESENTATIVES
            National security
         NASA representatives not experienced opticians
     Complacency

Figure 13.3 Groups, sub-groups, immediate and deeper causes

cause is stated: for example, inadequate training might be the immediate cause of people causing a problem. Sometimes this means rephrasing the words of the original list; before this is done, it is necessary to obtain confirmation that the rephrased words are true. Immediate causes are shown on the diagram by branch arrows leading either to a main group or to a sub-group of causes.

Finally, but only using the facts to hand, deeper causes are stated. Perhaps inadequate training was caused by lack of teachers caused by lack of investment and so on. Again, arrows are used to link these deeper causes successively to the immediate causes. Figure 13.3 illustrates immediate and deeper causes, and is one limb of a cause-and-effect diagram based on an article (in *Science*, 17 August 1990) that explored the reasons for an optical flaw which troubled the early days of the Hubble telescope.

The cause-and-effect diagram is a tool that helps focus on the causes of or reasons for success or failure and hence leads to activities that use quantitative methods. Statistics are necessary to assess priorities and the effect of particular actions.

■ 13.4 Fault tree analysis (FTA)

A fault tree analysis involves the construction of a *tree of causes*, which aims to portray *how* faults occur. It is distinct from the cause-and-effect diagram,

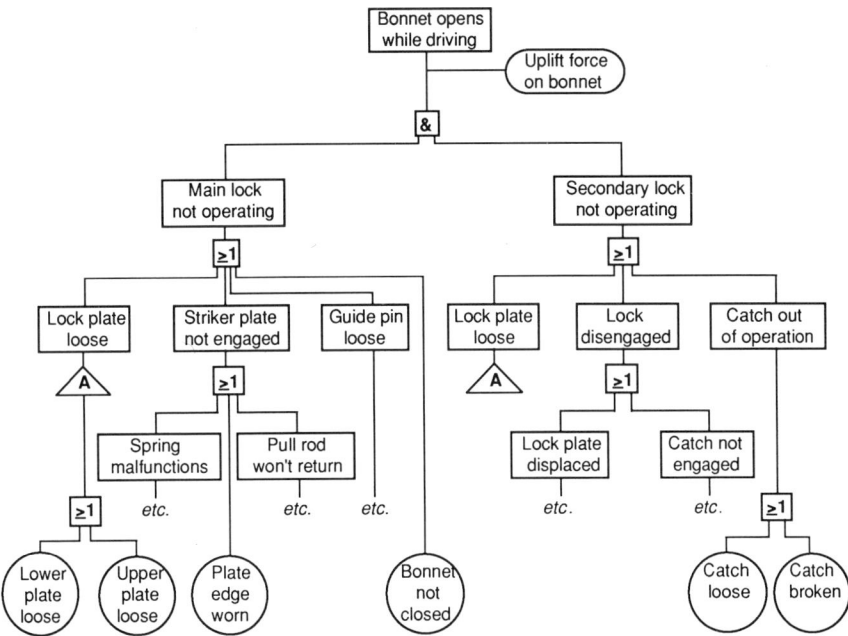

Figure 13.4 Tree of causes

which shows *what* caused the fault. A tree of causes breaks down a fault (or *top event*) into events which, singly or combined, cause the fault to occur. Each possible cause is then broken down again and again, until *basic events* are determined.

These basic events might become the subject of brainstorming and construction of a cause-and-effect diagram. The method is particularly useful when a problem has many causes and the initial investigation needs to isolate different avenues of approach. It is demonstrated by the example in Figure 13.4, where there is identification of basic events that might cause a vehicle's bonnet to open, when the vehicle is being driven.

The easily drawn symbols used in Figure 13.4 are defined in Figure 13.5 together with alternatives and some occasionally used refinements that might require a special template.

An application of the tree of causes leading to use of statistical techniques is shown in Figure 13.6. The problem in this example was productivity: an

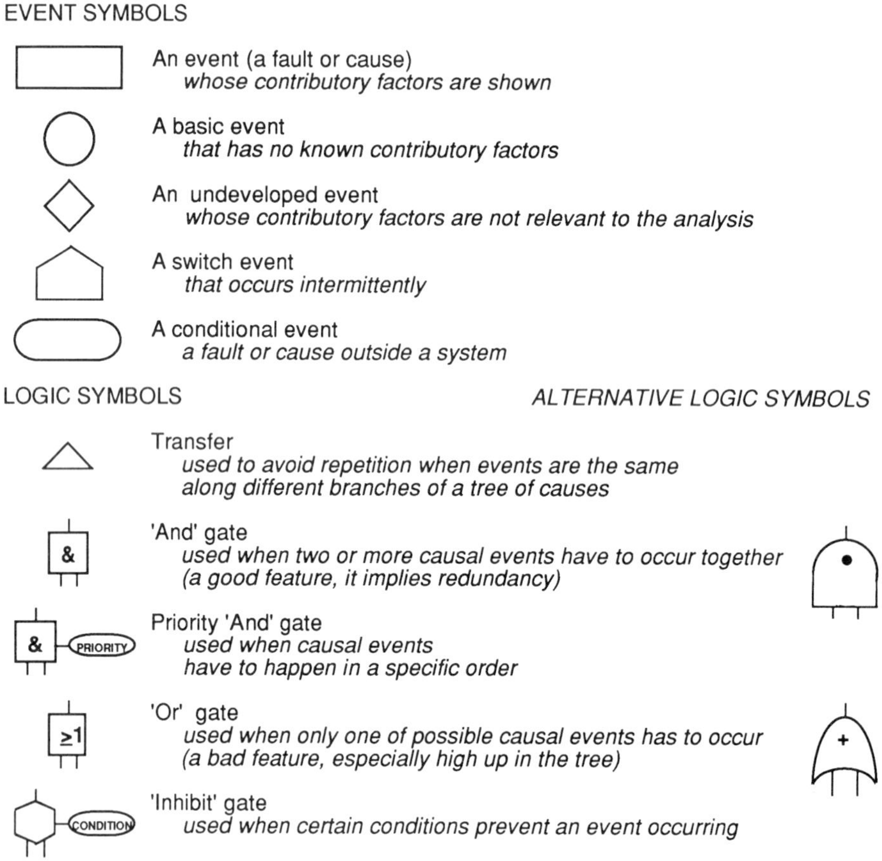

EVENT SYMBOLS

An event (a fault or cause)
whose contributory factors are shown

A basic event
that has no known contributory factors

An undeveloped event
whose contributory factors are not relevant to the analysis

A switch event
that occurs intermittently

A conditional event
a fault or cause outside a system

LOGIC SYMBOLS   *ALTERNATIVE LOGIC SYMBOLS*

Transfer
used to avoid repetition when events are the same
along different branches of a tree of causes

'And' gate
used when two or more causal events have to occur together
(a good feature, it implies redundancy)

Priority 'And' gate
used when causal events
have to happen in a specific order

'Or' gate
used when only one of possible causal events has to occur
(a bad feature, especially high up in the tree)

'Inhibit' gate
used when certain conditions prevent an event occurring

Figure 13.5   Symbols in the tree of causes

320

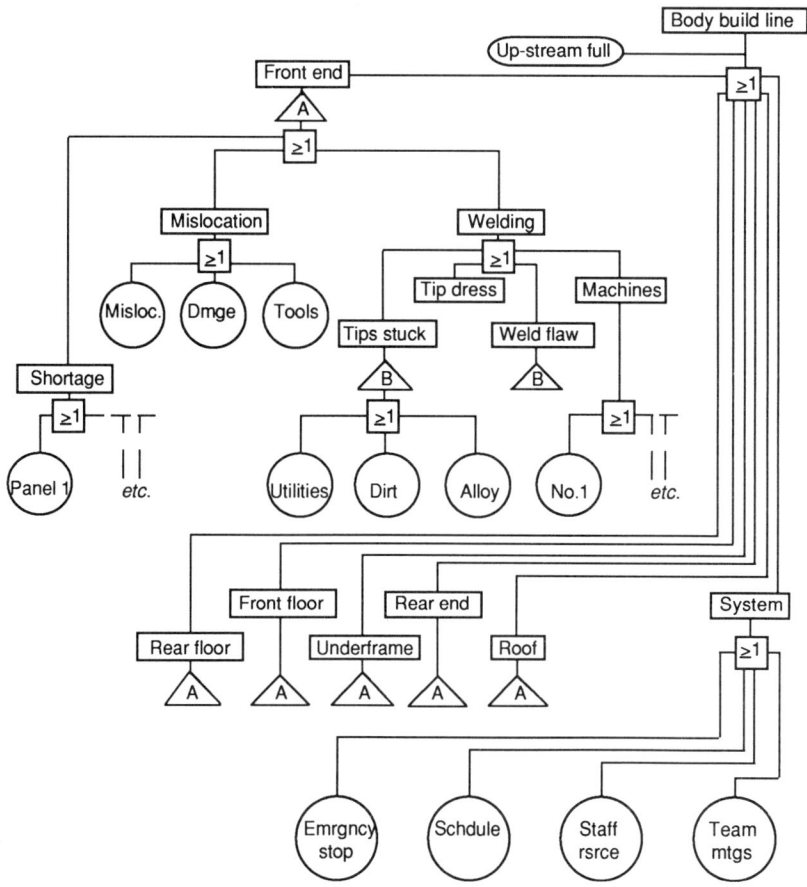

Figure 13.6 Application of fault tree analysis to process availability

automated machine did not produce its planned capacity. The tree of causes was constructed to help the identification of factors that affected machine availability.

These factors were measured, in terms of machine down-time, over several months. There are 77 factors summarized in Figure 13.6, most of which had an availability of better than 99.85%, a probability of failure of less than 0.0015. The theoretical machine availability, obtained simply by using the multiplication rule (say $0.9985^{77} = 0.8908$ or 89%), confirmed that merely driving the workforce would be unlikely to achieve the required productivity. Several possibilities were examined, such as duplicating the whole machine or splitting the machine and then either introducing a mid-way buffer stock or running one part of the machine faster than the other. A satisfactory solution was determined through statistical methods using parallel models (Section 4.12) and involving economics beyond the scope of this book.

■ 13.5 Failure mode, effect and criticality analysis (FMECA)

FMECA provides a disciplined approach to determining the seriousness of possible failures, and hence the nature of and priorities for preventive action. The method is applicable to components of, systems within and whole products and processes. It is a three-stage technique. All stages usually involve brainstorming and their acronyms summarize their content:

1. FMA – determine possible failure modes.
2. FMEA – then determine the effect of the possible failures.
3. FMECA – finally, assess the criticality of the failures.

Figure 13.7 illustrates FMECA applied to a simple component through a worksheet. Usually, the worksheet is first prepared at the outset of a project (before expense is incurred in pilot schemes or prototypes), but it is not a once-and-for-all document. If failure actually occurs at any time in the future, it is usual to interrogate the FMECA, determine the shortfall in design perception and hence advance the state of the art. Appropriately updated FMECA can be a basis for 'shelf engineering' or, in other words, a record of reliable products and processes.

The example was constructed by listing the components (the physical parts and their functions) of the system under consideration and then applying the three-stage FMECA technique to each component. Stages 1 and 2 are self-explanatory: possible failure modes and likely effects are listed (the list of failure causes is an extra that might not directly contribute to the technique). Stage 3 involves the calculation of a *criticality index*, C_F, which is the product of three subjective ratings ($C_F = P_o \times S \times P_d$) where
P_o is the subjective 'probability' of an occurrence (usually in a given time) when $1 =$ unlikely and $10 =$ definitely,
S is the subjective 'seriousness' of an occurrence when $1 =$ unimportant and $10 =$ a danger to life and limb, and
P_d is the subjective 'probability' of preventing failure by detection before it happens when $1 =$ will be found and $10 =$ will not be found.

With experience, the value of C_F can be interpreted as an indicator of necessary subsequent action, as indicated in the last column of the worksheet. Of course, at design and later stages, the target is continuously to reduce any value of C_F.

Figure 13.7 A prototype design FMECA worksheet

SYSTEM									
			Accelerator pedal						
COMPONENT		FAILURE CHARACTERISTICS			INDEX				ACTIONS
DSCRPTN	FUNCTION	MODE	EFFECTS	CAUSES	P_o	S	P_d	C_F	
Plate	Increases area	Loosens	Noise	Small size Wear	1	1	2	2	Screen
		Breaks	Loosens Discomfort	Damage Mat. flaw	1	2	1	2	Screen
Spring	Returns plate	Weakens	Slow return Discomfort	Mat. flaw Heat treatment	3	2	2	12	Process control
		Breaks	No return Discomfort	Fatigue	1	8	9	72	Review design
Stem	Transmits load	Distorts	Sticks	Small size Damage	2	8	3	48	Review design

■ 13.6 Quality function deployment

Quality function deployment (or QFD for short) is a planning technique that originated in Japan during the late 1960s. Its purpose is to focus product design, manufacturing design and production control on to the essentials of satisfying customers. It works in stages to translate sometimes vague statements of customer expectations into product concepts, then into detail features, capable manufacturing facilities and finally production process controls.

This might seem a simple and obvious sequence of events for any manufacturer. However, they represent a complex process that is rarely sufficiently well coordinated or communicated. The process is called QFD when particular charts of various types are put together and used methodically. Effective application of QFD requires extended use of engineer resources and considerable management commitment. The methodology described below was pioneered in the United States of America and is used within the motor industry.

At each stage of the process, the different charts are assembled into a single picture that is often called *the house of quality*. A typical picture is summarized in Figure 13.8, the purpose of which is to present all information that is relevant to the stage. It is not essential for every picture at every stage to contain all the detail shown in Figure 13.8 – the important parts will become apparent as the picture is explained. It is usual for the first stage, the conversion of sometimes vague customer requirements into an objective product concept specification, to have the most comprehensive picture.

The house of quality

The four areas (numbered 1, 3, 5 and 6) that are in bold print on Figure 13.8 should appear in the pictures at every stage of QFD. They represent the core information of customer requirements, design features, the relationships between them, and target values for the features. However, there is other information that might be available and of use. Figure 13.8 includes examples of this other information in areas which are numbered in the sequence that the information could be used.

1. Customer expectations are the starting point for the process. Obviously, the more definition there is, the greater the likelihood of a satisfactory product. Thorough market research and face-to-face contact are the most useful sources of the information, although in some situations a comprehensive document might suffice.
2. It is helpful if some level of importance can be attached to each expectation. Typically, the customer will be asked to rate the expectations, subjectively, perhaps on a scale where 5 = most essential and 1 = a cosmetic embellishment.

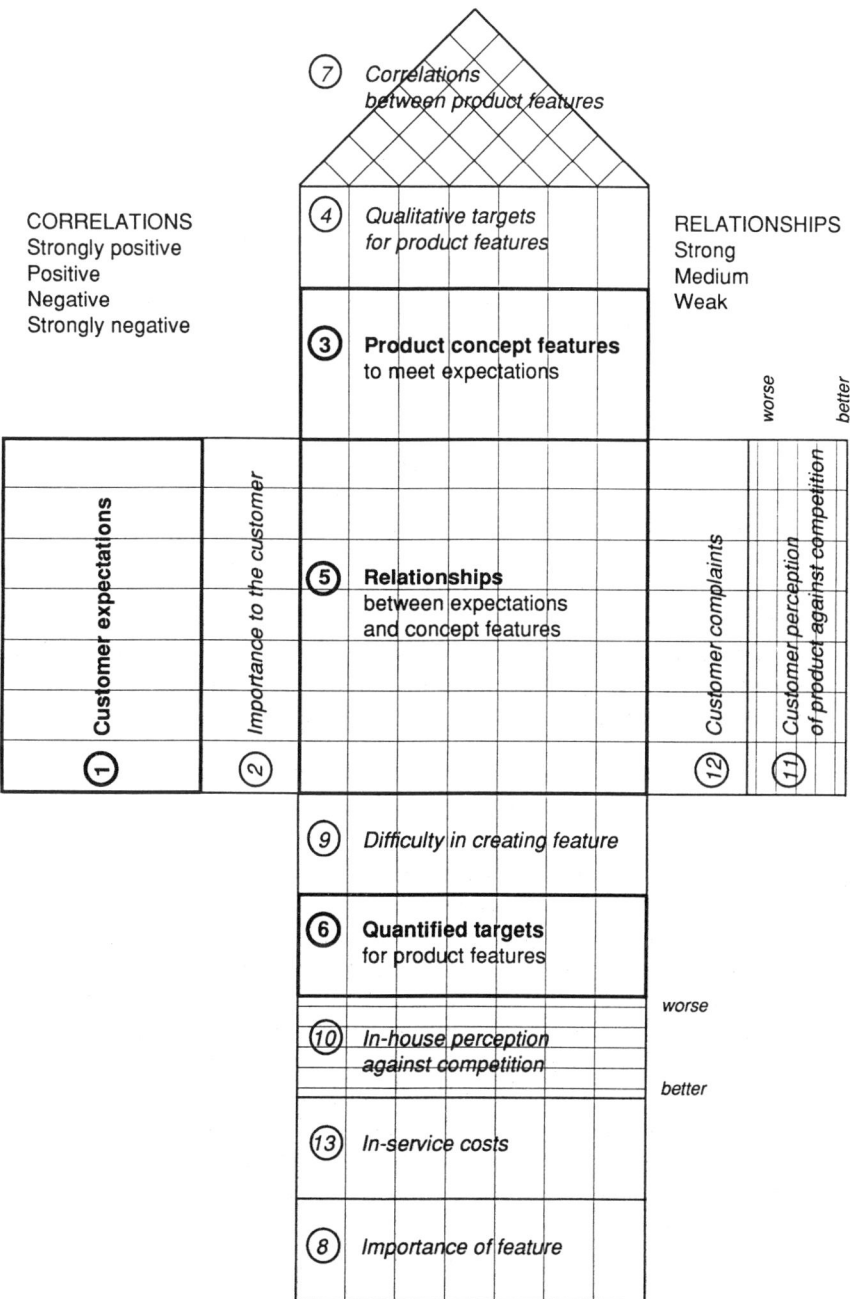

Figure 13.8 House of quality

3. With the expectations in mind, a concept is designed and its features are listed.

4. The designer might indicate broad targets for the features: for example, 'as soft as possible', which is a minimum target, or 'as hard as possible', which is a maximum target, or a precise target such as the height of a motor vehicle bumper, which is subject to regulations. On the picture, maximum, minimum and precise would be indicated respectively by symbols such as ↑, ↓ and ○.

5. The expectations will be delivered, to varying degree, by one or more product features. The matrix of relationships is drawn to show how this happens, whether strongly, weakly or in between. Again symbols are used on the picture to indicate the relationships: for example, strong, medium and weak could be indicated respectively by ■, ▲ and •.

   The matrix provides a check that all expectations are catered for, since any blank horizontal column indicates that something has been forgotten. Similarly, any blank vertical column suggests that a feature is surplus to requirements. At this time and later, the features list is revisited and perhaps redesigned to deliver more accurately what the customer expects.

6. Once the features list is settled, each feature must be quantified. For example, softness and hardness will be specified perhaps in terms of a shore hardness value.

7. Product features that deliver some expectations are sometimes helpful and sometimes counter-productive so far as other expectations are concerned. For example, a feature of heavy armour on a vehicle might increase weight and therefore reduce the effectiveness of features aimed at ease of handling. In other words, the heavy armour feature is negative so far as, say, the steering feature is concerned.

   The correlation matrix aims to highlight interactions between features. Symbols that are used include ↙ (strongly positive), + (positive), − (negative) and × (strongly negative).

8. Negative correlations suggest a need to revisit and perhaps respecify some features to reach a compromise among the expectations. Decision making is helped by information about the importance of affected features.

   There can be two approaches to assessing feature importance. The first is to consider matters not mentioned by the customer. This happens sometimes from ignorance and sometimes because the matter seems so obvious that it is not worth mentioning. The approach sweeps in the specialist technical knowledge of the designer, especially where regulations are concerned. For example, the colour of vehicle rear lights and the provision of car safety restraints are mandatory; they have the highest importance and would not be traded off against another feature.

   The second approach is to rate the features, using information already on the picture. If the relationships for each feature are weighted (from 5 above, say ■ = 9, ▲ = 3 and • = 1), multiplied by the customer rating (from 1 to 5 as in 2 above) and then added, each feature will have a subjective rating. The rating numbers are subjective and have no meaning except that they suggest relative importance. Any inconsistency with preconceived ideas about importance should be checked – perhaps errors have been made in the process so far.

9. While the concept is being developed, thoughts will have turned actually to making the product. The picture can include information about practicalities – perhaps resources will not support the concept. Another rating could be developed (1 = no difficulty to 5 = extreme difficulty), but the important matter is to channel effort into concepts that will succeed.

10. During development, information is needed as to how the concept will stand up in the market place. The picture can include comparison of its features list with those of its competitors: in other words, a technical assessment of the design.

11. A further assessment is needed before large-scale cash investment is committed. This one takes the customers' views of the concept into account. A typical method is to show the customers a prototype product alongside its competitors. The customers are asked to rank the fulfilment of their expectations by each on display.

    The designer should compare the results of the in-house technical assessment with the customer assessment. A feature might be superior to the same in a competitor, but the expectation with which the feature has a strong relationship might be better satisfied by the competitor. This would be a hint to go back to the drawing board!

12. Previous similar products can be a source of information that emphasizes the importance of some expectations. The level of customer complaints can indicate aspects of the concept that deserve particular effort. At worst, this information could be obtained after the event, when it would help debriefing with the future in mind.

13. Similar to customer complaints are the costs of ownership both to the customer and to the manufacturer. In the motor industry, these costs are grouped against product features. Again the information sources might be previous similar products or after the event. Whatever the source, the information is another trigger to revisit and perhaps redesign the product.

Deployment

The house of quality has been described in some detail for the first stage of the QFD process. The same principles apply in the later stages which are illustrated in Figure 13.9, but only the four principal areas of the picture are shown.

Figure 13.9 simply illustrates that the product feature targets determined at the concept stage (a) become the 'customer expectations' at the detail design stage (b); detail design targets are the 'customer expectations' for facility design (c); and facility targets are the 'customer expectations' for control (d).

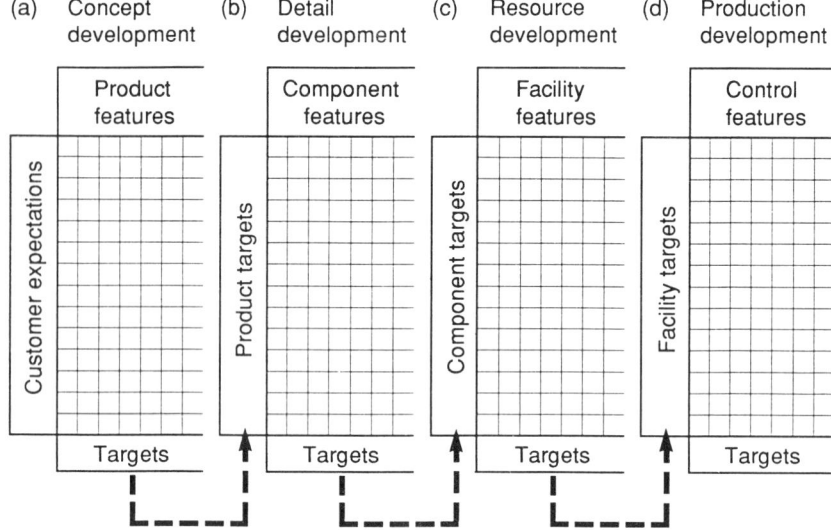

Figure 13.9   Stages in the QFD process

The illustration is of four stages, but this is not a hard-and-fast rule – deployment is taken through as many stages as are appropriate.

As described above for area 5 in the house of quality, the matrix at each stage provides a check that all expectations are catered for. Any blank horizontal column indicates that something has been forgotten. Similarly, any blank vertical column suggests that a feature is surplus to requirements. At all stages, the pictures are focused only on the expectations of the ultimate customer or user of the product.

Obviously, the pictorial representation of the deployment process could get bogged down in unnecessary detail. In practice, the pictures tend to be limited to matters that are new and particularly important or difficult to create.

■ 13.7 In conclusion

This chapter has outlined methods for analysing and testing processes and products. Some are subjective, others make use of several statistical techniques. They can indicate problems or solutions to problems and a course of further action. The effectiveness of that action should be measured.

The early chapters of this book discussed data collection and the way that real processes can be explained by probability models. The normal distribution and others are offered as models that have wide application in the management of processes and product quality.

To the casual reader, the choice of words in discussion and description might sometimes seem to lack the comfort of a prescriptive recipe. This is deliberate: it emphasizes that there may be several reasonable explanations for things that are observed in the real world. If a fairly simple model provides an adequate explanation, it can be used to make sensible decisions. Fitting more elaborate models usually involves collecting more data and doing more calculation, but it should result in making better decisions.

So to the final words, which are a warning! The application of statistical techniques must not be seen as some sort of magic cure or painless solution for the ills of industry. It is not. It is a practical and powerful tool to be used in diagnosis of the root causes of many of those ills.

■ References to further reading on the topics of Chapter 13

Feigenbaum (1983), Ishikawa (1978) and Juran and Gryna (1980) all discuss practical methods for planning, control and analysis of product and process quality.

See Bibliography for details of titles and publishers.

Bibliography

Barnett, V. (1975) 'Probability plotting methods and order statistics' *Applied Statistics*, vol. 24, no. 1, pp. 95–108.

Barnett, V., and Lewis, T. (1984) *Outliers in statistical data*, 2nd edition, Chichester: Wiley.

Bissell, A.F. (1990) 'How reliable is your capability index?', *Applied Statistics*, vol. 39, no. 3, pp. 331–40.

Cass, T. (1973) *Statistical Methods in Management 1*, 2nd edition, London: Cassell.

Chatfield, C. (1983) *Statistics for Technology*, 3rd edition, London: Chapman and Hall.

Cox, D.R. (1958) *Planning of Experiments*, New York: Wiley.

David, H.A., and Moeschberger, M.L. (1978) *The Theory of Competing Risks*, London: Griffin.

Davies, O.L. (1956) *Design and Analysis of Industrial Experiments*, 2nd edition, Edinburgh: Oliver and Boyd.

Deming, W.E. (1986) *Out of the Crisis*, Cambridge, Mass.: MIT Center for Advanced Engineering Study.

Draper, N., and Smith, H. (1981) *Applied Regression Analysis*, 2nd edition, New York: Wiley.

Feigenbaum, A.V. (1983) *Total Quality Control*, New York: McGraw-Hill.

Grant, E.L., and Leavenworth, R.S. (1988) *Statistical Quality Control*, 6th edition, New York: McGraw-Hill.

Hahn, G.J., and Shapiro, S.S. (1968) *Statistical Models in Engineering*, New York: Wiley.

Ishikawa, K. (1978) 'Quality control in Japan' Proceedings of 13th IAQ meeting, Kyoto, Japan.

Juran, J.M., and others (1974) *Quality Control Handbook*, New York: McGraw-Hill.

Juran, J.M., and Gryna, F.M. (1980) *Quality Planning and Analysis*, New York: McGraw-Hill.

Kao, J.H.K. (1959) 'A graphical estimation of mixed Weibull parameters in life-testing of electron tubes', *Technometrics*, vol. 1, no. 4, pp. 389–407.

Lawless, J.F. (1982) *Statistical Models and Methods for Lifetime Data*, New York: Wiley.

Logothetis, N. (1990) 'Box–Cox transformations and the Taguchi method', *Applied Statistics*, vol. 39, no. 1, pp. 31–48.

Mann, N.R., Schafer, R.D., and Singpurwalla, N.D. (1974) *Methods for Statistical Analysis of Reliability and Life Data*, New York: Wiley.

Mitra, A. (1993) *Fundamentals of Quality Control and Improvement*, New York: Macmillan.

Montgomery, D.C. (1985) *Introduction to Statistical Quality Control*, New York: Wiley.

Oakland, J.S. (1986) *Statistical Process Control: A practical guide*, Oxford: Heinemann.

Parzen, E. (1960) *Modern Probability Theory and Its Applications*, New York: Wiley.

Shewhart, W.A. (1931) *Economic Control of Quality of a Manufactured Product*, New York: Van Nostrand.

Taguchi, G. (1991) *Introduction to Quality Engineering*, White Plains, NY: Unipub/ Kraus International.

Walpole, R.F., and Myers, R.H. (1993) *Probability and Statistics for Engineers and Scientists*, 5th edition, New York: Macmillan.

Walton, M. (1989) *The Deming Management Method*, London: W.H. Allen.

Answers to Self-Assessment Questions

Chapter 3

1.   1.976, 2.424
2.   1, 2, 1.557
3. (a)   4.6, 4, 1.817
   (b)   4.5, 3.5, 3, 2.204
4.   0.071, 0.011, 0.00305
5.   41.10 μsecs, 11.19 μsecs
6.   46.0 μsecs
7.   41.7 μsecs
8. (b)   32.14 mins, 7.165 mins
   (c)   32.22 mins, 7.172 mins

Chapter 4

1. (a)   1/3
   (b)   (i) 1/12, (ii) 1/2
   (c)   (i) 1/84, (ii) 15/28
2. (a)   0.2
   (b)   0.04, 0.32
   (c)   0.008, 0.384
3. (a)   1/2
   (b)   1/12
   (c)   1/6
   (d)   1/36
   (e)   1/18
   (f)   125/216
4.   7/12

5. (a) 0.63
    (b) 0.37
6. (a) 1/12
    (b) 1/2
    (c) 7/12
7.   0.774
8. (a) 0.316
    (b) 0.422
9.   48
10.   5, 35, 45, 20, 4950, $n(n-1)/2$
11.   56
12. (a) 0.00000369
    (b) 0.719
    (c) 0.0000591
13. (a) £4.7
    (b) £47
14.   £418.75
15.   0.430
16. (a) 0.5685
    (b) 0.0244
    (c) 0.0698
       £1388
17. $f(1) = 0.6$, $f(2) = 0.2667$, $f(3) = 0.1$, $f(4) = 0.0286$, $f(5) = 0.0048$

Chapter 5

1. (a) 0.1075
    (b) 0.0102
    (c) 0.8980
    (d) 0.128
2. (a) 0.0228
    (b) 0.1492
    (c) 0.5948
    (d) 0.6639
       56.625 and 63.375
3. (a) 0.0317 (possibly none)
    (b) 54.8 (about 55)
    (c) 884.9 (about 885)
    (d) 999.313 (possibly all)
4. (a) 2.28%
    (b) 85.3%
    (c) 68.26%
5.   31.86 mpg
6.   2.44 g, 248.7 to 261.3 g

Chapter 6

1. (a) Mean = 67.5, standard deviation = 4.33;
      probability = $14/15 = 0.9330$
   (b) $P(z < 1.5) = 1 - 0.0668 = 0.9332$, so probability of completion within standard job time is effectively the same (approximately 93% certain) whether the distribution is uniform or normal.
2. (a) $p = 0.0228$
   (b) $f(0) = 0.6305$; $f(1) = 0.2942$; $P(x < 2) = 0.9247$
3. $\lambda = rt = 3.0$
   (a) 0.4232
   (b) 0.5768
   (c) 0.9161
4. $P(z > 5.5/5) = P(z > 1.1) = 0.1357$
5. $f(1) = 0.25$; $f(2) = 0.1875$; $f(3) = 0.1406$;
   $F(3) = 0.5781 = P(\text{less than 4 calls})$
6.

t	0	0.5	1.0	1.5	2.0	2.5
$h(t)$	0	0.96	2.20	3.58	5.05	6.61
$R(t)$	1	0.804	0.368	0.087	0.010	0.001

Chapter 7

1. $\hat{p} = 3/30 = 0.1$
2. $\hat{\lambda} = 6.6$
3. $\bar{x} = 10.5$; $\hat{r} = 0.095$
4. Confidence limits are 220 and 234 minutes (i.e. 3 hours 47 minutes ± 7 minutes).
5. Confidence limits for p are 0.87 ± 0.066, so for percentage they are 87 ± 6.6.
   About 74 more cars should be sampled.
6. Estimate from standard deviations is 5.9 and that from ranges is 5.8.

Chapter 10

1. (a) The process is capable.
   (b) $UCL_x = 0.3048$ mm, $LCL_x = 0.2953$ mm, $UCL_R = 0.0174$ mm, $LCL_R = 0$, is in control.
2. $UCL = 0.3059$ mm, $LCL = 0.2945$ mm, is in control.
3. $UCL_x = 509.6$ ml, $LCL_x = 497.9$ ml, $UCL_R = 24.1$ ml, $LCL_R = 0$, is in control.

4. (b) Variation reduced (the last 18 range values are all below the average range). The setting (average mean) does not appear to have changed. Using original average mean and new average range (of last 15 samples) gives $UCL_x = 505.7$ ml, $LCL_x = 501.8$ ml, $UCL_R = 8.2$ ml, $LCL_R = 0$.

5. $UCL = 0.136$, $LCL = 0$, is in control.

6. $UCL = 15.35$, $LCL = 0$, is not in control.

7. $UCL = 14.9$, $LCL = 0$, in control, slight improvement (8 of last 9 below average).

8. $UCL = 0.723$, $LCL = 0.465$, is in control.

9. $UCL_x = 317.1$, $LCL_x = 128.2$, $UCL_R = 116.0$, $LCL_R = 0$, is in control.

Chapter 11

1. $\chi^2 = 10.86$, $\nu = 4$, $P < 5\%$, significant, yes.

2. $\chi^2 = 2.25$, $\nu = 1$, $P > 10\%$, not significant, no.

3. $\chi^2 = 6.47$, $\nu = 1$, $P < 5\%$, significant, indicates difference between sprays.

4. $\chi^2 = 4.46$, $\nu = 6$, $P > 50\%$, not significant, no.

5. $\chi^2 = 2.1$, $\nu = 4$, $P > 50\%$, not significant, good fit, yes.

6. $F = 11.94$, $\nu_1 = 5$, $\nu_2 = 7$, 1-sided, $P < 1\%$, highly significant, yes.

7. $t = 2.27$, $\nu = 6$, 2-sided, $P > 5\%$, not significant, yes, but since $P < 10\%$ a larger sample may be taken.

8. (a) 1st $t = 3.41$, $\nu = 7$, 2-sided, $P < 5\%$, significant, evidence of bias. 2nd $t = 1.43$, $\nu = 7$, 2-sided, $P > 5\%$, not significant, no evidence of bias.

   (b) $F = 1.28$, $\nu_1 = 7$, $\nu_2 = 7$, 2-sided, $P > 5\%$, not significant, no.

9. $t = 2.50$, $\nu = 4$, 1-sided, $P < 5\%$, significant, yes.

10. $F = 2.1$, $\nu_1 = 4$, $\nu_2 = 6$, 2-sided, $P > 5\%$, not significant, use general t-test, $t = 1.98$, $\nu = 10$, 2-sided, $P < 5\%$, not significant.

11. (a) $t = 2.2$, $\nu = 9$, 1-sided, $P < 5\%$, significant, yes.

    (b) $t = 0.82$, $\nu = 18$, $P > 20\%$, not significant, no evidence of improvement.

12. $D = 0.246$, $n = 10$, $P > 5\%$, not significant, good fit to exponential distribution, suggests pumps are failing at random.

13. $C_s = 4.21$, $C_k = 3.30$, $C_s >$ tabulated value at 1% level, reject extreme value of 26 and repeat. $C_s = 1.47$, $C_k = 4.83$, $C_s >$ tabulated value at 1% level, $C_k >$ tabulated value at 5% level, reject extreme value of 24 and repeat, $C_s = 0.12$, $C_k = 2.15$, no more outliers.

Chapter 12

1. $F = 1.62$, $\nu_1 = 3$, $\nu_2 = 16$, $P < 1\%$, highly significant.
     There is a significant difference between machines.
2.   Antifreeze: mean square $<$ residual mean square, not significant.
     No difference between types of antifreeze.
     Inhibitor: $F = 1.18$, $\nu_1 = 3$, $\nu_2 = 6$, not significant.
     No evidence of a difference.
3. (a) $r = -0.66$, $n = 12$, $|r| >$ tabulated value of 0.576, indicates a relationship.
   (b) $y = 46.01 - 0.686x$, 22.0
4. (b) $y = 15.26 + 0.418x$
   (d) $r = 0.968$
   (e) $n = 8$, $r >$ tabulated value, indicates a relationship.
   (f)   19.0
5. (b) $y = 44.9 + 0.50x$
   (c)   44.85, 49.85, 54.85, etc.
   (d) -2.01, -0.72, 0.32, etc.
   (e)   0.999, 99.7%

Appendix A
Glossary of Symbols

■ The Greek alphabet

A	α	alpha	a
B	β	beta	b
Γ	γ	gamma	g
Δ	δ	delta	d
E	ε	epsilon	e
Z	ζ	zeta	z
H	η	eta	ee
Θ	θ, ϑ	theta	th
I	ι	iota	i
K	$\varkappa$	kappa	k
Λ	λ	lambda	l
M	μ	mu	m
N	ν	nu	n
Ξ	ξ	xi	x
O	o	omicron	o
Π	π	pi	p
P	ρ	rho	r
Σ	σ	sigma	s
T	τ	tau	t
Υ	υ	upsilon	u
Φ	ϕ, φ	phi	ph
X	χ	chi	ch
Ψ	ψ	psi	ps
Ω	ω	omega	oo

■ Special uses of Greek letters

Greek letters tend to be used to denote distribution *parameters*.

α	tail area of probability distribution
	also significance level
β	Weibull shape parameter
γ	Weibull location parameter
δ	infinitely small increment
η	Weibull characteristic life
λ	mean of Poisson distribution
μ	population mean
	also micro (10^{-6})
	also micron $(10^{-4}$ cm$)$
μm	nano (10^{-9})
$\mu\mu$	pico (10^{-12})
ν	degrees of freedom (see also df)
π	a constant $\simeq 3.142$
ρ	coefficient of correlation
σ	population standard deviation
σ^2	population variance
σ_e	standard error of mean
Σ	sum of
ϕ	degrees of freedom (not preferred, see ν)
$\phi(z)$	pdf of standard normal distribution
$\Phi(z)$	cdf of standard normal distribution
χ^2	a quantity used in significance tests

■ Special uses of English letters

English letters tend to be used to denote *statistics* obtained from data. Letters that denote variable quantities are sometimes printed in italics.

C_k	coefficient of kurtosis
C_m	machine or short-term capability index
C_p	process capability index
C_{pk}	process setting index
C_s	coefficient of skewness
D	a quantity used in significance tests
df	degrees of freedom (see also DF and ν)
DF	degrees of freedom (see also df and ν)
e	a constant $\simeq 2.7183$
f	frequency
$f(x)$	pmf of discrete x, pdf of continuous x

F	cumulative frequency
	also a quantity used in significance tests
$F(x)$	cumulative probability function of x
$h(t)$	hazard function or hazard rate
H_A	alternative hypothesis
H_0	null hypothesis
$H(t)$	cumulative hazard function
k	number of samples (see also m)
m	number of samples (see also k)
M	median (not preferred, see $\tilde{x}$)
n	sample size
N	population size
p	probability of success
P	probability
q	probability of failure
r	sample correlation coefficient
	also average rate for Poisson process
R	range
$R(t)$	reliability
s	standard deviation of a sample
s^2	variance of a sample
t	a quantity used in significance tests and confidence limits
w	range (not preferred, see R)
z	standardized variable, standardized deviate

■ Common abbreviations

AMM	average movement of mean
ARL	average run length
cdf	cumulative distribution function
LAL	lower action line
LCL	lower control line
LWL	lower warning line
MTBF	mean time between failures
pdf	probability density function
pmf	probability mass function
UAL	upper action line
UCL	upper control line
UWL	upper warning line

■ Statistical symbols

x, y and z below, can be substituted by other English or Greek letters.

x	a feature
y	another feature
x_i	the ith value of feature x
$\bar{x}$	mean of several x values (usually within a sample)
$\bar{\bar{x}}$	mean of several $\bar{x}$ values (usually minimum 25)
$\tilde{x}$	median of several x values (usually within a sample)
$\bar{\tilde{x}}$	mean of several $\tilde{x}$ values (usually minimum 25)
$\hat{x}$	estimate of x
$\hat{x}$	mode of several x values (not a standard symbol)
$x \mid y$	x is conditional upon y
$\lvert x \rvert$	treat $\pm\ x$ as $+$
$x \sim y$	the range x to y
$x \rightarrow y$	x approaches y
$x : y$	the ratio of x to y
$x : y :: y : z$	x is to y as y is to z (not in common use)
$x \propto y$	x is proportional to y
$=$	equals
$\neq$	does not equal
$\equiv$	equivalent to
$\approx$	approximates to
$\Rightarrow$	it implies that
$\cup$	or
$>$	greater than
$\geq$	greater than or equal to
$\ngtr$	not greater than
$<$	less than
$\leq$	less than or equal to
$\nless$	not less than
$\therefore$	therefore
$\because$	because (not in common use)
∞	infinity
$\sqrt{}$	square root
$\int$	integral
$!$	factorial

Appendix B
Standard Normal Probability
Table

$$\alpha = \frac{1}{\sqrt{2\pi}} \int_{z_i}^{\infty} exp\left[-\frac{1}{2} z^2\right] dz$$

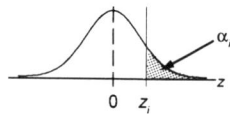

Proportions (α values) of the standard normal distribution that lie to the right of selected positive z values

z	.00	.01	.02	.03	.04	.05	.06	.07	.08	.09
0.0	.5000	.4960	.4920	.4880	.4840	.4801	.4761	.4721	.4681	.4641
0.1	.4602	.4562	.4522	.4483	.4443	.4404	.4364	.4325	.4286	.4247
0.2	.4207	.4168	.4129	.4090	.4052	.4013	.3974	.3936	.3897	.3859
0.3	.3821	.3783	.3745	.3707	.3669	.3632	.3594	.3557	.3520	.3483
0.4	.3446	.3409	.3372	.3336	.3300	.3264	.3228	.3192	.3156	.3121
0.5	.3085	.3050	.3015	.2981	.2946	.2912	.2877	.2843	.2810	.2776
0.6	.2743	.2709	.2676	.2643	.2611	.2578	.2546	.2514	.2483	.2451
0.7	.2420	.2389	.2358	.2327	.2296	.2266	.2236	.2206	.2177	.2148
0.8	.2119	.2090	.2061	.2033	.2005	.1977	.1949	.1922	.1894	.1867
0.9	.1841	.1814	.1788	.1762	.1736	.1711	.1685	.1660	.1635	.1611
1.0	.1587	.1562	.1539	.1515	.1492	.1469	.1446	.1423	.1401	.1379
1.1	.1357	.1335	.1314	.1292	.1271	.1251	.1230	.1210	.1190	.1170
1.2	.1151	.1131	.1112	.1093	.1075	.1056	.1038	.1020	.1003	.0985
1.3	.0968	.0951	.0934	.0918	.0901	.0885	.0869	.0853	.0838	.0823
1.4	.0808	.0793	.0778	.0764	.0749	.0735	.0721	.0708	.0694	.0681
1.5	.0668	.0655	.0643	.0630	.0618	.0606	.0594	.0582	.0571	.0559
1.6	.0548	.0537	.0526	.0516	.0505	.0495	.0485	.0475	.0465	.0455
1.7	.0446	.0436	.0427	.0418	.0409	.0401	.0392	.0384	.0375	.0367
1.8	.0359	.0351	.0344	.0336	.0329	.0322	.0314	.0307	.0301	.0294
1.9	.0287	.0281	.0274	.0268	.0262	.0256	.0250	.0244	.0239	.0233
2.0	.0228	.0222	.0217	.0212	.0207	.0202	.0197	.0192	.0188	.0183
2.1	.0179	.0174	.0170	.0166	.0162	.0158	.0154	.0150	.0146	.0143
2.2	.0139	.0136	.0132	.0129	.0125	.0122	.0119	.0116	.0113	.0110
2.3	.0107	.0104	.0102	.00990	.00964	.00939	.00914	.00889	.00866	.00842
2.4	.00820	.00798	.00776	.00755	.00734	.00714	.00695	.00676	.00657	.00639
2.5	.00621	.00604	.00587	.00570	.00554	.00539	.00523	.00508	.00494	.00480
2.6	.00466	.00453	.00440	.00427	.00415	.00402	.00391	.00379	.00368	.00357
2.7	.00347	.00336	.00326	.00317	.00307	.00298	.00289	.00280	.00272	.00264
2.8	.00256	.00248	.00240	.00233	.00226	.00219	.00212	.00205	.00199	.00193
2.9	.00187	.00181	.00175	.00169	.00164	.00159	.00154	.00149	.00144	.00139

z	α
3.0	.00135
3.1	.000968
3.2	.000687
3.3	.000483
3.4	.000337
3.5	.000233
3.6	.000159
3.7	.000108
3.8	.0000723
3.9	.0000481
4.0	.0000317
4.1	.0000207
4.2	.0000133
4.3	.00000854
4.4	.00000541
4.5	.00000340
4.6	.00000211
4.7	.00000130
4.8	.000000793
4.9	.000000479
1.2816	.1
2.3263	.01
3.0902	.001
3.7190	.0001
4.2649	.00001
1.6449	.05
2.5758	.005
3.2905	.0005
3.8906	.00005
4.4172	.000005

Appendix C
Binomial Table

- The tabulated function is the cumulative distribution function

$$F(x) = f(0) + f(1) + \ldots + f(x) \quad \text{where} \quad f(x) = \frac{n!}{x!(n-x)!}\, p^x\,(1-p)^{n-x}$$

- For $x > 0$, values of $f(x)$ can be found by subtraction: $f(x) = F(x) - F(x-1)$.

- In all cases, $F(n) = 1$ exactly. Other omitted entries at top or bottom of any column are 0 or 1 respectively, to four decimal places.

x	p (Proportion defective)						
	0.05	0.10	0.15	0.20	0.30	0.40	0.50
n (Sample size) = 5							
0	.7738	.5903	.4437	.3277	.1681	.0778	.0313
1	.9774	.9185	.8352	.7373	.5282	.3370	.1875
2	.9988	.9914	.9734	.9421	.8369	.6826	.5000
3		.9995	.9978	.9933	.9692	.9130	.8125
4			.9999	.9997	.9976	.9898	.9688
n (Sample size) = 10							
0	.5987	.3487	.1969	.1074	.0282	.0060	.0010
1	.9139	.7361	.5443	.3758	.1493	.0464	.0107
2	.9885	.9298	.8202	.6778	.3828	.1673	.0547
3	.9990	.9872	.9500	.8791	.6496	.3823	.1719
4	.9999	.9984	.9901	.9672	.8497	.6331	.3770
5		.9999	.9986	.9936	.9527	.8338	.6230
6			.9999	.9991	.9894	.9452	.8281
7				.9999	.9984	.9877	.9453
8					.9999	.9983	.9893
9						.9999	.9990
n (Sample size) = 20							
0	.3585	.1216	.0388	.0115	.0008		
1	.7358	.3917	.1756	.0692	.0076	.0005	
2	.9245	.6769	.4049	.2061	.0355	.0036	.0002
3	.9841	.8670	.6477	.4114	.1071	.0160	.0013
4	.9974	.9568	.8298	.6296	.2375	.0510	.0059
5	.9997	.9887	.9327	.8042	.4164	.1256	.0207
6		.9976	.9781	.9133	.6080	.2500	.0577
7		.9996	.9941	.9679	.7723	.4159	.1316
8		.9999	.9987	.9900	.8867	.5956	.2517
9			.9998	.9974	.9520	.7553	.4119
10				.9994	.9829	.8725	.5881
11				.9999	.9949	.9435	.7483
12					.9987	.9790	.8684
13					.9997	.9935	.9423
14						.9984	.9793
15						.9997	.9941
16							.9987
17							.9998
18							
19							

F(x)

x

- The tabulated function is the cumulative distribution function
  $F(x) = f(0) + f(1) + ... + f(x)$ where $f(x) = \dfrac{n!}{x!(n-x)!}\, p^x\,(1-p)^{n-x}$

- For $x > 0$, values of $f(x)$ can be found by subtraction: $f(x) = F(x) - F(x-1)$.

- In all cases, $F(n) = 1$ exactly. Other omitted entries at top or bottom of any column are 0 or 1 respectively, to four decimal places.

x	*p* (Proportion defective)						
	0.01	0.02	0.03	0.04	0.05	0.10	0.15
	n (Sample size) = 50						
0	.6050	.3642	.2181	.1300	.0769	.0052	.0003
1	.9106	.7358	.5553	.4005	.2794	.0338	.0029
2	.9862	.9216	.8108	.6767	.5405	.1117	.0142
3	.9984	.9822	.9372	.8609	.7604	.2503	.0460
4	.9999	.9968	.9832	.9510	.8964	.4312	.1121
5		.9995	.9963	.9856	.9622	.6161	.2194
6		.9999	.9993	.9964	.9882	.7702	.3613
7			.9999	.9992	.9968	.8779	.5188
8				.9999	.9992	.9421	.6681
9					.9998	.9755	.7911
10						.9906	.8801
11						.9968	.9372
12						.9990	.9699
13						.9997	.9868
14						.9999	.9947
15							.9981
16							.9993
17							.9998
18							.9999
19							

F(x) — x

Appendix D
Poisson Table

x	Expected mean (λ)																	
	0.5	1.0	1.5	2.0	2.5	3.0	4.0	5.0	6.0	7.0	8.0	9.0	10.0	11.0	12.0	13.0	14.0	15.0
0	.6065	.3679	.2231	.1353	.0821	.0498	.0183	.0067	.0025	.0009	.0003	.0001						
1	.9098	.7358	.5578	.4060	.2873	.1991	.0916	.0404	.0174	.0073	.0030	.0012	.0005	.0002	.0001			
2	.9856	.9197	.8088	.6767	.5438	.4232	.2381	.1247	.0620	.0296	.0138	.0062	.0028	.0012	.0005	.0002	.0001	
3	.9982	.9810	.9344	.8571	.7576	.6472	.4335	.2650	.1512	.0818	.0424	.0212	.0103	.0049	.0023	.0010	.0005	.0002
4	.9998	.9963	.9814	.9473	.8912	.8153	.6288	.4405	.2851	.1730	.0996	.0550	.0293	.0151	.0076	.0037	.0018	.0009
5		.9994	.9955	.9834	.9580	.9161	.7851	.7029	.4457	.3007	.1912	.1157	.0671	.0375	.0203	.0107	.0055	.0028
6		.9999	.9991	.9955	.9858	.9665	.8893	.7622	.6063	.4497	.3134	.2068	.1301	.0786	.0458	.0259	.0142	.0076
7			.9998	.9989	.9958	.9881	.9489	.8666	.7440	.5987	.4530	.3239	.2202	.1432	.0895	.0540	.0316	.0180
8				.9998	.9989	.9962	.9786	.9319	.8472	.7291	.5925	.4557	.3328	.2320	.1550	.0998	.0621	.0374
9					.9999	.9989	.9919	.9682	.9161	.8305	.7166	.5874	.4579	.3405	.2424	.1658	.1094	.0699
10						.9997	.9972	.9863	.9574	.9015	.8159	.7060	.5830	.4599	.3472	.2517	.1757	.1185
11						.9999	.9991	.9945	.9799	.9466	.8881	.8030	.6968	.5793	.4616	.3532	.2600	.1848
12							.9997	.9980	.9912	.9730	.9362	.8758	.7916	.6887	.5760	.4631	.3585	.2676
13							.9999	.9993	.9964	.9872	.9658	.9261	.8645	.7813	.6815	.5730	.4644	.3632
14								.9998	.9986	.9943	.9827	.9585	.9165	.8540	.7720	.6751	.5704	.4657
15								.9999	.9995	.9976	.9918	.9780	.9513	.9074	.8444	.7636	.6694	.5681
16									.9998	.9990	.9963	.9889	.9730	.9441	.8987	.8355	.7559	.6641
17									.9999	.9996	.9984	.9947	.9857	.9678	.9370	.8905	.8272	.7489
18										.9999	.9994	.9976	.9928	.9823	.9626	.9302	.8826	.8195
19											.9997	.9989	.9965	.9907	.9787	.9573	.9235	.8752
20											.9999	.9996	.9984	.9953	.9884	.9750	.9521	.9170
21												.9998	.9993	.9977	.9939	.9859	.9712	.9469
22												.9999	.9997	.9990	.9970	.9924	.9833	.9673
23													.9999	.9995	.9985	.9960	.9907	.9805
24														.9999	.9993	.9980	.9950	.9888
25															.9997	.9990	.9974	.9938
26															.9999	.9995	.9987	.9967
27																.9998	.9994	.9983
28																.9999	.9997	.9991
29																	.9999	.9996
30																		.9998
31																		.9999
32																		

$F(x)$

x

- The tabulated function is the cumulative distribution function
  $F(x) = f(0) + f(1) + \dots + f(x)$ where $f(x) = \dfrac{\lambda^x\, e^{-\lambda}}{x!}$.

- For x > 0, values of f(x) can be found by subtraction: $f(x) = F(x) - F(x-1)$.

- Omitted entries at top or bottom of any column are 0 or 1 respectively, to four decimal places.

Appendix E
t-Table

Percentage points of the *t*-distribution for a 1-sided test

double P for 2-sided tests

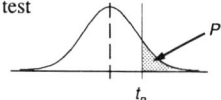

ν	Probability (*P* %)								
	25%	**10%**	**5%**	**2.5%**	**1%**	**0.5%**	**0.25%**	**0.1%**	**0.05%**
1	1.00	3.08	6.31	12.7	31.8	63.7	127	318	637
2	0.816	1.89	2.92	4.30	6.97	9.92	14.1	22.3	31.6
3	0.765	1.64	2.35	3.18	4.54	5.84	7.45	10.1	12.9
4	0.741	1.53	2.13	2.78	3.75	4.60	5.60	7.17	8.61
5	0.727	1.48	2.01	2.57	3.37	4.03	4.77	5.89	6.87
6	0.718	1.44	1.94	2.45	3.14	3.71	4.32	5.21	5.96
7	0.711	1.42	1.89	2.36	3.00	3.50	4.03	4.79	5.41
8	0.706	1.40	1.86	2.31	2.90	3.36	3.83	4.50	5.04
9	0.703	1.38	1.83	2.26	2.82	3.25	3.69	4.30	4.78
10	0.700	1.37	1.81	2.23	2.76	3.17	3.58	4.14	4.59
11	0.697	1.36	1.80	2.20	2.72	3.11	3.50	4.03	4.44
12	0.695	1.36	1.78	2.18	2.68	3.05	3.43	3.93	4.32
13	0.694	1.35	1.77	2.16	2.65	3.01	3.37	3.85	4.22
14	0.692	1.35	1.76	2.14	2.62	2.98	3.33	3.79	4.14
15	0.691	1.34	1.75	2.13	2.60	2.95	3.29	3.73	4.07
16	0.690	1.34	1.75	2.12	2.58	2.92	3.25	3.69	4.01
17	0.689	1.33	1.74	2.11	2.57	2.90	3.22	3.65	3.96
18	0.688	1.33	1.73	2.10	2.55	2.88	3.20	3.61	3.92
19	0.688	1.33	1.73	2.09	2.54	2.86	3.17	3.58	3.88
20	0.687	1.33	1.72	2.09	2.53	2.85	3.15	3.55	3.85
21	0.686	1.32	1.72	2.08	2.52	2.83	3.14	3.53	3.82
22	0.686	1.32	1.72	2.07	2.51	2.82	3.12	3.51	3.79
23	0.685	1.32	1.71	2.07	2.50	2.81	3.10	3.49	3.77
24	0.685	1.32	1.71	2.06	2.49	2.80	3.09	3.47	3.74
25	0.684	1.32	1.71	2.06	2.49	2.79	3.08	3.45	3.72
26	0.684	1.32	1.71	2.06	2.48	2.78	3.07	3.44	3.71
27	0.684	1.31	1.70	2.05	2.47	2.77	3.06	3.42	3.69
28	0.683	1.31	1.70	2.05	2.47	2.76	3.05	3.41	3.67
29	0.683	1.31	1.70	2.05	2.46	2.76	3.04	3.40	3.66
30	0.683	1.31	1.70	2.04	2.46	2.75	3.03	3.39	3.65
40	0.681	1.30	1.68	2.02	2.42	2.70	2.97	3.31	3.55
60	0.679	1.30	1.67	2.00	2.39	2.66	2.91	3.23	3.46
120	0.677	1.29	1.66	1.98	2.36	2.62	2.86	3.16	3.37
∞	0.674	1.28	1.64	1.96	2.33	2.58	2.81	3.09	3.29

Appendix F
Control Chart Constants

For Shewhart (USA system) variables charts

Sample size	x̄ & R means chart control lines	x̃ & R medians chart control lines	x̄ & s means chart control lines	s chart lower control line	s chart upper control line	Correction factor for s	Correction factor for R	R chart lower control line	R chart upper control line
n	A_2	$\widetilde{A_2}$	A_3	B_3	B_4	c_4	d_2	D_3	D_4
2	1.880	1.880	2.659	0	3.267	0.798	1.128	0	3.267
3	1.023	1.187	1.954	0	2.568	0.886	1.693	0	2.574
4	0.729	0.796	1.628	0	2.266	0.921	2.059	0	2.282
5	0.577	0.691	1.427	0	2.089	0.940	2.326	0	2.114
6	0.483	0.548	1.287	0.030	1.970	0.952	2.534	0	2.004
7	0.419	0.508	1.182	0.118	1.882	0.959	2.704	0.076	1.924
8	0.373	0.433	1.099	0.185	1.815	0.965	2.847	0.136	1.864
9	0.337	0.412	1.032	0.239	1.761	0.969	2.970	0.184	1.816
10	0.308	0.362	0.975	0.284	1.716	0.973	3.078	0.223	1.777
11	0.285		0.927	0.321	1.679	0.975	3.173	0.256	1.744
12	0.266		0.886	0.354	1.646	0.978	3.258	0.283	1.717
13	0.249		0.850	0.382	1.618	0.979	3.336	0.307	1.693
14	0.235		0.817	0.406	1.594	0.981	3.407	0.328	1.672
15	0.223		0.789	0.428	1.572	0.982	3.472	0.347	1.653
16	0.212		0.763	0.448	1.552	0.984	3.532	0.363	1.637
17	0.203		0.739	0.466	1.534	0.985	3.588	0.378	1.622
18	0.194		0.718	0.482	1.518	0.985	3.640	0.391	1.608
19	0.187		0.698	0.497	1.503	0.986	3.689	0.403	1.597
20	0.180		0.680	0.510	1.490	0.987	3.735	0.415	1.585
21	0.173		0.663	0.523	1.477	0.988	3.778	0.425	1.575
22	0.167		0.647	0.534	1.466	0.988	3.819	0.434	1.566
23	0.162		0.633	0.545	1.455	0.989	3.858	0.443	1.557
24	0.157		0.619	0.555	1.445	0.989	3.895	0.451	1.548
25	0.153		0.606	0.565	1.435	0.990	3.931	0.459	1.541

Appendix G
Chi-Squared Table

Percentage points of the χ^2 distribution

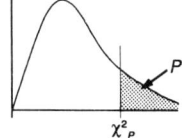

χ^2_P

Values of chi-squared (χ^2) for $P\%$

v	99.5%	99%	97.5%	95%	90%	75%	50%	25%	10%	5%	2.5%	1%	0.1%
1	4×10^{-5}	2×10^{-4}	1×10^{-3}	4×10^{-3}	0.02	0.10	0.46	1.32	2.71	3.84	5.02	6.63	10.83
2	0.01	0.02	0.05	0.10	0.21	0.58	1.39	2.77	4.61	5.99	7.38	9.21	13.82
3	0.07	0.12	0.22	0.35	0.58	1.21	2.37	4.11	6.25	7.81	9.35	11.34	16.27
4	0.21	0.30	0.48	0.71	1.06	1.92	3.36	5.39	7.78	9.49	11.14	13.28	18.47
5	0.41	0.55	0.83	1.15	1.61	2.68	4.35	6.63	9.24	11.07	12.83	15.09	20.52
6	0.67	0.87	1.24	1.64	2.20	3.46	5.35	7.84	10.64	12.59	14.45	16.81	22.46
7	0.99	1.24	1.69	2.17	2.83	4.26	6.35	9.04	12.02	14.07	16.01	18.48	24.32
8	1.34	1.65	2.18	2.73	3.49	5.07	7.34	10.22	13.36	15.51	17.53	20.09	26.13
9	1.74	2.09	2.70	3.33	4.17	5.90	8.34	11.39	14.68	16.92	19.02	21.67	27.88
10	2.16	2.56	3.25	3.94	4.87	6.74	9.34	12.55	15.99	18.31	20.48	23.21	29.59
11	2.60	3.05	3.82	4.57	5.58	7.58	10.34	13.70	17.28	19.68	21.92	24.73	31.26
12	3.07	3.57	4.40	5.23	6.30	8.44	11.34	14.85	18.55	21.03	23.34	26.22	32.91
13	3.57	4.11	5.01	5.89	7.04	9.30	12.34	15.98	19.81	22.36	24.74	27.69	34.53
14	4.08	4.66	5.63	6.57	7.79	10.17	13.34	17.12	21.06	23.68	26.12	29.14	36.12
15	4.60	5.23	6.26	7.26	8.55	11.04	14.34	18.25	22.31	25.00	27.49	30.58	37.70
16	5.14	5.81	6.91	7.96	9.31	11.91	15.34	19.37	23.54	26.30	28.85	32.00	39.25
17	5.70	6.41	7.56	8.67	10.09	12.79	16.34	20.49	24.77	27.59	30.19	33.41	40.79
18	6.27	7.01	8.23	9.39	10.86	13.68	17.34	21.61	25.99	28.87	31.53	34.81	42.31
19	6.84	7.63	8.91	10.12	11.65	14.56	18.34	22.72	27.20	30.14	32.85	36.19	43.82
20	7.43	8.26	9.59	10.85	12.44	15.45	19.34	23.83	28.41	31.41	34.17	37.57	45.32
21	8.03	8.90	10.28	11.59	13.24	16.34	20.34	24.94	29.62	32.67	35.48	38.93	46.80
22	8.64	9.54	10.98	12.34	14.04	17.24	21.34	26.04	30.81	33.92	36.78	40.29	48.27
23	9.26	10.20	11.69	13.09	14.85	18.14	22.34	27.14	32.01	35.17	38.08	41.64	49.73
24	9.89	10.86	12.40	13.85	15.66	19.04	23.34	28.24	33.20	36.42	39.36	42.98	51.18
25	10.52	11.52	13.12	14.61	16.47	19.94	24.34	29.34	34.38	37.65	40.65	44.31	52.62
26	11.16	12.20	13.84	15.38	17.29	20.84	25.34	30.43	35.56	38.89	41.92	45.64	54.05
27	11.81	12.88	14.57	16.15	18.11	21.75	26.34	31.53	36.74	40.11	43.19	46.96	55.48
28	12.46	13.56	15.31	16.93	18.94	22.66	27.34	32.62	37.92	41.34	44.46	48.28	56.89
29	13.12	14.26	16.05	17.71	19.77	23.57	28.34	33.71	39.09	42.56	45.72	49.59	58.30
30	13.79	14.95	16.79	18.49	20.60	24.48	29.34	34.80	40.26	43.77	46.98	50.89	59.70

Appendix H
F-Table

Percentage points of the *F*-distribution for a 1-sided test

double P for 2-sided tests

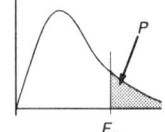

ν_2	P%	1	2	3	4	5	6	7	8	9	10	15	∞
1	10	39.9	49.5	53.6	55.8	57.2	58.2	58.9	59.4	59.9	60.2	61.2	63.3
	5	161	199	216	225	230	234	237	239	241	242	246	254
	2.5	648	800	864	900	922	937	948	957	963	969	985	1018
	1	4052	4999	5403	5625	5764	5859	5928	5982	6022	6056	6157	6366
2	10	8.53	9.00	9.16	9.24	9.29	9.33	9.35	9.38	9.30	9.39	9.42	9.49
	5	18.5	19.0	19.2	19.2	19.3	19.3	19.4	19.4	19.4	19.4	19.4	19.5
	2.5	38.5	39.0	39.2	39.3	39.3	39.3	39.4	39.4	39.4	39.4	39.4	39.5
	1	98.5	99.0	99.2	99.2	99.3	99.3	99.4	99.4	99.4	99.4	99.4	99.5
3	10	5.54	5.46	5.39	5.34	5.31	5.28	5.27	5.25	5.24	5.23	5.20	5.13
	5	10.1	9.55	9.28	9.12	9.01	8.94	8.89	8.85	8.81	8.79	8.70	8.53
	2.5	17.4	16.0	15.4	15.1	14.9	14.7	14.6	14.5	14.5	14.4	14.3	13.9
	1	34.1	30.8	29.5	28.7	28.2	27.9	27.7	27.5	27.3	27.2	26.9	26.1
4	10	4.54	4.32	4.19	4.11	4.05	4.01	3.98	3.95	3.94	3.92	3.87	3.76
	5	7.71	6.94	6.59	6.39	6.26	6.16	6.09	6.04	6.00	5.96	5.86	5.63
	2.5	12.2	10.7	10.0	9.6	9.4	9.2	9.1	9.0	8.9	8.8	8.7	8.3
	1	21.2	18.0	16.7	16.0	15.5	15.2	15.0	14.8	14.7	14.5	14.2	13.5
5	10	4.06	3.78	3.62	3.52	3.45	3.40	3.37	3.34	3.32	3.30	3.24	3.10
	5	6.61	5.79	5.41	5.19	5.05	4.95	4.88	4.82	4.77	4.74	4.62	4.36
	2.5	10.0	8.4	7.8	7.4	7.1	7.0	6.9	6.8	6.7	6.6	6.4	6.0
	1	16.3	13.3	12.1	11.4	11.0	10.7	10.5	10.3	10.2	10.1	9.72	9.02
6	10	3.78	3.46	3.29	3.18	3.11	3.05	3.01	2.98	2.96	2.94	2.87	2.72
	5	5.99	5.14	4.76	4.53	4.39	4.28	4.21	4.15	4.10	4.06	3.94	3.67
	2.5	8.8	7.3	6.60	6.23	5.99	5.82	5.70	5.60	5.52	5.46	5.27	4.85
	1	13.7	10.9	9.78	9.15	8.75	8.47	8.26	8.10	7.98	7.87	7.56	6.88
7	10	3.59	3.26	3.07	2.96	2.88	2.83	2.78	2.75	2.72	2.70	2.63	2.47
	5	5.59	4.74	4.35	4.12	3.97	3.87	3.79	3.73	3.68	3.64	3.51	3.23
	2.5	8.07	6.54	5.89	5.52	5.29	5.12	5.00	4.90	4.82	4.76	4.57	4.14
	1	12.2	9.55	8.45	7.85	7.46	7.19	6.99	6.84	6.72	6.62	6.31	5.65
8	10	3.46	3.11	2.92	2.81	2.73	2.67	2.62	2.59	2.56	2.54	2.46	2.29
	5	5.32	4.46	4.07	3.84	3.69	3.58	3.50	3.44	3.39	3.35	3.22	2.93
	2.5	7.57	6.06	5.42	5.05	4.82	4.65	4.53	4.43	4.36	4.30	4.10	3.17
	1	11.3	8.65	7.59	7.01	6.63	6.37	6.18	6.03	5.91	5.81	5.52	4.86

ν_1 (corresponding to the greater sample variance)

P%	v_2	1	2	3	4	5	6	7	8	9	10	15	∞
10	9	3.36	3.01	2.81	2.69	2.61	2.55	2.51	2.47	2.44	2.42	2.34	2.16
5		5.12	4.26	3.86	3.63	3.48	3.37	3.29	3.23	3.18	3.14	3.01	2.71
2.5		7.21	5.72	5.08	4.72	4.48	4.32	4.20	4.10	4.03	3.96	3.77	3.33
1		10.6	8.02	6.99	6.42	6.06	5.80	5.61	5.47	5.35	5.26	4.96	4.31
10	10	3.29	2.92	2.73	2.61	2.52	2.46	2.41	2.38	2.35	2.32	2.24	2.06
5		4.96	4.10	3.71	3.48	3.33	3.22	3.14	3.07	3.02	2.98	2.85	2.54
2.5		6.94	5.46	4.83	4.47	4.24	4.07	3.93	3.86	3.78	3.72	3.52	3.08
1		10.0	7.56	6.55	5.99	5.64	5.39	5.20	5.06	4.94	4.85	4.56	3.91
10	12	3.18	2.81	2.61	2.48	2.39	2.33	2.28	2.24	2.21	2.19	2.10	1.90
5		4.75	3.89	3.49	3.26	3.11	3.00	2.91	2.85	2.80	2.75	2.62	2.30
2.5		6.55	5.10	4.47	4.12	3.89	3.73	3.61	3.51	3.44	3.37	3.18	2.73
1		9.33	6.93	5.95	5.41	5.06	4.82	4.64	4.50	4.39	4.30	4.01	3.36
10	15	3.07	2.70	2.49	2.36	2.27	2.21	2.16	2.12	2.09	2.06	1.97	1.76
5		4.54	3.68	3.29	3.06	2.90	2.79	2.71	2.64	2.59	2.54	2.40	2.07
2.5		6.20	4.77	4.15	3.80	3.58	3.42	3.29	3.20	3.12	3.06	2.86	2.40
1		8.68	6.36	5.42	4.89	4.56	4.32	4.14	4.00	3.89	3.80	3.52	2.87
10	16	3.05	2.67	2.46	2.33	2.24	2.18	2.13	2.09	2.06	2.03	1.94	1.72
5		4.49	3.63	3.24	3.01	2.85	2.74	2.66	2.59	2.54	2.49	2.35	2.01
2.5		6.12	4.69	4.08	3.73	3.50	3.34	3.22	3.13	3.05	2.99	2.79	2.32
1		8.53	6.23	5.29	4.77	4.44	4.20	4.03	3.89	3.78	3.69	3.41	2.75
10	24	2.93	2.54	2.33	2.19	2.10	2.04	1.98	1.94	1.91	1.88	1.78	1.53
5		4.26	3.40	3.01	2.78	2.62	2.51	2.42	2.36	2.30	2.25	2.11	1.73
2.5		5.72	4.32	3.72	3.38	3.16	3.00	2.87	2.78	2.70	2.64	2.44	1.94
1		7.82	5.61	4.72	4.22	3.90	3.67	3.50	3.36	3.26	3.17	2.89	2.21
10	60	2.79	2.39	2.18	2.04	1.95	1.87	1.82	1.77	1.74	1.71	1.60	1.29
5		4.00	3.15	2.76	2.53	2.37	2.25	2.17	2.10	2.04	1.99	1.84	1.39
2.5		5.29	3.93	3.34	3.01	2.79	2.63	2.51	2.41	2.33	2.27	2.06	1.48
1		7.08	4.98	4.13	3.65	3.34	3.12	2.95	2.82	2.72	2.63	2.35	1.60
10	∞	2.71	2.30	2.08	1.94	1.85	1.77	1.72	1.67	1.63	1.60	1.49	1.00
5		3.84	3.00	2.60	2.37	2.21	2.10	2.01	1.94	1.88	1.83	1.67	1.00
2.5		5.02	3.69	3.12	2.79	2.57	2.41	2.29	2.14	2.11	2.05	1.83	1.00
1		6.63	4.61	3.78	3.32	3.02	2.80	2.64	2.51	2.41	2.32	2.04	1.00

v_1 (corresponding to the greater sample variance)

Appendix I

Table of the

Kolmogorov—Smirnov

Statistic

Percentage points for D
in the Kolmogorov-Smirnov one-sample test

Sample size (n)	Probability (P %)				
	20%	**15%**	**10%**	**5%**	**1%**
6	0.410	0.436	0.470	0.521	0.618
7	0.381	0.405	0.438	0.486	0.577
8	0.358	0.381	0.411	0.457	0.543
9	0.339	0.360	0.388	0.432	0.514
10	0.322	0.342	0.368	0.410	0.490
11	0.307	0.326	0.352	0.391	0.468
12	0.295	0.313	0.338	0.375	0.450
13	0.284	0.302	0.325	0.361	0.433
14	0.274	0.292	0.314	0.349	0.418
15	0.266	0.283	0.304	0.338	0.404
16	0.258	0.274	0.295	0.328	0.392
17	0.250	0.266	0.286	0.318	0.381
18	0.244	0.259	0.278	0.309	0.371
19	0.237	0.252	0.272	0.301	0.363
20	0.231	0.246	0.264	0.294	0.356
25	0.21	0.22	0.24	0.27	0.32
30	0.19	0.20	0.22	0.24	0.29
35	0.18	0.19	0.21	0.23	0.27
Over 35	$\dfrac{1.07}{\sqrt{n}}$	$\dfrac{1.07}{\sqrt{n}}$	$\dfrac{1.07}{\sqrt{n}}$	$\dfrac{1.07}{\sqrt{n}}$	$\dfrac{1.07}{\sqrt{n}}$

Index